The Soari
Pilot's Ma

The Soaring Pilot's Manual

SECOND EDITION

Ken Stewart

Illustrated by
Mark Taylor

Airlife

First published in 2000 by
Airlife Publishing, an imprint of
The Crowood Press Ltd
Ramsbury, Marlborough
Wiltshire SN8 2HR

www.crowood.com

This impression 2018

British Library Cataloguing-in-Publication Data
A catalogue record for this book is available from the British Library.

ISBN 978 1 84797 044 2

Disclaimer
The information contained in this book is true and complete to the best of our knowledge. All recommendations are made without any guarantee on the part of the Publisher, who also disclaims any liability incurred in connection with the use of this data or specific details.

Typeset by Rowland Phototypesetting Ltd, Bury St Edmunds, Suffolk
Printed and bound by CPI Group (UK) Ltd.

Acknowledgements

Any book on soaring has to delve, to a large degree, into meteorology. I would like to thank Tom Bradbury for his support, the time he has given to checking the text of this book, and the many suggestions he has made. Without his specialist knowledge, I am certain that the contents would not be as comprehensive and that their accuracy would have suffered.

Also, I would like to thank Peter Disdale and Diana Bartlett who spent many hours checking and correcting the text, and the many others who have applied their expertise to specific sections.

I am indebted to John Williamson for allowing me to use the JSW Final Glide Calculator in illustrations.

Lastly, I would like to thank all those who, over the years, have passed on their hard-gained knowledge to improve others. Without them, this book would never have been written.

Ken Stewart

Units Used

Let us start with a short discussion on the units used in gliding and throughout this book. If there is any one item that is designed to confuse someone beginning gliding, it is the continual hopping between the multitudes of different units used by glider pilots in a way that would be pounced upon by any decent mathematics teacher.

So, in an attempt to clarify why metric, statute, and even nautical units are often used in the same sentence, here are some reasons.

In the UK and several other countries, the instruments in gliders are calibrated to indicate airspeed and rate of climb (or descent) in nautical miles per hour (knots). Altimeters are calibrated in feet. Fortunately, one knot is equal to almost exactly 100 feet per minute and so rate of climb and gain in altitude are easily cross-checked.

Glider instruments in continental European countries are calibrated in metric units (kilometres per hour for airspeed, metres per second for rate of climb, and metres for altitude).

However, as gliding awards and records are recognised worldwide, they have been standardised by the Fédération Aéronautique Internationale using metric units.

Therefore, it is not uncommon to hear a British glider pilot say something like, 'I was 20 miles from home on the last leg of a 300 kilometre flight when I found a 6 knot thermal which took me to 6,000 feet and I was then able to glide home at 90 knots'.

Throughout this book, I will attempt to standardise as follows:

* Distances for navigational purposes will be quoted in nautical miles (nm)
* Airspeeds will be given in knots (kts)
* Rates of climb or descent will be given in knots
* Heights and altitudes will be given in feet
* Only the length of tasks, average speeds achieved, and heights for international or nationally recognised awards or records will be given in metric units.

I hope the reader will appreciate that these eccentricities are not of the author's choosing but that he promises to try to keep the various international standards separated as far as possible in the text.

Preface

Having gained the judgement and learned the skills that are necessary to fly a glider safely, most glider pilots wish to progress to become soaring pilots. This book assumes that these basic skills have been mastered, and covers the knowledge and the many different skills that must be gained if a pilot is to be successful at soaring.

Gliding has developed into a sport with a large number of levels of achievement and many different goals. For instance, many pilots are content just to soar locally within range of their own airfield, while others favour cross-country flying, measuring their achievement by the distance covered. Many prefer cross-country speed tasks, flown either individually or against others in competitions. Some are more satisfied by high-altitude flying. Because of all this variety, this book has been written in an attempt to cover all the known variations of soaring. It starts with basic soaring, continues through the various levels, and attempts to cover most of the avenues of soaring that are practised today.

Over the years, the basic training a glider pilot receives has become well-structured and well-organised. Unfortunately, after a pilot's first solo flight, the learning of soaring techniques is often left to the individual. This means that each new soaring pilot has to negotiate all of the pitfalls that more experienced pilots negotiated previously. Ideally, their experience would be passed on by some form of training – but alas, it seldom is. This manual attempts to offer some of the information that the author has learned, either from his own experiences or from the many other experienced pilots who have been willing to share the secrets of their success.

Some parts of this book refer to weather patterns and phenomena. It would be an impossible task to detail the weather aspects in all of the countries where soaring takes place. This has led to a more localised emphasis, dealing mainly with British soaring weather, with references to conditions in other countries. However, most of the general aspects will be pertinent in many countries.

Contents

Section 1

Basic Soaring

Chapter 1

The Principle of Soaring

The wing of an aircraft, be it an aeroplane or a glider, must move through the air if it is to produce the lift force which is necessary to balance the aircraft's weight. On a powered aircraft, the engine can produce thrust to move the aircraft through the air, thus causing airflow past the wing. This airflow past the wing creates the lift force that allows an aeroplane to maintain level flight.

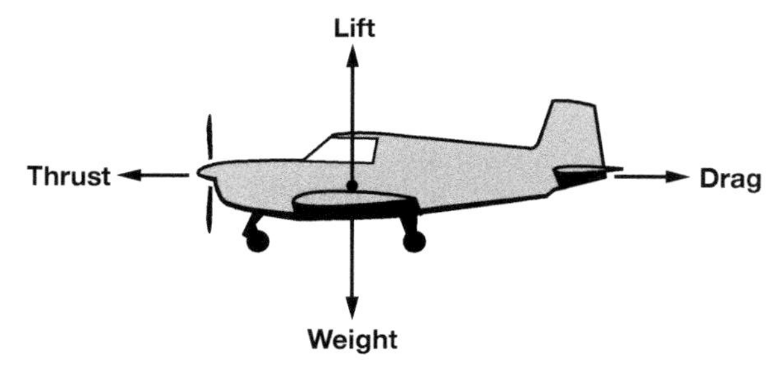

Fig 1.1 Forces in balance. In an aeroplane, lift normally balances weight, and thrust normally balances drag.

A glider, on the other hand, not having the luxury of an engine, relies on the force of gravity to propel it forward through the air. A suitable analogy is a ball rolling down a gentle slope. The ball's forward motion is due to gravity pulling it downward. In fact, if it were not for the surface of the slope, the ball would fall vertically under the influence of gravity. Similarly, gravity would make a glider descend vertically if it were not for the lift force produced by the wing, which, being tilted forward, produces the 'slope' down which the glider flies (figure 1.2).

If a glider tries to fly level, then the lift force will no longer be tilted forward, giving no forward component to propel the glider through the air. It will lose airspeed and the reduced airflow past the wing will cause a reduction in the lift force, eventually resulting in the wing stalling.

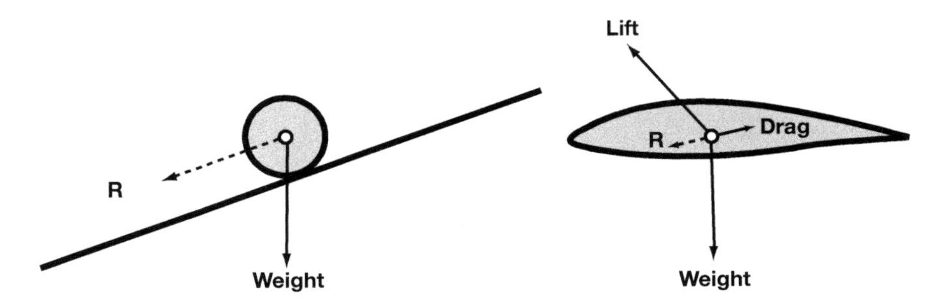

Fig 1.2 Lift and weight versus drag. The resultant of lift and weight 'propels' a glider forward.

Therefore a glider must continually fly 'down a slope' to maintain flight. Put another way, it must always descend *relative to the air mass in which it is flying.*

If the air mass has no updraughts or downdraughts, a glider is said to be flying in STILL AIR. The rate at which a glider descends in still air is dependent on its airspeed – in normal flight, the greater its airspeed, the greater its rate of descent through the air.

As a glider will always be descending relative to the air in which it is flying, in order to maintain height it must be flown in air which is rising at a rate equal to its still air rate of descent. For instance, if a glider is flying in still air at a steady airspeed and is descending at one knot (100 feet per minute), then it will have to encounter an area of air which is ascending at a rate of one knot just to arrest its descent. This would be similar to your walking down an ascending escalator at the same rate as it ascends.

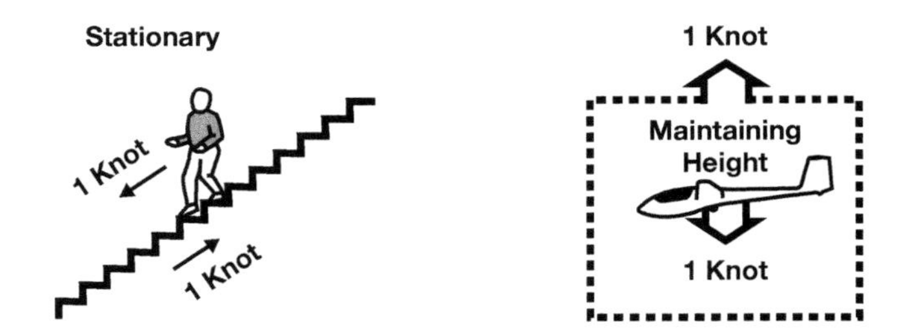

Fig 1.3 Rising air arresting a glider's descent. A glider will constantly descend unless it is flown in air that is rising at a rate equal to its still air descent rate. This is similar to walking down an escalator at the same rate as it ascends.

To gain height, the air mass in which the glider is flying will need to be ascending at a rate greater than the glider's rate of descent in still air – that is, in this example, greater than one knot. (Now the escalator has increased its ascent rate.)

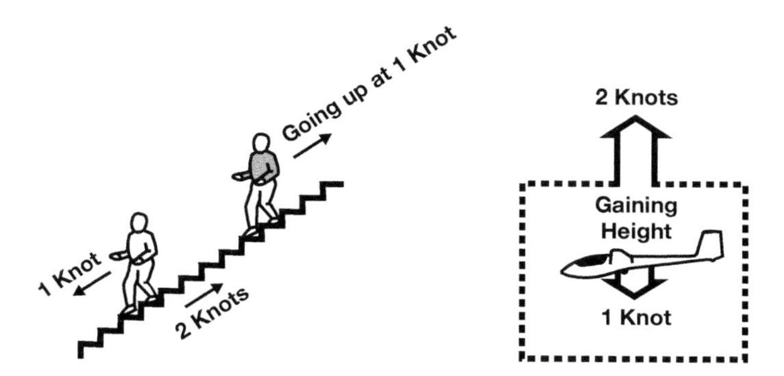

Fig 1.4 Rising air causing a glider to gain height. Air rising at a rate greater than a glider's still air descent rate will result in an increase in the glider's height. This is similar to walking down an escalator which is ascending faster than your walking speed.

Successful soaring depends on your finding and using such areas of rising air. The first task, finding a suitable updraught, can be challenging enough, depending on the nature of the up-current. Often both experienced and inexperienced glider pilots will fortuitously stumble into an area of rising air. This is often the way in which many pilots achieve their first soaring flights. However, becoming a good soaring pilot depends on learning to consistently seek out rising air. Once a suitable up-current is found, you will need to be able to fly accurately in order to climb efficiently. This will require many of the handling skills that you have gained during your basic training.

NOTE: *Often when discussing rising air currents, glider pilots refer to these simply as LIFT. This expression, which will be used extensively throughout the text, should not be confused with the aerodynamic force called 'lift', which will be used to a much lesser extent. The context in which the word is used will, hopefully, make its meaning obvious. The expression that is commonly used to describe descending air is SINK.*

Various atmospheric phenomena cause air to rise in quantities large enough and at a fast enough rate to keep a glider airborne. The main causes of this lift are:

* THERMALS, which are parcels of warm air which rise much as a hot air balloon rises
* HILL LIFT, where the wind is deflected upwards when it meets the face of a hill or mountain
* MOUNTAIN LEE WAVES, which are caused by the deflection of the air mass after it has flowed over a line of hills or mountains
* CONVERGENCE LIFT, where two air masses meet, causing air to be forced upwards

All of these are discussed in the chapters that follow.

Chapter 2

Thermals

The earth's atmosphere

The earth is surrounded by an envelope of gases, which we call the atmosphere. We live in that atmosphere. Not only do we live in it, we live under it. By living on the earth's surface, virtually all of the atmosphere is above us. The atmosphere is retained around the earth by the pull of the earth's gravity, which gives weight to the mass of air aloft.

As an object ascends through the atmosphere, the pressure acting on it reduces. Most people who have ascended or descended quickly in an aircraft have sensed this change of pressure on their eardrums. If the object has no firm sides (such as a child's balloon or a bubble of air), when it is relieved of some atmospheric pressure, it will expand. When air expands it cools. When air descends, it is compressed and as a result warms up. This latter fact can be experienced when using a bicycle pump – compressing the air heats up the pump barrel.

At first, these facts may seem academic, but as we go on to discuss thermals, and indeed air movement in general, remembering them will help you understand the reason why the air behaves as it does.

The environmental lapse rate

The sun emits large amounts of energy, some of which reaches the earth. Some of this energy is scattered and filtered by the atmosphere, but much of it reaches the earth's surface and causes it to warm up. The energy that passes through the atmosphere does not heat the air directly, as it is in the form of short-wavelength radiation, which is not absorbed by the air. However, when it reaches the earth's surface, this energy is absorbed by the land and seas, and surface heating occurs. In a sense, the atmosphere acts like the glass of a greenhouse, in that the glass itself does not get warm, but the non-transparent contents of a greenhouse do.

When the earth's surface warms up, it re-emits some of this energy as longer-wavelength radiation, and much of this is absorbed by the atmosphere. The result is that the air close to the surface is heated due to

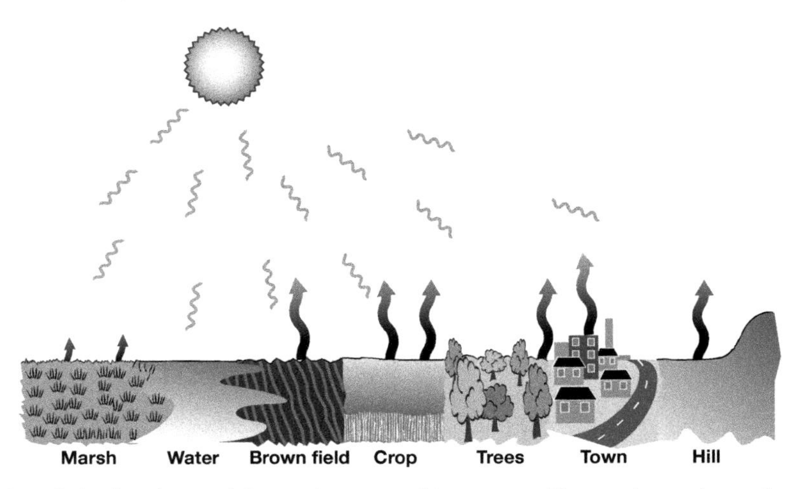

Fig 2.1 Solar heating and the environmental lapse rate. The sun heats the surface of the ground without significantly heating the atmosphere. The warm ground then heats the air close to the ground. Therefore, the temperature of the atmosphere generally reduces as altitude increases.

its contact with the warm ground, and not directly by the sun.

Therefore, air temperature decreases as altitude is increased. With minor local variations, this general rule holds good for altitudes up to around 36,000 feet. The rate at which the temperature decreases with altitude on any one day is known as the ENVIRONMENTAL LAPSE RATE (ELR).

The ELR will not only vary from day to day but will also vary as altitude increases. In fact, over certain altitude bands the temperature may stop decreasing with height, giving what is called an ISOTHERMAL LAYER. Often, the temperature trend may reverse and *increase* with altitude for a time. This increase of temperature with height is called a TEMPERATURE INVERSION, or simply an INVERSION.

How thermals form

When the sun's energy strikes the earth, different surfaces will heat up at different rates, as will the air coming into contact with them. This differential heating will mean that adjacent areas of air may vary in temperature by several degrees.

As the temperature of a parcel of air increases, it becomes lighter and will want to rise. If the temperature difference between this parcel of air

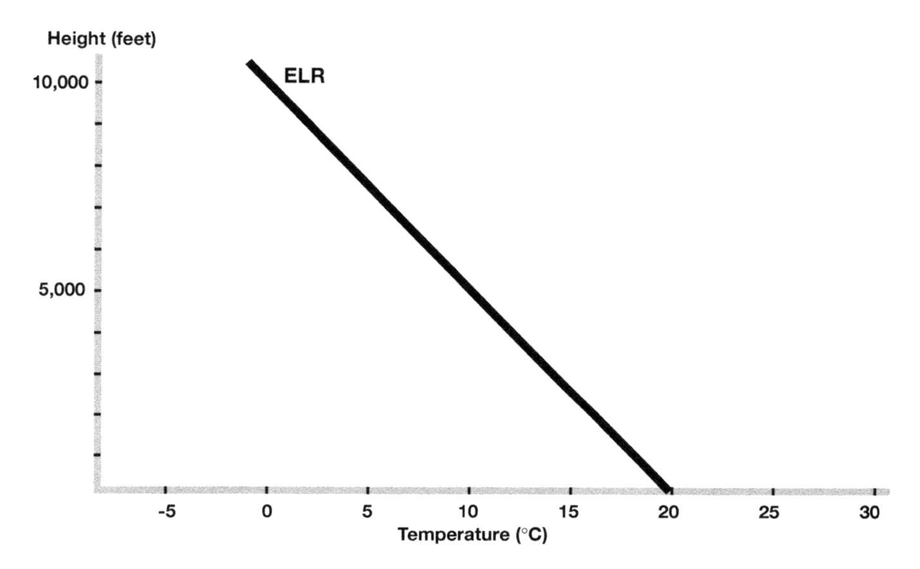

Fig 2.2 The environmental lapse rate. Air temperature normally decreases as altitude increases.

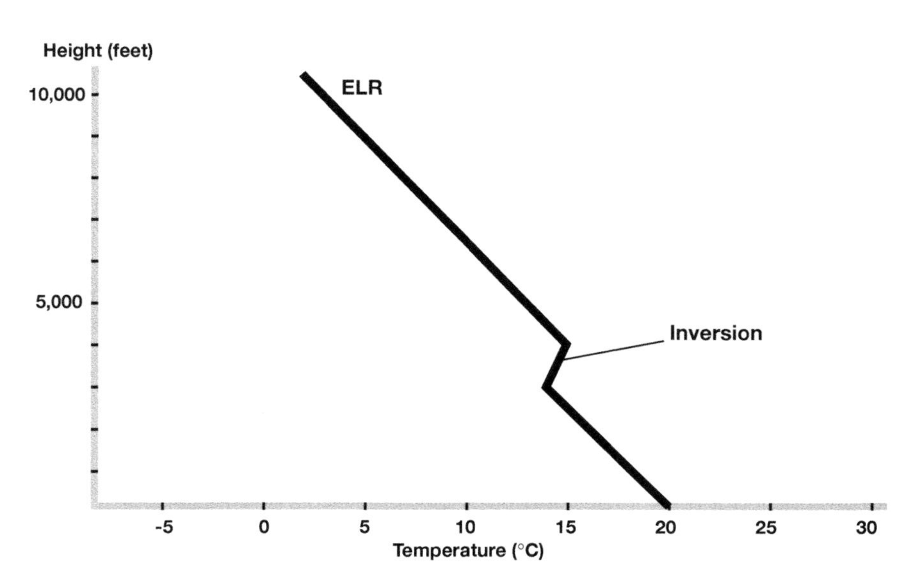

Fig 2.3 An inversion is where air temperature increases as altitude increases.

and the air surrounding it is great enough, then, should it become dislodged from the surface, it will ascend. This rising parcel of air is what is known as a THERMAL. As it ascends, the atmospheric pressure upon it decreases, allowing it to expand. As it expands, it cools. This cooling occurs at a rate of 3°C per 1,000 feet. This figure is known as the DRY ADIABATIC LAPSE RATE (DALR).

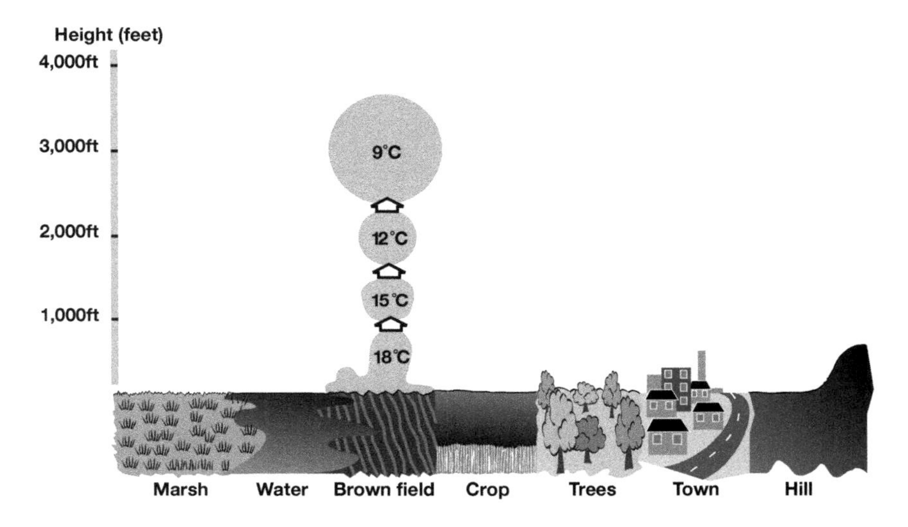

Fig 2.4 Dry adiabatic lapse rate. As a parcel of air rises, it cools at the dry adiabatic lapse rate.

If the DALR is greater than the ELR, then the thermal will eventually reach a height at which its temperature is the same as its surroundings. In theory, when the thermal reaches this height it will stop ascending. (In practice, as the thermal may contain as much as 50,000 tons of air rising at possibly 1,000 feet per minute, its momentum will carry it some height above this temperature equilibrium level. This is one reason why it is not uncommon to find thermals bursting through an inversion, to temporarily leave either a haze dome or a cumulus cloud showing above the inversion.)

Up to this height, this air mass is said to be UNSTABLE. Above this height, STABILITY is said to exist.

When a thermal reaches an inversion, it is entering a layer where the air becomes warmer with height. The rising thermal soon finds itself colder

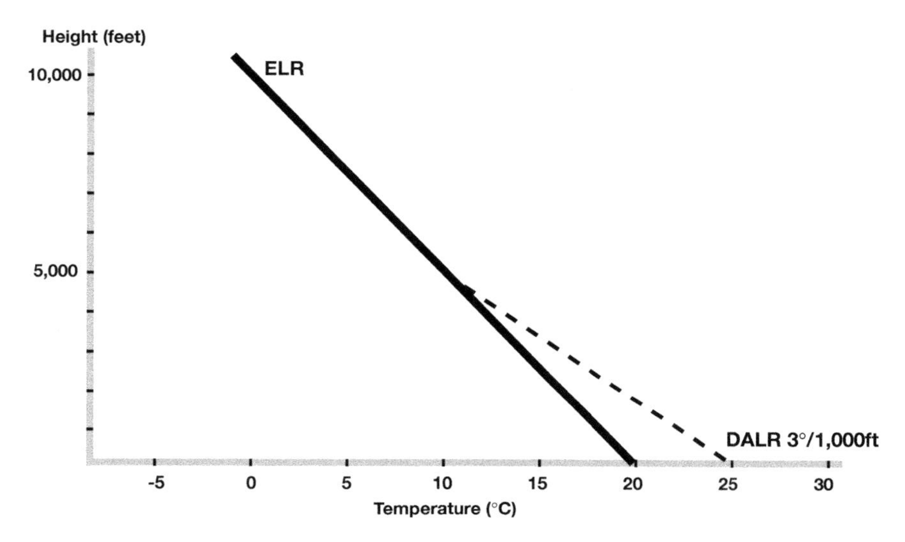

Fig 2.5 DALR greater than ELR. When the DALR is greater than the ELR, a thermal will eventually reach a height at which its temperature is the same as the surrounding air and it will stop ascending.

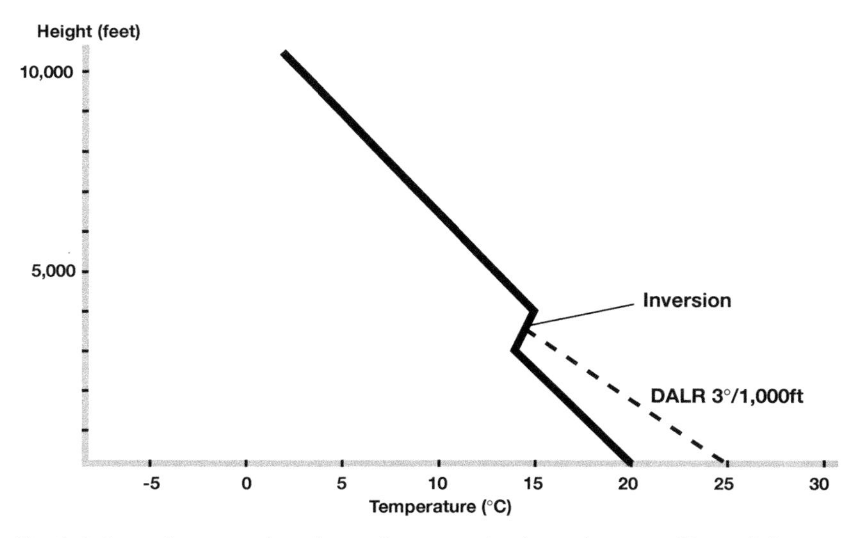

Fig 2.6 Inversion stopping thermal ascent. An inversion acts like a lid, stopping further ascent of thermals.

than its environment, loses buoyancy and starts to sink. The inversion acts like a lid, limiting the top of convection.

An inversion close to the ground may even prevent the formation of thermals, until the surface heating is great enough to warm the air in contact with the surface to a degree where it is not only more buoyant than its surroundings, but also warm enough to break through the inversion.

Figure 2.7 shows two inversions, one just above the surface and one between 3,000 and 4,000 feet. This is a common situation early in the morning. As the sun's heating increases the ground temperature, the inversion close to the surface will break down, allowing thermals to start.

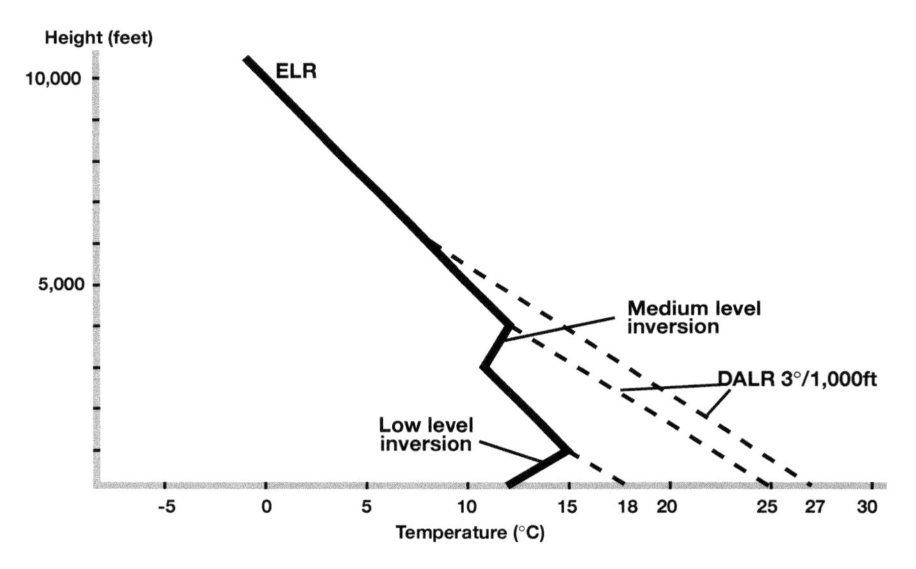

Fig 2.7 Surface inversion. Early in the morning, there is often an inversion near the surface as well as one higher up.

Cumulus clouds

If a thermal contains a reasonable amount of water vapour, then the above characteristics change somewhat, often to the advantage of the glider pilot searching for a thermal.

The amount of water vapour that any parcel of air can contain depends on its temperature – the higher the temperature, the more water that can be contained as a vapour. Because the air's temperature decreases as it ascends, it becomes more saturated, until eventually the parcel of air can no longer contain all of its water as vapour. At this temperature, known as the DEW POINT, the water vapour will condense into water droplets and form cloud.

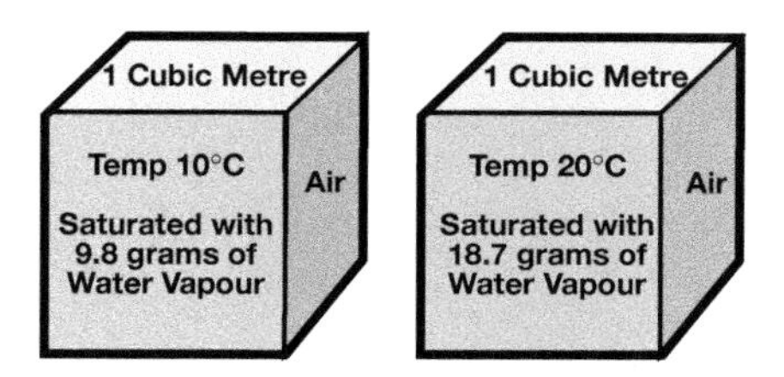

Fig 2.8 Relative humidity. The amount of water vapour that a parcel of air can hold decreases as its temperature is reduced.

If a parcel of air is still rising when it reaches its dew point, a cloud, known as a CUMULUS CLOUD, will form. These clouds are often called 'fair weather clouds' and are recognisable by their 'cauliflower-like' appearance.

The height at which these clouds start to form is known as the CONDENSATION LEVEL. Cumulus clouds mark the position where there is, or perhaps has been, a thermal.

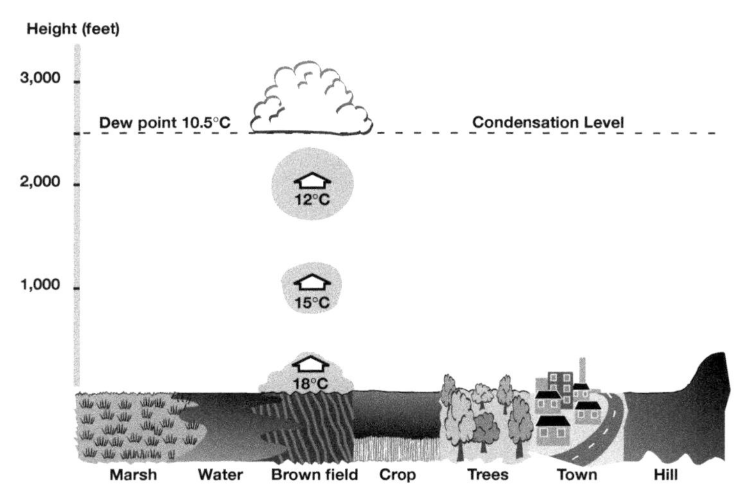

Fig 2.9 Formation of cumulus cloud. When a thermal reaches the condensation level, a cumulus cloud will form.

The likely height of the condensation level, and therefore the base of any cumulus clouds, can be determined by comparing the forecast maximum air temperature at the surface and the dew point temperature. The difference between the two is known as the DEW POINT DEPRESSION. For each degree Celsius difference between these two temperatures, the condensation level rises approximately 400 feet. Therefore, if the maximum temperature forecast is 20°C, and the dew point temperature remains at 10°C (a difference of 10°C), then the base of any cumulus can be expected to reach 4,000 feet. (The maximum temperature expected is normally given on television and radio weather forecasts, while the actual temperature and the dew point temperature are given in airfield reports on VOLMET and ATIS transmissions. Unless some general change in the weather is likely, the dew point will not change by a significant amount during the day.)

The condensing out of the water vapour into cloud releases heat (known as the LATENT HEAT OF CONDENSATION) into our parcel of air, giving it an added boost. This means that the air within the cloud will be capable of ascending faster, and possibly further, than it would have done if it had been too dry for cloud to form.

From the point at which cloud starts to form, the rising air cools at the SATURATED ADIABATIC LAPSE RATE (SALR) which is only around 1.5°C per 1,000 feet.

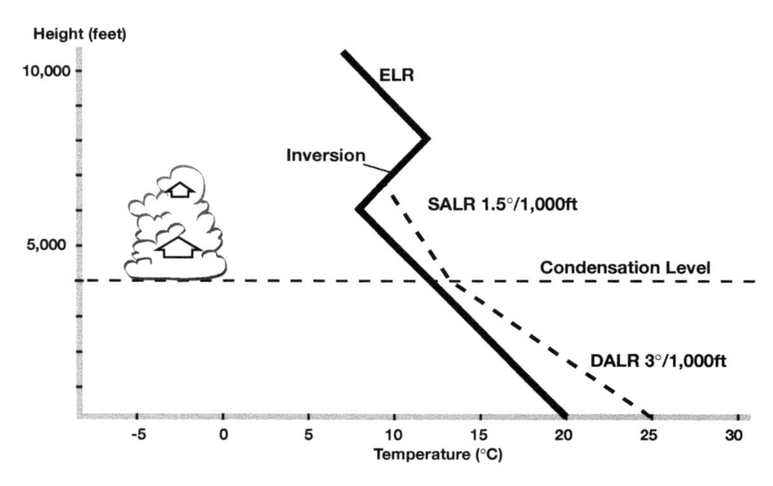

Fig 2.10 Saturated adiabatic lapse rate. When cloud forms, the air in the thermal cools at the SALR.

If the ELR is such that air continues to rise within the cloud, the small cumulus cloud can build into a huge, towering CUMULONIMBUS cloud which could result in heavy rain, hail, thunder and lightning.

Fig 2.11 Cumulonimbus cloud. If the air within the cumulus cloud continues to rise, then a large cloud called a cumulonimbus cloud may form.

Alternatively, if the growing cloud reaches an inversion, the cloud's vertical growth may be halted. If the thermal is still feeding the cloud, the moist air will continue to arrive at the inversion level and, not having enough energy to burst through the inversion, will spread out horizontally. The result may be that the spreading cumulus may join with other cumulus clouds that have suffered the same fate, creating clouds known as STRATOCUMULUS. On some days, this stratocumulus may become widespread and deep enough to stop much of the sun's energy from reaching the ground. If this happens, thermals may not be as strong (or may even stop developing completely), until some of the stratocumulus evaporates and disappears, thus allowing renewed solar heating of the ground. This cycle may continue for much of the day or as long as the offending inversion exists.

Days when such spreadout is likely are often heralded by cumulus starting much earlier than usual – raising false hopes about the soaring potential of the day. Another sign is the tendency of cumulus clouds to develop rapidly into narrow turrets of cloud instead of growing slowly. If small lenticular caps form at the top of growing cumulus clouds, this is a sign that the air at the level of the inversion is moist and that the formation of stratocumulus is a possibility.

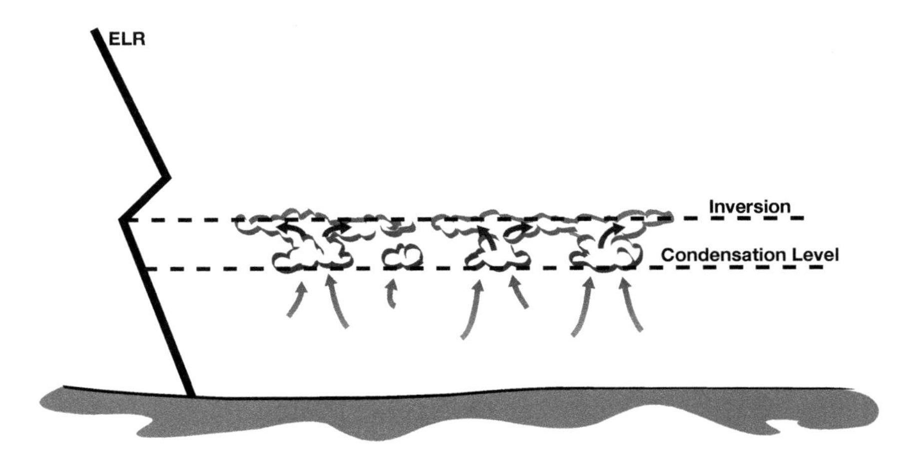

Fig 2.12 Stratocumulus cloud. Stratocumulus cloud may form when an inversion prevents the vertical growth of cumulus clouds.

To summarise, some of the basic rules of the atmosphere, which affect the glider pilot, are listed below.

1. Air that is warmer than its surroundings will want to rise.
2. Air that is cooler than its surroundings will descend.
3. As air rises it cools.
4. As air descends it is heated.
5. Air will stop rising when (or shortly after) it cools to the same temperature as the surrounding air.
6. If the air is moist enough, cloud will form when the air rises to its condensation level.
7. If the condensation level is below a strong inversion, a layer of stratocumulus cloud may form.

Blue thermals

The fact that on any given day cumulus clouds may not form does not mean that there are no thermals. It may be that the air is too dry for cloud to form. Remember that a cumulus cloud, when present, is a product of the thermal, and not vice versa!

If the air in the thermal does not contain enough water vapour, the thermal may reach the inversion before it cools to its dew point, therefore no cloud will form. On such days, the inversion can often be seen by a layer of haze. This haze layer is formed by dust, which has been carried up by thermals to the level of the inversion.

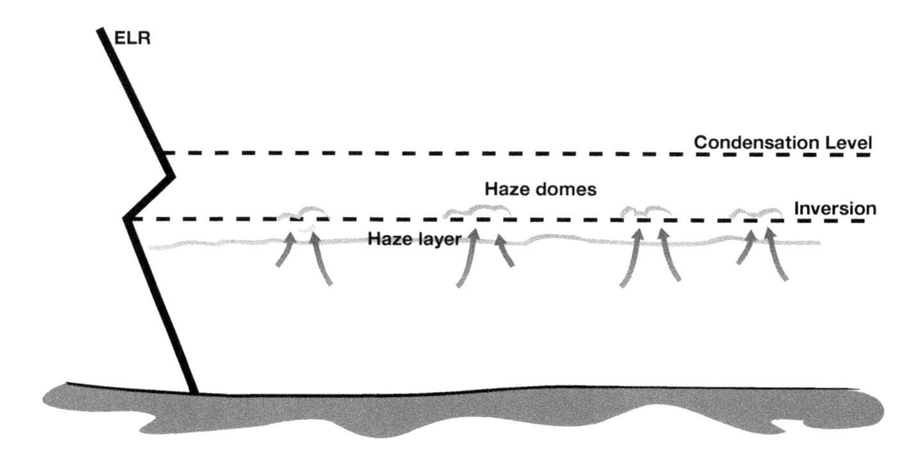

Fig 2.13 Blue thermals. If an inversion exists below the condensation level, thermals may stop rising before they cool enough to form cloud.

These DRY THERMALS are known as BLUE THERMALS and the days when they occur are called BLUE DAYS.

Often, a day which started off with cumulus cloud, will see a reduction of the amount and depth of such cloud and, as the day progresses, cumulus cloud may disappear completely. This is often the result of the surface temperature increasing during the day, to the extent that when a thermal breaks away from the ground it is warm enough to reach the inversion before it cools to its dew point. When this situation occurs no cloud will form.

A similar situation occurs when an area of HIGH PRESSURE (known as an ANTICYCLONE, or RIDGE) is approaching the region. Anticyclones tend to lower the inversion. If the inversion becomes lower than the condensation level, then thermals may not go high enough to cool to their dew point.

The opposite situation may occur on some days. Late in a day that has had only blue thermals, small cumulus clouds may start to form. This occasionally occurs on hot days and is a result of a gradual lifting of the inversion to a height above the condensation level. This lifting of the inversion is a result of thermals reaching and pushing into the inversion, causing mixing of the warmer air above the inversion with the air below the inversion. This increases the temperature of the air below the inversion, and eventually the height of the inversion itself.

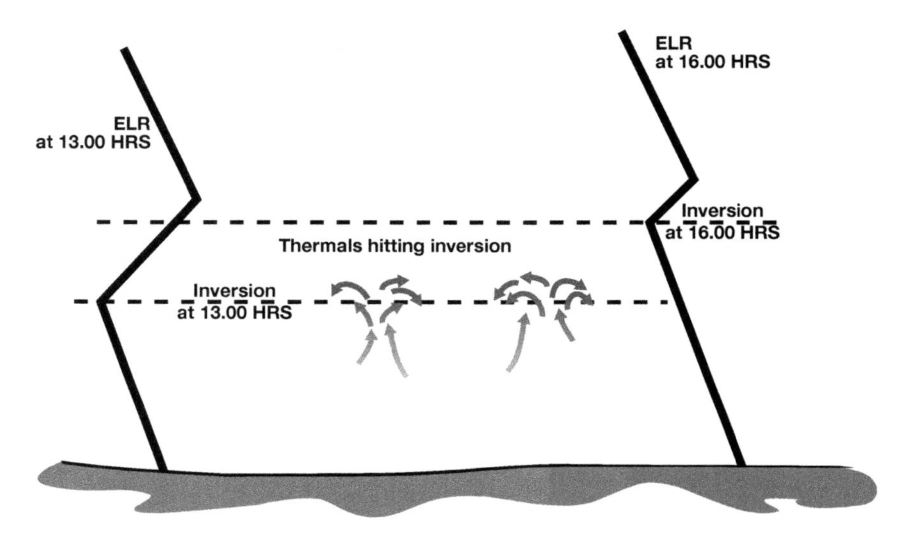

Fig 2.14 Inversion level rising. As the day progresses, the mixing of air as thermals push into the inversion may cause its level to rise.

Not all thermals require a large volume of air to be heated by a 'hot spot' on the ground. Any air that is warmed enough to become buoyant will rise. Cumulus clouds and cloud streets observed far out over a sea or ocean are an example of streams of rising air that develop above a surface of uniform temperature. Such rising air is unlikely to keep a glider airborne, except perhaps at cloud base. For this reason, the descriptions of 'thermals' in the remainder of this book refer to thermals of a size and strength likely to be usable by a glider pilot – what could be called 'soarable thermals'.

The source of a thermal

Normally, for a thermal to form, the air must be heated as a result of its contact with the ground. It therefore follows that the warmer the ground, the warmer and more buoyant the air will become. It is also necessary for the air to heat up differentially; that is, for neighbouring areas to heat up at different rates. Fortunately, the surface of the earth varies immensely and so every surface has its own characteristics as a thermal source, ranging from poor to excellent.

Poor thermal sources would include marshy areas, wetland, lakes or generally areas which have a large water content. This is because, as the sun's energy starts to heat such areas, much of the energy goes into evaporating moisture, or is conducted deeper into the ground or water, rather than warming up the surface. As a result, the surface may never become warm enough to heat the air above it to produce a reasonable thermal. Another problem of wet areas is that there will be a certain amount of reflection as opposed to absorption of the sun's energy.

Surfaces which tend to be good thermal sources therefore tend to be those which are dry, and are often of a colour which will absorb the sun's energy more readily. For instance, a bare, brown field will display totally different thermal producing qualities when it is dry from when it is waterlogged.

Another criterion that will affect the quality of a thermal is how well its source area retains the parcel of air next to the warm ground so that heating of the air can occur. The longer the air is stationary over the thermal source, the warmer it will become. For instance, in strong winds, a parcel of air may only be over an area of warm ground for a short time before it is blown past it. Some areas will trap air for considerable periods, allowing it to increase in temperature. Crop fields and areas protected from the wind by hills or buildings may allow the air to remain in contact with the surface just long enough to create a better than average thermal.

No matter how good the thermal source, it still needs the all-important solar heating to raise its temperature. Areas under cloud shadow or in the

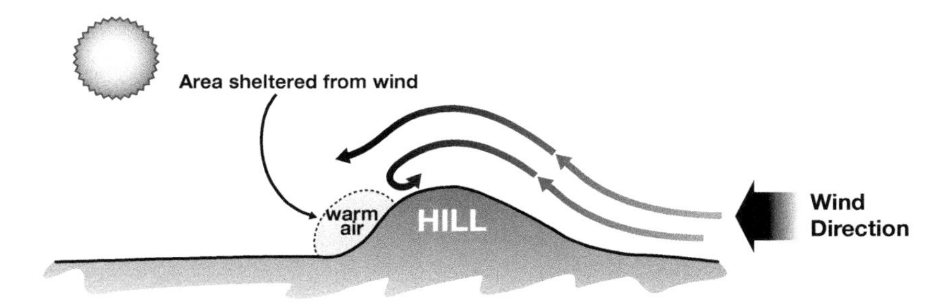

Fig 2.15 Wind shadow. Areas sheltered from the wind by a hill will allow air to remain in contact with the warm ground for longer, increasing the chances of a good thermal forming.

shadow of a hill or mountain will not warm up as much as land that is receiving direct sunlight.

On the other hand, if the surface is inclined in such a way as to receive the sun's rays at a more direct angle, then the surface will be a much more efficient thermal source. Hillsides and mountain slopes facing the sun will provide such sources.

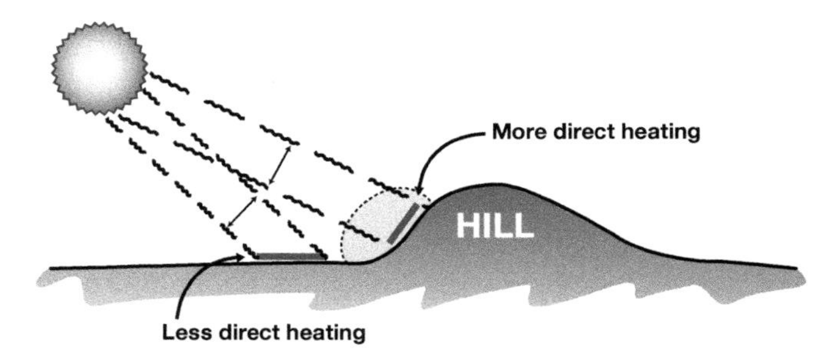

Fig 2.16 Sun angle. The angle of the sun's rays. The more directly the sun's rays strike the ground, the greater the surface heating.

Triggers

Once our parcel of air is heated by the thermal source, it will not necessarily break away from the ground immediately. It needs something to dislodge it. The ascent may be caused by the parcel of air becoming so warm that its buoyancy is too great to allow it to remain close to the ground. Alternatively, some external influence (known as a TRIGGER) may cause the beginning of a thermal.

Many of these triggers are man-made. Anything that disturbs the air can kick off a thermal. Cars, trains, launching cables and tow-planes, even combine harvesters and tractors, can all stir up the air enough to start a thermal on its way.

It might be hard to believe that something as small as a car or tractor may cause a large enough disturbance to trigger a thermal. However, remember that the trigger is not forming the thermal, only stirring it into action. The trigger only needs to start a small amount of air moving, the result of which is movement of surrounding air, which may be enough to disturb our incipient thermal.

Moisture thermals

Water vapour (containing a large proportion of hydrogen) is less dense than dry air. It is therefore more buoyant. It follows that a parcel of moist air surrounded by drier air will want to rise. Whether or not this would create a usable thermal is doubtful, but the effect on a thermal of having a large moisture content will be to enhance the temperature differential between the thermal and the surrounding air. In theory, this should make for a better thermal.

The formation and structure of a thermal

To talk about a typical thermal is difficult, as any meteorological phenomenon is blessed with the ability to be infinitely variable in its structure, shape and size. The best that can be achieved here is to describe an idealised thermal, which will give a general picture of what a thermal would look like, if only the air of which it is composed were visible.

The beginning of our ideal thermal can be regarded as a volume of air next to a part of the ground which is hotter than the neighbouring surfaces.

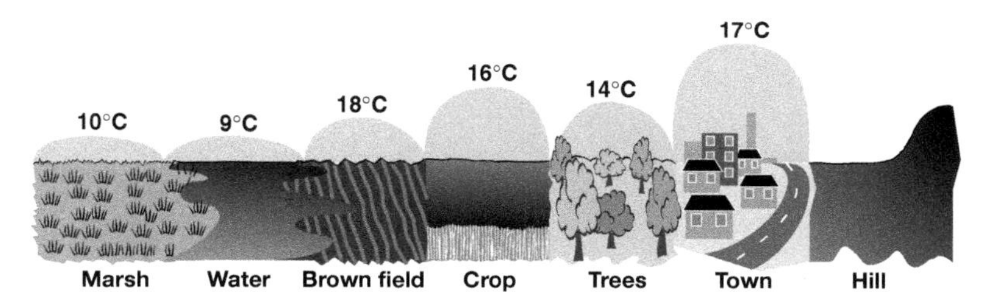

Fig 2.17 Thermal sources. A thermal starts as a volume of air being heated by an adjacent 'hot spot' on the ground.

As the surface heats this local air mass, the air will become increasingly buoyant until, due to either some independent disturbance (a trigger) or its own degree of buoyancy, it overcomes its tendency to adhere to the ground and breaks away.

As this thermal 'bubble' rises, it will be ascending to levels where the atmospheric pressure upon it is less. Therefore, it will expand as it rises. It will also encounter some drag from the air through which it is passing. This drag will affect the outer areas of the thermal. There will also be some degree of mixing with, and entrainment of, the air through which it is passing. The result of these latter events is that the outer air of

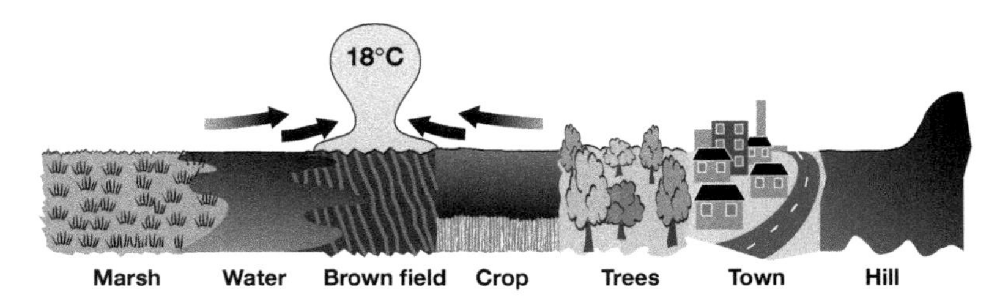

Fig 2.18 Birth of a thermal. Eventually the air will break away from the ground and rise.

the thermal bubble will be cooling and descending relative to the centre of the thermal bubble (the CORE). The accepted shape of a thermal is therefore one of a three-dimensional vortex ring, with air rising up the centre and spilling downwards on the periphery.

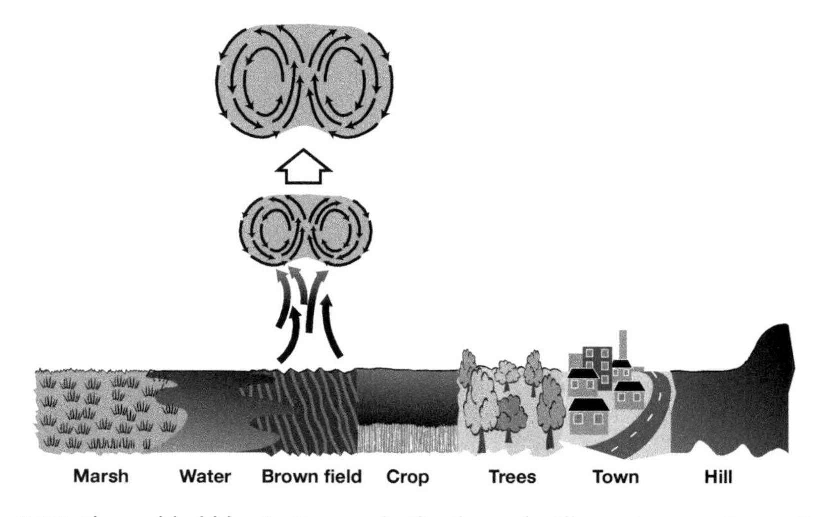

Fig 2.19 Thermal bubble. As it ascends, the thermal will acquire a vortex motion, with its core rising faster than its outer edges.

This entrainment and mixing with the outside air not only leads to the thermal expanding but also adds to its cooling. This cooling will reduce the thermal's buoyancy and may eventually stop its ascent. As a result, smaller thermals, which will suffer more from this cooling effect, may not last as long as their bigger brothers.

This vortex structure will assume some horizontal movement of the air at the top and bottom of the thermal bubble, with an inflow at the bottom and an outflow at the top.

Two important points should be realised from this model of a thermal.

1. Despite the fact that the air on the outside is descending relative to the core, the whole bubble is rising relative to the ground.
2. From the above, it can be visualised that in the core of a thermal, the air will be ascending faster than the thermal bubble as a whole.

Whether all thermals take the form of a bubble being released in this way, or whether some are continuous streams of rising air forming a column, has been a point of debate for many years. Most thermals probably begin as a column of lift that is still being fed from its ground source, but as the reservoir of warm air at the source is exhausted, the thermal breaks away and forms a rising bubble.

If conditions are favourable, a good thermal source will fire off regular thermals, and if these are frequent enough, then the result may well give the impression of one continuous column of rising air, as each bubble released follows behind its predecessor. The spiralling nature of a dust devil may suggest that some thermals at least form the shape of a column rather than a bubble of rising air. Straw stubble and heath fires supply a continuous source of heat while they are alight, and such a thermal source is likely to form a continuous column of rising air through which will rise pulses of faster rising air.

However, all too often, it is possible to find oneself searching unsuccessfully for lift under other climbing gliders. This would suggest that the thermal bubble has ascended past your level. In other words, 'you have missed the bubble'. Once the warm air close to the ground has ascended, it will take time for the warm ground to heat the cooler air that has replaced it. How long this takes, will depend on the prevailing conditions. If you are lucky, the next thermal bubble may be on its way up. If you are unlucky, you may watch from the field below, while your colleagues climb in this later thermal bubble. Who knows, your landing may even have disturbed the air enough to trigger this new thermal!

Another fact which supports this bubble theory is that gliders contacting a new (lower) bubble often find that they can catch up with gliders which are hanging around in the weak lift near the top of the thermal. This is often observed in the club environment where inexperienced pilots often jealously guard their hard-won height.

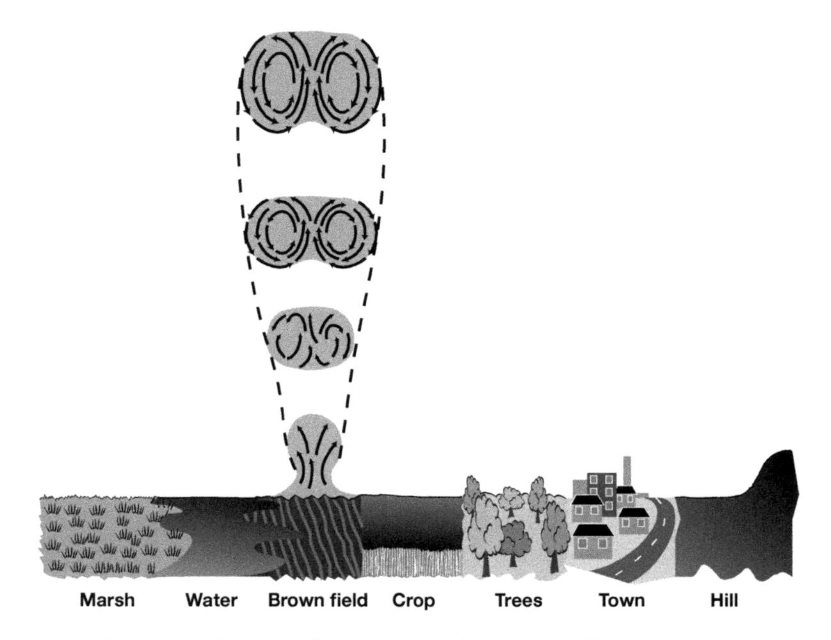

Fig 2.20 Thermal column. Often a thermal source will set off regular thermal bubbles, giving the impression of a continuous column of lift.

The dimensions and rate of ascent of any one thermal are as hard to pin down as an average shape. Figures of around 1,000 feet (300 metres) in width have been suggested for 'typical' British thermals, with glider pilots achieving average rates of climb of 4 knots regarding themselves as having found a good thermal. On average, the height which a British thermal might reach before forming cloud would be around 4,000 feet, with an exceptional day being 6,000 feet or higher. (The greatest reported height reached by a thermalling glider in the UK was 11,000 feet during the legendary heat wave of 1976!)

These figures would be regarded as poor by glider pilots from hotter countries such as Australia, South Africa and some parts of the USA. In such countries, climb rates of over 10 knots and much wider thermals rising to well in excess of 10,000 feet may be considered the norm.

These descriptions of thermal shapes are an over-simplification to help visualise these normally invisible entities. Thermals may have two or more cores. Several thermals may feed the same cloud. These points are often manifested when two gliders, which are initially thermalling in neighbouring thermals, find themselves in overlapping circles or even in the same core. (One wonders whether thermals are like raindrops running down the glass of a window, in that they occasionally merge and increase their speed.)

Thermal streets

In certain conditions, the wind and the circulation caused by rising thermals will result in a phenomenon known as THERMAL STREETING. This is most obvious in moderate to strong winds when cumulus clouds form long lines in the sky, called CLOUD STREETS. Under such streets can be found long lines of lift while under the clear areas between them, you can expect to find large amounts of descending air. Figure 2.21 shows the circulation which causes such conditions.

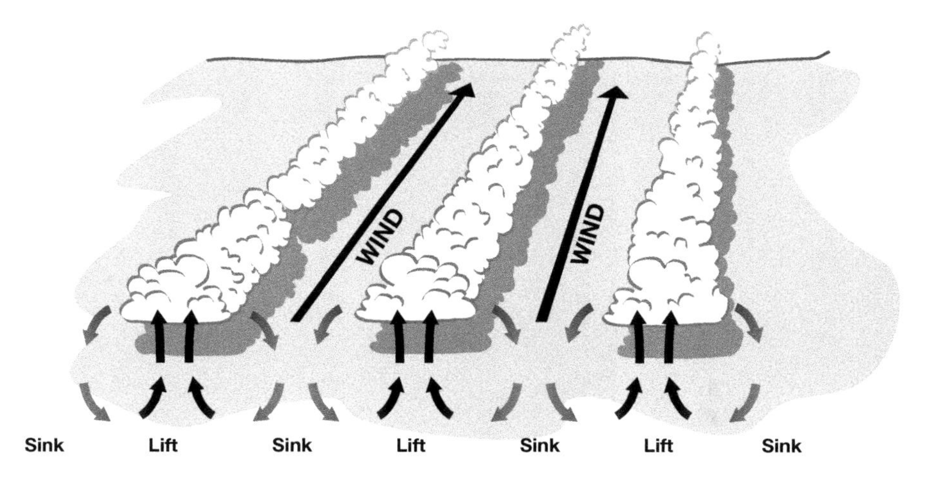

Fig 2.21 Cloud streets. In windy conditions cloud streets will often form, indicating the presence of long lines of lift.

These lines of cumulus will be aligned more or less parallel to the wind direction at cloud base. The distance between the streets has been found to be around three times the height of the cloud tops. Such streeting is most likely to occur when the wind speed is increasing with height between the surface and cloud base.

Occasionally, a single 'cloud street' will form which differs from 'true' cloud streets in that it is not a result of the general circulation just mentioned. Such a line of cumulus may be caused by a thermal source which persistently fires off thermals more often than the surrounding land, or perhaps by two differing air masses converging and causing thermals to be triggered along a line.

It should be noted that thermal streeting is not restricted to days when cumulus clouds are present to show the streets. Blue thermal streets are also common, and the lack of cloud in these situations can add to the challenge of thermal finding.

Chapter 3

Thermal Soaring

It is out there somewhere but it is invisible. It is not solid; you cannot touch it but you can feel it. It changes shape, changes speed and sometimes it changes direction. It has character and, more often than not, it is antisocial. *Above all, it has energy which you need to tap.*

No, this is not a description from a Star Trek movie. It is the description of a thermal. 'Character? Antisocial?' All thermals have characteristics and many, once you enter them, will do their utmost to get rid of you by tipping your wing or apparently disappearing temporarily or changing position.

Finding a thermal

Finding a thermal is the first task that you will face when you embark on a thermal soaring flight. How difficult or easy this may be will depend very much on the local weather conditions on the day. It will also depend on how you read the signs which give clues to a thermal's presence, and on the way in which you search for a thermal.

Ideally, you will launch into a thermal by aerotow or even from a wire launch. Therefore, the time to start looking for a thermal is while awaiting launch. Unfortunately, the average club environment does not always allow you to select the exact moment of launch. Often your glider will be in a queue and you will be forced to launch as soon as there is a tow-plane or a launch cable available. Alternatively, whichever of these you have chosen to use may not be forthcoming at the time when you consider it would be ideal to launch.

Despite these setbacks, your spare time awaiting a launch should still be spent thinking about the position of possible thermals. Watch the clouds. Which ones are forming? Which ones are collapsing? Observe how well or how badly other gliders are doing. The knowledge you gain from these observations will help you decide where to search after you release the winch cable, or where to tell the tow-plane pilot to take you.

Do not waste the time immediately before launch. Doing so may cost you another launch fee or even that badge flight you were after.

If you launch by aerotow, there is every chance that you will be towed through some thermals. Learn to read the variometer during the aerotow, and get used to the rate of climb it will display during most of the climb in still air. Only by having some idea of what is the normal rate of climb will you be able to judge whether you are being towed through a worthwhile thermal. At the higher airspeeds of an aerotow, even weak thermals can seem quite turbulent and, as a result, can be mistaken for good lift. This often causes pilots to release from aerotow early, only to find that the 'thermal' they thought was a good one was too weak to use. This also is often the reason for a pilot landing back on the airfield within minutes of launch.

If you do think you are going through a good thermal while on aerotow, do not release the towrope. Mark the spot and when you reach your intended launch height, go back to the place where the thermal was. To mark the spot, first glance upwards to note which cloud you were passing under, if any. Secondly, or if there are no clouds, fix the spot by reference to ground features. After releasing the towrope, fly directly to that position (unless of course you have found something better). The chances are, if the thermal you sensed is worth using, that it will still be there; and, as you will probably enter it higher up, it will be better formed and easier to use than it would have been lower down. These tactics are especially important in countries such as the UK where thermals are often weak. In countries where thermals are strong and wide, there is less likelihood of misjudging whether or not the glider is in reasonable lift during an aerotow.

If you are launched by winch or car, there is still a chance that you will launch into or through a thermal. Depending on the thermal strength, this may be indicated by a surge or an airspeed increase. Often, a strong thermal will break the weak link of the cable assembly or even the cable itself. If this occurs well up the launch, you may not have to plan a premature circuit and landing. Instead, you may be able to soar away from the tangle of cable which your colleagues on the ground will have to sort out.

Skyreading

If cumulus clouds are present, you will probably find a thermal beneath or close below such a cloud. You might wrongly assume that directly under every cumulus cloud will be a thermal. Remember, however, that the cloud is the end product of a thermal and that the thermal which produced the cloud may have stopped feeding the cloud some time ago. If this is the case, the cloud may well be decaying, and under such a cloud the air may be descending rather than rising. A cumulus cloud therefore has various stages in its life depending on the degree of activity of the thermal or thermals supplying it; there may be more than one thermal feeding a cloud.

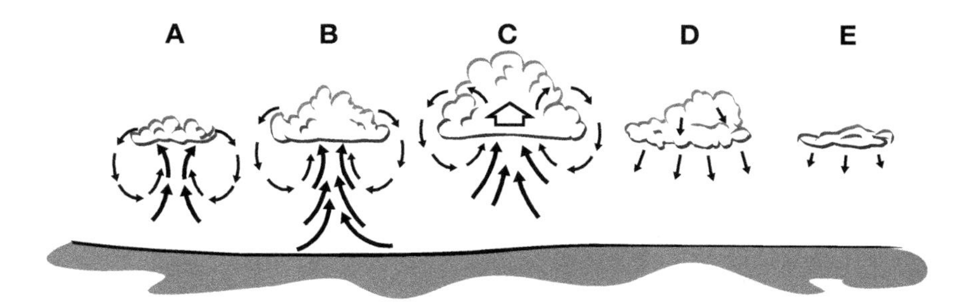

Fig 3.1 Life of a cumulus cloud. A cumulus cloud starts as a thin wisp, and builds into a well-defined cloud before collapsing and dissipating.

At first, as the thermal reaches the condensation level, the cumulus will appear as a small wisp of cloud (figure 3.1A). As more moisture condenses out, the cloud will build into a more recognisable cumulus cloud, with a firm base and well-defined sides (figures 3.1B and C). If the thermal feeding the cloud ceases, then the cloud will start to decay. It will lose its well-defined appearance and start to break up and subside (figures 3.1D and E).

Next time you fly in a commercial aircraft on a day with small, well-spaced cumulus clouds (the sort of day when you wish you were at the gliding club), look down on the clouds. From above, those which are building will be obvious. They will appear to have a 'bulge' (indicating a growing cell) on their tops, whereas decaying clouds will appear flat. The next thing to observe is how many of the clouds are apparently still growing. You will probably find that it is only about one in four at best. Identifying these active cumulus clouds from below cloud base is a much more difficult task.

Short-lived cumulus

The first stage of a cumulus, the formation of thin wisps of cloud, can often be very difficult to see. A good pair of sunglasses is a must, to enhance the visibility of the cloud. Finding the correct type of sunglasses is difficult, as only use in soaring conditions will prove your choice was correct. Like so much of soaring, preferences differ from pilot to pilot. Polarised sunglasses seem to be ideal. Some sunglasses darken the sky too much, giving the impression that the day is gloomier than it actually is, especially under a temporary overcast. Others, such as the 'orange' lens types, while showing up clouds well, seem to over-brighten the sky and also give a colour hue which may be deceiving when assessing types of crops in fields. Whichever sunglasses you eventually choose, you would be wise to make sure that they meet all of the nationally set standards for protection against ultraviolet rays.

On some days, identifying this early stage of a cloud's existence can be critical. Unfortunately, it can be difficult to distinguish from the wisps left after a cloud has decayed. The only way to reduce the chances of charging into an area of sink left by a decaying cloud is to monitor the sky constantly, watching for wisps forming where no cloud has been and noting which clouds are collapsing and their position.

The day appeared good from the start with small, well-separated cumulus clouds at around 2,200 feet. I headed off cross-country shortly after launch. After a short climb in a thermal under one of these small cumulus, I headed towards another cloud. As I approached it, it evaporated. The next cumulus looked too far away to reach safely. On turning to retreat back to the cloud I had just left, it too was gone. Fortunately, another cumulus cloud started to form just within reach and I used its thermal to save me from an early outlanding.

Having only just survived this first trial, I had realised that to stay airborne during the early part of this day, I could not afford to climb to the top of every thermal for as long as the lift was good, but instead would have to leave for the next thermal the moment I saw its cloud begin to form.

One learns a lot about skyreading in conditions such as these!

On days when the air is relatively dry, small cumulus clouds may form and last for a very short period. In these conditions, you may find that any delay in heading for a newly formed cloud results in it dissipating before your eyes while you are still trying to reach it. On your arrival where the cloud had been, you will probably find sink. This situation often occurs early in the morning of good soaring days when cloud base is still low, which makes your decisions more critical as the available height to search for another thermal is limited.

Small and moderate-sized cumulus

Once thermals start to form recognisable cumulus clouds, the problem of identifying a growing cumulus (and therefore one that is being fed by a thermal) moves on a stage. When individual cumulus clouds are small, they will tend to be shorter lived than larger clouds. This is because smaller cumulus clouds suffer more from entrainment and the cooling effects of the surrounding air. This means that their shape will change faster, and therefore more visibly, than larger cumulus clouds.

When looking at a cloud from a distance, look at its structure. If it is ragged around the edges, if it has a base which is poorly defined or is narrow compared with the upper part of the cloud, or if the cloud is showing signs of splitting, then avoid it like the plague. It is likely to be decaying; that is, not being fed by a thermal. As the cloud evaporates, the air will cool, which will cause it to descend. Underneath it you are likely to find sink.

If, on the other hand, a cloud has firm sides and a good firm base which is either flat or concave, it is probably still being fed by one or more thermals. It may even be possible to see the cloud growing as you watch it. There may be tendrils of cloud which visibly change shape, hanging down from the base of the cloud. These wisps of cloud show an area where moisture is condensing out of a thermal.

Another indication of a cloud's state is given by its shadow on the ground. If this too is firm and distinct with no gaps where sunlight is shining through holes or clefts in the cloud, this, combined with the above signs, will indicate a healthy cloud and an active thermal. (Cloud shadows are also useful in judging the distance to a cloud. They can also be used to give an idea of where other cumulus clouds are when the sky ahead is obscured by a nearby cloud, as is the case when you are close to cloud base.)

Occasionally even a cloud which is decaying will be fed by another thermal and find a new lease of life. This will manifest itself by a new cell building through the old decaying cloud. This may be enough to make it worth looking for a thermal under that cloud, but you may have to run the gauntlet of increased areas of sink from the decaying parts of the cloud before finding the lift.

Small cumulus, providing they are being fed by a thermal, do not present much of a problem when looking for their lift because they have a relatively small area of base under which to search. If the cloud is of moderate size, and the base is wide enough that a decision about where to look for the lift has to be made, then look for any wisps of cloud condensing out at the base, or failing this, an area where the base looks firmer or darker. If none of these signs are seen, then search for lift under the centre of the cloud. If this fails, search for the thermal on the upwind side of the cloud.

On some occasions cumulus clouds can be seen with bases which are at different levels. In such instances, the best lift is usually found under the part of the cloud which has the higher base.

Often, especially when the glider is at a much lower level than the cloud, the thermal will not be found directly under the cloud. Instead it may be displaced to one side. As the thermal ascends, it will be passing through different air masses, which will be subject to differing wind strengths and directions. It will also be subject to wind shears caused by neighbouring vertical and horizontal air movements. All of this may have a tilting effect on the thermal's path. As a result, at lower levels the thermal may be found some distance upwind from the cumulus cloud it is

Fig 3.2 Sloping thermal ascent. If any wind is present a thermal may follow a sloping path.

feeding. If the wind is strong, the thermal source may be a considerable distance upwind.

Once the relationship of the thermals to their clouds is established on a particular day, it will probably remain the same for several hours, or maybe even the whole of the day, unless wind or soaring conditions change. As a general rule, if lift cannot be found directly beneath a cloud, try searching upwind of the cloud. The stronger the wind and the lower the glider, the further upwind you should search for the thermal. You will also increase your chances of finding the thermal faster if you plan your flight path so that you approach the area where you expect to find the thermal by flying directly into wind or downwind.

Soaring in a northerly wind, it was possible to climb in good lift under cumulus clouds over the sea, some five miles or so from England's south coast. What had begun as a ridge-running flight along the South Downs had become a flight of fascination as our gaggle of four gliders enjoyed the unusual view of sea below and the beaches of Brighton and Worthing to the north.

Large cumulus

As cumulus clouds grow larger, they may develop several cells. Their base will be larger, as will be their vertical depth. Occasionally, a cumulus top will power up well above the height of its companions, either as an individual cloud or as a cell within a cloud. Here is a thermal with ambition. It has more energy than the other thermals which formed the clouds with tops at a lower level. If you happen to be in the right place at the right time, you may well have a good rate of climb in such a thermal. However, it is unhealthy for a cloud to be up on its own like this. Unless it has a continuous supply from a thermal source (such as a power station), the chances are that as soon as the thermal feeding it ceases, the cloud, which is now stuck up into drier air, will quickly evaporate. If this happens, then energy will be taken out of the cloud (the latent heat of evaporation) and the cloud will quickly cool, causing the air to descend rapidly. Beneath this type of cloud is certainly not the place to be when this happens.

As a cloud builds, any increase of wind with height will tend to make it

Fig 3.3 'Rocket cloud'. A narrow cloud which rises well above the normal cloud tops may have a dynamic but short life.

lean towards the downwind side. In windy conditions, new thermals may feed the upwind side of the cloud, while old decaying cells of the cloud subside on the downwind side. Subsequently, the place to look for a thermal is near the windward side of the cloud, as strong sink will probably be found under the collapsing side.

Fig 3.4 Toppling cumulus. Stronger winds aloft may cause clouds to topple downwind as they decay. Any lift will be towards the upwind side.

Once a cumulus starts to develop multiple cells and grows both horizontally and vertically, the problems of finding the thermal under the cloud are somewhat more difficult. Larger cumulus clouds may have several thermals feeding them; this means not only that there may be more places where lift may be found, but also that there may be more areas of sink.

Searching for lift may therefore require covering more sky (and flying through more sink) than would be necessary under a smaller cloud. In order to increase the chances of finding the best lift more quickly, and without the need to wander around in sink, it is necessary to approach the cloud with a mental picture of the cloud as seen from afar. The fact that you have selected this particular cloud probably meant that you 'liked the look of it'. Ask yourself why? Was it its shape? Was it the firmness or the darkness of its base? Perhaps it was the fact that there appeared to be a particular cell powering up at the top of the cloud.

Identify the feature which made you pick this cloud – whatever it was – and note its position relative to the centre or an edge of the cloud. Do this early in your flight towards the cloud. The reason for doing this early is that, as you get closer to the cloud, it will no longer be in front of you but increasingly above you. Hence, you will not be able to see the feature of the cloud which previously indicated the position of the suspected thermal.

From the moment you have selected the feature, fly towards it; and as the perspective of the cloud changes, try to keep heading for the same area under the cloud.

Once under it, the cloud will appear relatively featureless. However, there will be darker areas where moisture is condensing out from thermals. There may also be wisps of cloud – again where moisture is condensing out. The cloud may be concave in places, giving a higher base where the air is warmer. If the base has two levels, lift is more likely to be found just in front of the step up to the higher level, below the higher part of the cloud base. Hopefully your initial attack on the cloud will have put you in lift. If not, moving towards one of these signs should find you a thermal.

Cumulonimbus cloud

Once a cumulus cloud builds into a cumulonimbus cloud, we enter a totally different ball game. Cumulonimbus clouds vary in size from the tropical giants to the less large cumulonimbus clouds more common in temperate latitudes. However, in the summer, even countries such as

Britain can find large cumulonimbus clouds forming when warm, humid air arrives from the south. The air in the cumulonimbus clouds commonly seen in temperate regions tends to be colder, and therefore is not capable of holding as much moisture as the warmer air found in tropical cumulonimbus clouds. As it is the condensing out of water vapour into cloud, and the consequent release of latent heat, which gives the cloud its energy, temperate latitude cumulonimbus clouds tend not to be as large as their tropical cousins. However, in terms of size, both types of cumulonimbus are in a different league from even large cumulus clouds, with cloud tops of 25,000 feet being typical for the 'smaller' cumulonimbus clouds and nearer 50,000 feet for tropical cumulonimbus.

Cumulonimbus clouds may also cover a large area horizontally, with a base several miles across and an anvil at the top of the cloud stretching for many miles downwind. The cloud may comprise several small cells, each breeding from older decaying cells; or it may consist of one gigantic cell, known as a super-cell, which self-generates with its own complicated circulation of air. The common denominator is rain. Each cloud will produce copious amounts. According to theory, when the rain starts so does the sink – and in the case of the smaller type of cumulonimbus, a cloud from which rain is falling is a decaying cloud. (This may be an over-simplification, as during the British National Championships in 1981, which was plagued by cumulonimbus clouds, at least one leading pilot announced that his new LS4 seemed to climb very well in the heaviest of rain. Fact – or fiction to out-psych the other competitors? In the event, the answer to that question did not matter – after ten days we had not managed a single contest day in the Standard Class, mainly thanks to cumulonimbus clouds.)

In normal situations, most glider pilots would choose to avoid these ferocious giants. Cumulonimbus clouds, although thermic powerhouses and therefore containing plenty of lift both in and below the cloud, tend to cause such disruptive effects that for the purpose of ordinary thermal soaring they are bad news. Yes, there is a huge amount of lift below the cloud. Unfortunately, there are also huge downdraughts which cover large areas, often suppressing normal thermic activity for miles around the storm. These areas of rapidly sinking air can cause gust fronts to radiate out from under the cloud. Heavy rain or hail may be encountered under or near the cloud, reducing visibility.

Big cumulonimbus clouds, such as the type which develop in hot countries, can produce exceptionally powerful downdraughts known as 'downbursts', or 'microbursts', in which the air may be descending at

60 knots (6,000 feet per minute) or more. When this descending air hits the ground it spreads out horizontally, giving a gust which may be as strong as the downdraught itself. Microbursts have been cited as the cause of accidents to several large airliners which have been trying to either take off or land. The associated gust can be strong enough to blow down trees – just one good reason why gliders and tow-planes are better in the hangar when large cumulonimbus clouds are reported or observed.

Lightning may occur between cloud and ground, between clouds, or within the cloud itself. Cloud base is often stepped down, becoming very low on occasions.

Should you decide to dabble with a cumulonimbus, the following notes might help you do so with more success and safety.

1. Flying in the vicinity of a cumulonimbus should only be attempted by experienced pilots.
2. Stay well away from large cumulonimbus clouds, especially if thunder has been heard or lightning observed.
3. Lift is often found under the cloud towards its leading edge (downwind side) or just in front of the rain band, or in front of any step down of cloud base (under the higher part of the base). For a short period at least, you may be able to climb well with half of your circle in rain and half outside it. Although some pilots would rather keep the glider's wings dry, the rate of climb may be worth the performance loss incurred by temporarily wet wings.
4. Lift can be very strong, often sucking the glider up into the cloud, despite the airbrakes being open and airspeed being increased.
5. Sink can be equally strong and accompanied by rain or hail which could damage the glider.
6. Visibility is normally poor near cloud base and visual contact with the ground can easily be lost, requiring instrument flying techniques in what may be turbulent air.
7. Should a landing have to be made near a cumulonimbus, watch out for wind direction changes (possibly even reversal of the wind direction) and allow for large increases in the wind strength. The strongest gusts often come from the direction of the cumulonimbus.
8. Take care not to enter a cumulonimbus unintentionally, and only enter it deliberately if you are in practice at cloud flying and

legally allowed to do so. Your glider must be suitably instrumented and cleared for cloud flying.

9. Oxygen equipment is likely to be required.

NOTE: *In some countries, cloud flying in gliders is prohibited.*

> *In April 1999, near Dunstable, England, an ASK 21 was struck by lightning while flying at about 2,500 feet in the vicinity of a large cloud. The glider disintegrated. Both pilots parachuted to safety.*

Despite these warnings, dabbling with the lift under smaller cumulonimbus clouds can be fun and educational. Should such a cumulonimbus be seen approaching your airfield, then it is often worth taking a launch (not a wire launch – because of the lightning risk) to attempt to soar beneath or ahead of it, safe in the knowledge that you can scamper back to the airfield if you get caught in the sink or the rain.

Make sure you decide to retreat early, so that you do not risk landing in gusty conditions or have the airfield blotted out by rain or low cloud before you can reach it. If this happens, go for a field landing well away from the rain area. DO NOT try to dive back to the airfield in bad visibility or through rain. Such cloud and rain can result in zero visibility down to ground level. Remember also that your return to the airfield may be in strong and constant sink, and that modern gliders perform very badly with wet wings.

> *During the British Standard Class National Championships of 1985, about a dozen pilots reported receiving electric shocks from their control columns while taking high climbs in cumulonimbus clouds. All were awarded rubber gloves at the next morning's briefing.*

Stratocumulus

When cumulus cloud has spread out to form stratocumulus, it will prevent the sun reaching certain areas of the ground and, as a result, reduce the amount of heating these receive. Thermal sources under this cloud layer may no longer warm up enough to give off thermals.

When this occurs, there are likely to be only two places where thermals might be found. The first is near cloud base, where thermals which were triggered earlier may still be feeding the clouds. The second is at the edge of, or at any breaks in, the stratocumulus layer where the sun is still getting through to the ground below. The rules about looking for wisps of cloud at cloud base or darker areas of cloud base still apply. In any case, you can expect thermals to be much weaker than those found earlier when more sunlight reached the ground. The most important tactic, if you find yourself under such a sheet of stratocumulus, is to avoid getting low where there is less likelihood of contacting a thermal. With any luck, the stratocumulus will break and decent thermals will return before you have to land.

The importance of skyreading

Skyreading is a skill that takes time to acquire, but once acquired it is one of the biggest advances which any soaring pilot can make towards improving his/her ability. Becoming aware of what the clouds are indicating and recognising their potential or warning signs is an essential part of soaring. Fortunately, it is a skill which one can practise not only while airborne (although that is the ideal time to see if you were right about a cloud's promise of a thermal), but any time there are cumulus clouds in the sky. So next time you are sunbathing on the beach, a passenger in a car or train, or waiting for your syndicate partner's 'come and retrieve me' call, PRACTISE SKYREADING.

Ground reading

On blue days, when no cumulus clouds are present to aid your search for a thermal, or when you are too far vertically from cloud base for the clouds to assist in accurately pinpointing a thermal, you will have to use other clues to find lift.

In the absence of any other visual clues, identifying a likely thermal source and flying in the air above or near it may be your only hope of finding lift. Most gliding clubs have known, local features where pilots regularly find their first climb. When searching for a thermal near these features, remember to take into account any wind effect.

Thermal sources

Remember the two characteristics which make a surface a good thermal source. Firstly, it must heat up more than the neighbouring surfaces. Secondly, the longer it retains the air in contact with the surface, the more the air will warm up and the more buoyant it will become.

When low, try to visualise which area or surface within reasonable range might best fulfil these two requirements. Areas that may be sheltered from the wind, allowing pockets of air to be undisturbed while they warm, are usually easy to identify. Identifying surfaces which will heat up well can be somewhat more difficult.

Let us look at various surfaces and study their merits or otherwise as thermal sources.

Brown fields

Early in the British soaring season, the surface of many fields is either young, winter-planted crop or ploughed or harrowed brown earth. Even the fields of young crop will tend to show as much brown as green, as the crops will be no more than shoots. The surface heating of such a field will depend greatly on whether the field is dry or waterlogged. The latter is always a danger in February and March in the UK, reducing a field's potential as a thermal source. If the field is very dry, the dark colour of the soil will tend to absorb the sun's energy and it will have the potential to be a reasonably good thermal source. Whether or not the air will be stationary over the field will depend on whether there is any significant wind, or, if there is, on what sort of shelter the field gets from hedges, trees or adjacent hills. Later in the season, fields with crops such as potatoes and turnips still show much brown soil and have the same characteristics, although there may be more moisture content in the broad leaves of these crops to 'waste' some of the sun's energy.

Green crop fields

As cereal crops (wheat, barley and the like) grow, less soil is exposed to direct sunlight and therefore the surface heating reduces. Again, there will be more moisture content in the crop. However, what the source field loses in terms of solar heat absorption is probably compensated for by the retention of air among the stalks for a longer period.

Ripe fields

Crops such as wheat and barley, when ready for harvest, are very good thermal sources. The crops are long, retaining vast quantities of air close to the surface. The golden crop is relatively dry and therefore little of the sun's energy is used evaporating moisture. In fact, the temperature of the air among the stalks of a wheat field is measurably higher than that of the air outside the crop.

Dark soil fields

Darker surfaces absorb energy more than lighter-coloured surfaces because there is less wastage of the sun's energy due to reflection. Darker fields, such as areas where stubble (or heath) has been burnt, or simply darker soil types, often produce good thermals.

Grass fields

Whether or not a grass or pasture field will be a good thermal source will depend on how dry and how sheltered it is. The surrounding landscape should give a clue as to how dry the field is likely to be. If the field is on high ground then the chances are that it is fairly dry. If on the other hand it is near a meandering river or a lake, then expect it not to be such a good bet for a thermal. Pasture fields are often fields which are too small or too undulating to bother cultivating. This means that hedges, trees or slopes may be abundant and may protect the air from being disturbed while it is being heated.

Cut grass (silage or hay) fields

Newly cut grass is fairly wet and uses up a lot of the sun's energy evaporating moisture. Once the cut grass is lifted and the ground dried out, then the surface of the field may heat more quickly, thus warming the air above it. Dried hay will also warm well and can be considered a reasonable thermal source.

Pig farms

Fields where pigs are snuffling around tend to lack any form of vegetation. When the weather has been hot and dry, the bare soil warms up well and can be regarded as a reasonable place to find a thermal. (Whether or not any added heat from these hot, sweaty creatures contributes to the thermals often found when low above such fields is simply the author's conjecture – but he has yet to land among the pigs!)

Moorland

Moorland varies its qualities as a thermal producer, depending on whether it is marshy or dry. Even areas of high moorland can be soggy. Areas of high moorland which are dry or parched seem to give off good thermals. As moorland tends to be exposed, any areas which are sheltered by hollows or hillocks may produce better thermals.

Forests

Forested areas have mixed qualities as thermal sources. All forests have the ability to shelter large volumes of air, so this is not a problem. The type of forest will give some clue as to the type of soil. Deciduous trees prefer richer-quality land with more moisture available, whereas coniferous trees are happy with drier soils. Such drier soils may produce better thermals. Deciduous trees have much larger leaf areas than coniferous trees and give off more moisture. They therefore have a larger water content. This may mean that deciduous forests take longer to warm up than coniferous forests.

On the whole, forests generally do not rate highly as thermal sources, often only giving off thermals late in the day. These thermals may take the form of large areas of weak lift. (Much of this rising air may be due to a process called transpiration, during which trees release large amounts of water vapour.) However, most large forests have cleared land such as firebreaks and areas where trees have been felled for timber. These areas often consist of bare soil or scrub, which will heat up well.

Cities, towns and villages

Concrete areas, such as towns, have good warming characteristics and will often contribute by having their own heat sources, such as factories, car emissions and heating (and cooling) systems in buildings.

Airfields

Airfields can often be regarded as good thermal sources. How good depends on their situation, size and surface.

A large airfield, of the type built by the Ministry of Defence, will tend to be built on the highest ground in the immediate area. This will probably mean that it is well drained. At any rate, if it has concrete runways, these will be designed against flooding, allowing water to run off and away from the area. The concrete itself will provide an excellent surface, which will heat up and act as a good thermal source. These larger airfields will also have dispersals, buildings and hangars (or the remains of) which will also provide dry stonework which will warm up well. These buildings will provide 'dead' areas where the air can linger while it is being heated.

Grass airfields may also present a well-drained surface which, although not as good as concrete as a thermal source, may still give rise to good thermals. Unfortunately, many grass airfields tend to be on poor quality land, which is why it is not being used for farming. This may be because of the soil type but may also be due to the fact that the land suffers from waterlogging. If this is the case, then the airfield may not make a good thermal source in anything but the driest of weather. When looking at an airfield as a potential thermal source, study the lie of the land and in particular its proximity to rivers, streams and lakes.

Another problem associated with airfields when looking for a thermal is that of controlled airspace. Larger active airfields normally have some type of airspace protecting them. Know the rules and whether your presence is welcome at the airfield in question, and you will be less likely to cause an air traffic problem. Such knowledge will enable you to thermal away with impunity. You may even have found a convenient landing area if you do not find a thermal.

Motorways

As mentioned earlier, concrete surfaces heat up well. Motorways are no exception. Often the problem with motorways is that they are fairly exposed to wind and therefore the air is disturbed before larger thermals can form. However, there are areas where motorways are sheltered by woods or embankments. Providing the disturbance caused by traffic is not too great, these may allow the air to be in contact with the surface long enough to form a good thermal. Service areas provide air traps among the buildings and parked vehicles as well as supplying additional heat to the air from heating systems and vehicle engines.

Railway yards

Blots on the landscape such as large railway shunting yards and railway depots usually have a surface that consists of stones or concrete as well as metal rails and sleepers. There are often other minerals present, either for use in the yard or in transit, such as coal or sand. These surfaces will heat up better than surrounding farmland and will probably be dry. There will also be vehicles contributing to the local heating, as well as buildings or embankments, which will offer some shelter to the air while it warms up.

Power stations

Many power stations are recognised thermal sources, producing much steam from cooling towers and often smoke from the main chimney. Whether the lift associated with these stations comes from the towers and chimneys, or the complex in general, is open to debate. One thing is for certain: when these sources are producing smoke and thermals, you can rely on a smelly ascent in the choking smoke which often marks the lift.

Factories

Large factories usually provide large areas of brick or concrete that will heat up admirably and therefore be good thermal sources. Car plants and steel works, or any plants where huge amounts of heat are produced, will add to the area's potential as a thermal source. These industrial sites also consist of lots of buildings between which air can be trapped while its temperature increases.

Hills and mountains

Hills and mountains tend to provide many good thermal sources, providing they are not snow-covered. Slopes of scree and bare rock warm up well, especially if they are facing the sun. Such slopes tend to be well drained. Outcrops and high valleys harbour air from the wind while it is being warmed. Air that is warmed on lower slopes will tend to hug the slope as it ascends. This lengthened contact with the surface will add to the heating of the parcel of air. When the thermal eventually does break away from the slope, the temperature differential between it and the ambient air will be greater. In mountains this tends to make for stronger thermals that go higher.

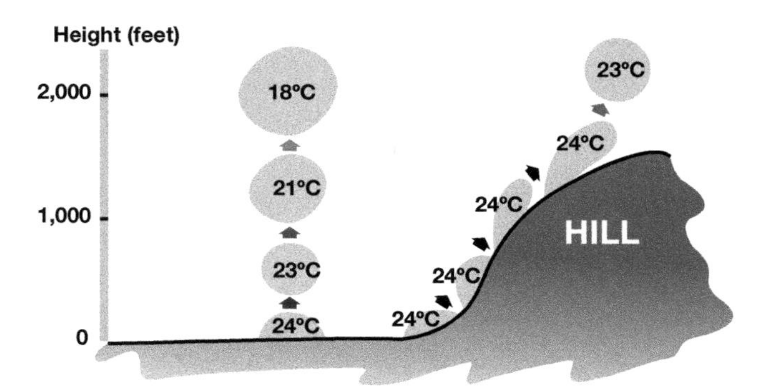

Fig 3.5 Slope thermals. If a thermal remains in contact with warm ground while it rises, it will benefit from surface heating during this part of its ascent.

Sandy areas
Sandy soils seem to have mixed blessings as thermal sources, with some areas acting as reasonable sources and others proving disappointing.

Lakes, wide river valleys and wetlands
These should be avoided as they generally prove to be poor thermal sources for the reasons given in the previous chapter; that is, the problems pertaining to their high water content.

Snowfields
Areas covered by snow reflect large amounts of the sun's energy rather than heating the surface and are therefore poor thermal sources.

When looking for a thermal by ground reading, study the land below and try to visualise the hot spots. Do not look at any one detail in isolation. Two fields may have the same crop but the one on the slope facing the sun will be a better bet. However, will that have more potential as a thermal source than the village along the ridge, which is sheltered by the hill from the wind and is still in sunshine? To make your choice easier, any areas obscured from the sun by clouds for any length of time, or lying in the shadow of a hill, can be ruled out. Any cloud shadows passing over or sitting upon an area will reduce or even cancel its thermal producing qualities.

Triggers

If you can identify a good thermal source which is accompanied by a suitable trigger feature, then your chances of finding a thermal will be increased considerably. Some of these triggers are visible from the air (although it would be difficult to prove that any one trigger was responsible for the birth of a particular thermal). Other triggers are invisible, and whether or not one of these activates a thermal in your search area is largely a matter of luck.

Meteorological triggers

Heat
If the temperature of the air in contact with the thermal source becomes high enough, the parcel of air may become so buoyant that it breaks away from the surface without any external influence.

Wind

The wind may disturb the layers of air close to the ground sufficiently to cause a thermal to break away. However, on windy days, the air may not remain over a source area long enough to develop into a reasonable thermal. Hence, on such days thermals may be broken and hard to use. This will add to the problems caused by the wind in general.

On very windy, unstable days, turbulence can cause mixing of the lower layers of the air. In this instance, thermals may be the result of warmer air being displaced upwards into the cooler air above. Once elevated into this situation, a parcel of air will have overcome its initial reluctance to ascend and may form a narrow, turbulent thermal. This compounds the problem of trying to identify thermal sources in windy conditions – and indeed such thermals may never have had a fixed source.

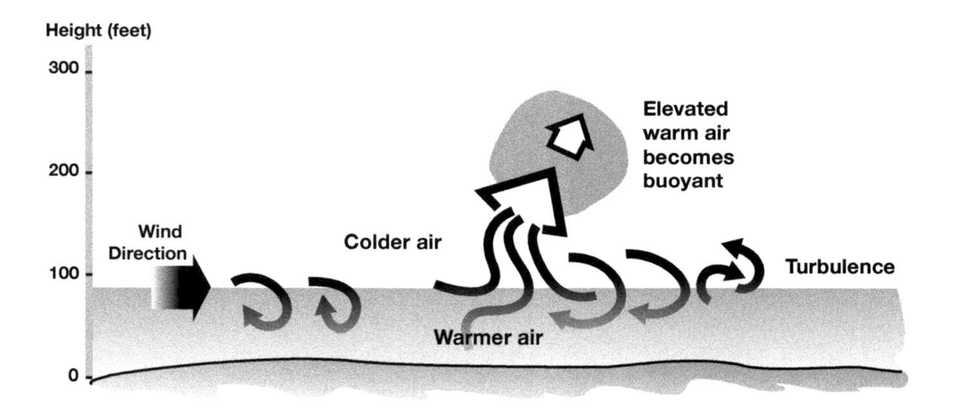

Fig 3.6 Turbulence as a trigger. In windy conditions, mixing of air at low levels may cause the initial triggering of a thermal.

Other thermals

How many times have you sat in the garden on a calm, hot day only to find a sudden breeze has started blowing papers around? More than likely a thermal has just kicked off somewhere nearby. As it went, it caused air to flow horizontally across the surface, taking your papers with it. It may only have been a very small thermal but the air movements that it caused may be enough to trigger off a larger, more useful thermal close by. Such gusts are quite common on thermic days, and once thermals become plentiful, the descending air from thermic activity may descend to the surface, where it spreads out fast enough to trigger other thermals.

Downdraughts from clouds

As clouds decay the sinking air may act in a similar way to the downdraughts described above. With cumulonimbus clouds, a gust front may be set up ahead of the cloud as the huge downdraughts reach the ground and are deflected forward away from the cloud. It is not uncommon to see a line of smaller, yet sizeable cumulus clouds sprouting ahead of the main storm cell. These will occur where the gust front (consisting of colder air) undercuts the warmer air ahead of it, thus kicking off thermals.

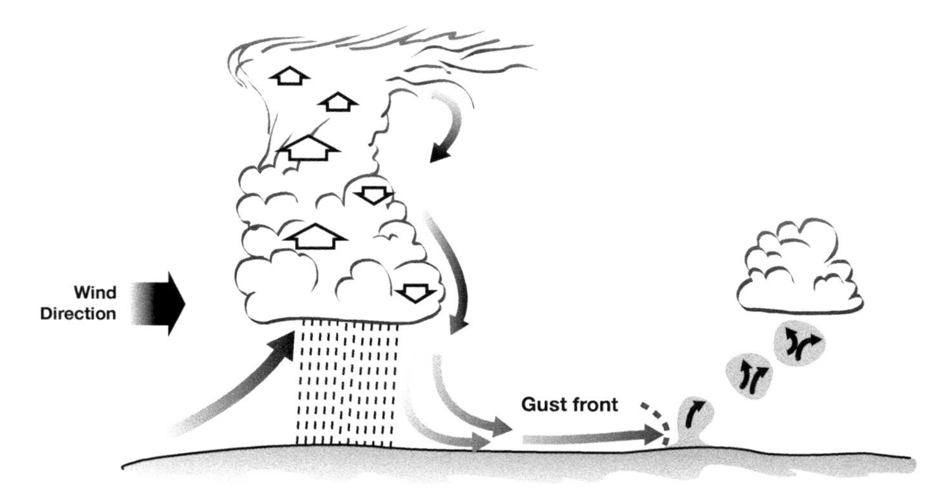

Fig 3.7 Downdraught triggering. Downdraughts from large clouds (especially cumulonimbus clouds) can cause other thermals to be triggered.

Convergence of air masses

In any situation where a colder, more dense air mass meets a warmer, less dense air mass there will be a tendency for the more dense to undercut the less dense air mass, potentially dislodging any incipient thermals. One condition where this occurs is where sea air encroaches over warmer land, often triggering a line of thermals, or in extreme cases, thunderstorms. (This is displayed on a grand scale when a cold front is formed.)

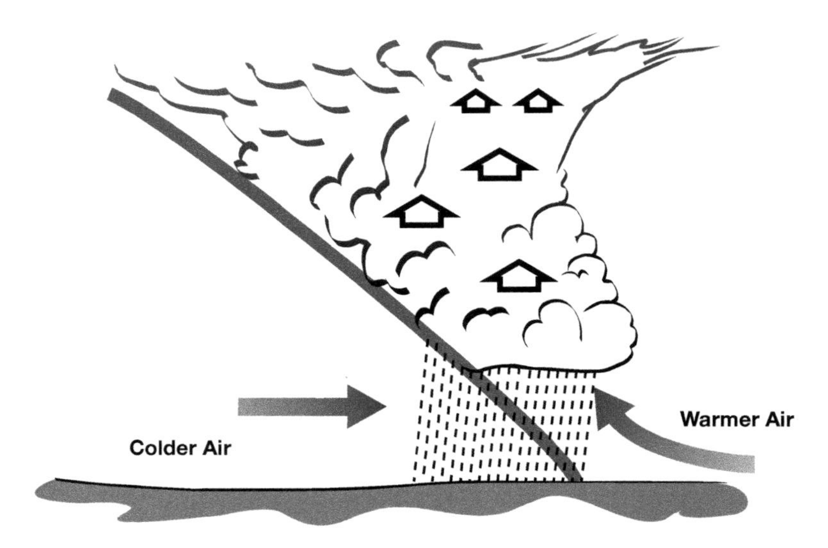

Fig 3.8 Convergence triggering. When two air masses which have different temperatures (and therefore densities) converge, the colder, denser air mass acts like a wedge, pushing the other upwards, as occurs on a cold front.

Cloud shadows

Another form of convergence triggering occurs when cloud shadows drift over the surface. When a cloud passes between you and the sun when sunbathing, the temperature change is noticeable. In light winds, as this cooler air drifts over the surface, it is possible for it to drive a wedge under any pockets of warm air, thus starting a thermal on its way.

Topographical triggers

Thermals will tend to break away from the surface from the highest point of the thermal source. This means that promontories on the landscape will tend to act as the point where thermals will be triggered. Hilltops, ridges, escarpment edges and mountain peaks are all favourite places for thermals to break away from the surface, especially if these exhibit a sharp edge.

When crossing a wide river valley or flood plain, the only thermals to be found may be above small areas of higher ground such as hillocks, as these are the only relatively dry ground in an otherwise soggy environment.

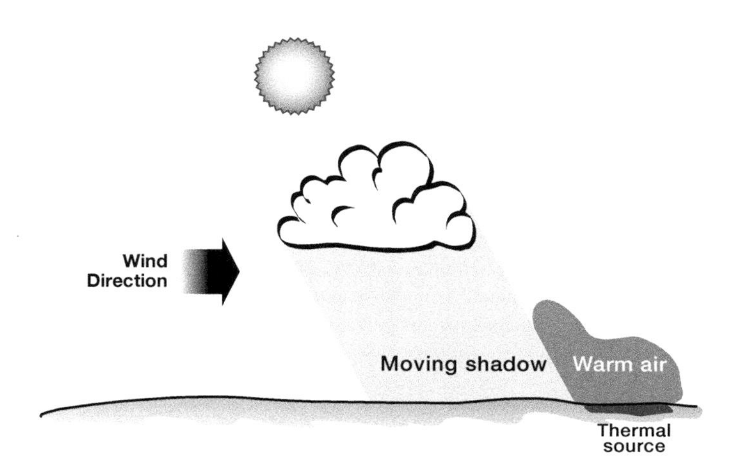

Fig 3.9 Cloud shadow triggering. The cooler air moving in the shadow of a cloud may trigger a thermal.

Manmade triggers

Cars and trucks

The next time you follow a car down a road where fallen leaves are lying, observe the amount of air which is being disturbed in its wake. Any disturbance such as this can dislodge potential thermals.

Railway trains

Because of their high speeds, railway trains push and drag along huge quantities of air, which will influence air all around their path. Warm air lying in cuttings and in other sheltered areas may be dislodged, giving rise to reasonable thermals.

Tractors and combine harvesters

Although slower than road vehicles or trains, vehicles working in fields are often working in an area where there is a lot of trapped air (e.g. in standing crops). If this trapped air has reached a temperature where it is already buoyant, then all that may be needed is a small disturbance to trigger a thermal. Even these slow vehicles may be enough to provide this.

Wire launches

Have you ever noticed that after a fruitless search for a thermal following a wire launch, you encounter lift on base leg or final approach – when you are too low to risk using it? 'Murphy's Law', some would say, but just look at what has happened. You took off into a light wind. Your take-off roll involved stirring up the air close to the hot airfield surface. The launch cable also whistled through the air – and if the launch was by car or reverse pulley launch, then the launch vehicle itself would disturb the air. All this kicks off a thermal, which drifts slowly downwind. Four minutes later, you meet that thermal as you are on approach – about to land! Next time, if traffic and local rules permit, try flying downwind immediately after launch. This way you might still have enough height to use the thermal you have just paid for as part of your launch fee.

Aerotow activity

Tow-planes and other powered aircraft have the same effect, although, as well as disturbing the air as a result of the drag they produce, they also create their own 'wind' in the form of their propeller wash.

Fires

A small fire, such as a bonfire, probably will not create enough warm air to form a usable thermal – but even the smallest fires need to draw in air if they are to survive. As well as creating air movements, they will add some heat to the neighbouring air. Larger fires, such as the burning of straw stubble or heath, will act as creator and trigger for their own powerful thermals.

Giving thermals time to rise

When watching for a trigger you will need to allow some time for the thermal it activates to reach the glider's height. Assuming that a thermal is weak, say rising at 2 knots, it will take 5 minutes to reach 1,000 feet. If the thermal is a strong one, it will take much less time to gain height. For this reason, it is essential that you look for triggers when you are looking for sources. Ideally, once you have spotted a source and trigger combination, by the time you reach the source area, the thermal will have reached your height.

The day was a very poor one by Australian standards. A strong inversion had prevented thermals from forming until late afternoon. When thermals did begin, they reached around 2,800 feet but were blue, weak and far apart.

Three of us set off on a 112 kilometre goal-and-return task from Benalla to Yarrawonga and back. We reached Yarrawonga airstrip where the Super Arrow landed. Conditions were poor and as I struggled to stay airborne, I watched the club's tow-plane arrive to retrieve the Super Arrow, thinking that soon I might also need the tow-plane pilot's services. It was then that I heard on the radio that the Jantar pilot had landed 15 kilometres south of Yarrawonga on the way home.

Getting low and with the air still feeling very dead, I decided that if a landing was imminent then I would land on the Yarrawonga airstrip, after the tow-plane and Super Arrow had got airborne. This would make my retrieve easier.

As I overflew the airstrip before joining a circuit pattern to land, I hit the first definite thermal I had encountered in the last 20 minutes. I climbed to 2,500 feet in steady lift and having lost the lift started gliding homeward.

As I approached the area where the Jantar had landed, I saw the same Super Arrow being aerotowed from a 'paddock'. The Jantar had been aerotowed out of this field, but the Super Arrow pilot, feeling adventurous, had released from aerotow to soar home. However, his brave intentions had landed him in another paddock.

In air that was still, I decided to test the theory that the tow-plane's activity had triggered the thermal that had saved me at Yarrawonga. After watching the Super Arrow launch for the second time, I flew over its launch point and, sure enough, there was another perfectly usable thermal. Another climb and a cautious glide on track followed.

Monitoring the Super Arrow pilot's conversation with the tow-plane pilot informed me that the Super Arrow was needing another aerotow 'retrieve' from yet another paddock.

Arriving overhead this new paddock in time to see the tow-plane begin another launch, I was rewarded by another thermal which provided me with enough height for a cautious final glide in air that was still lifeless.

After my Mosquito and the tow-plane had been returned to the hangar, and as we were heading for the clubhouse, the tow-plane pilot was called to the office and asked if he could do another aerotow retrieve. The Super Arrow pilot had landed in a fourth paddock!

Other visual clues

Haze domes

Even on cloudless days, thermals may produce slight domes in the otherwise flat haze layer. These domes are caused when a thermal hits the inversion layer, pushing it upwards slightly. They are best seen when the glider is near the inversion and the pilot is wearing polarising sunglasses.

Soaring birds

Soaring birds have a greater instinct for finding and using lift than the most experienced glider pilot. Vultures and eagles are classic examples of experts at thermal soaring. Larger birds save energy by using thermals to support their considerable weight rather than by flapping their wings. Other more common species such as buzzards, hawks and gulls can also teach us a thing or two about thermalling and, indeed, about soaring in general.

Therefore, when birds are seen circling or flying without flapping their wings for any length of time, this can be taken as an indication of the presence of lift. Indeed, it is not only an indication that lift is present, but also usually where the best lift is to be found. Providing you have sufficient height margin to safely reach the bird's thermal, on joining it you will usually be rewarded with a good rate of climb. Conversely, a soaring bird will not hesitate to join your thermal if it thinks that you have found better lift than it has. This can be taken as the ultimate compliment.

If the air is suddenly full of swifts darting around your glider as they chase insects, you can be assured that locally, lift is plentiful. This sight more commonly occurs near cloud base.

When crops are very young or when fields appear mainly brown, flocks of birds (gulls and the like) settle on the fields to feed. If a thermal rises from the field, a flock of birds can often be seen taking to the air to feed on insects being carried aloft by the rising air. From above, initially the birds appear to be hovering just above the ground, but if observed for longer, you will realise that they are getting closer as their thermal gradually reaches your level. Look out for birds on or near the ground just as much as those soaring high above you. They may save you from an outlanding!

Flying a Janus B, we were thermalling just north of the Murray River. The eagle had been with us for the last five miles and was now in our thermal. I had soared with birds before but this was different. Firstly, the eagle had shared more than one thermal with us and secondly, it was not outclimbing us. In fact, it was deliberately staying at our level and on occasions came in beside our wing to within a few feet of our cockpit to inspect us further.

After what seemed a long time in a close encounter with this magnificent beast, it moved off and rapidly outclimbed us. I pointed this out to Sam in the front seat who was doing the flying. He understood my intentions and straightened up towards the eagle's position and the apparently better lift. Almost immediately, I realised why the bird had at last climbed above us. It was now diving at our cockpit – talons first.

Instinctively, I pushed forward on the control column (there was little time to discuss control handover) and, as I did this, I immediately regretted my reflex action, realising that if this huge creature hit the tailplane then we would be wishing we had the parachutes we lacked.

Whether the manoeuvre or the change of closing speed made it alter its plans, I know not. The eagle folded its wings and when I last saw it, it was a ball of brown feathers with a beak and eyes (still fixed on us) hurtling under our left wing.

Two somewhat shaken pilots flew at 100 knots into the distance, not stopping to circle until we had put some distance between us and what I assume was that eagle's territory.

Other gliders

Sometimes, just as helpful, although often less reliable than thermalling birds, are circling gliders. Unlike birds, gliders circle for reasons other than to maintain or gain height (for instance when practising turns, for other training purposes, or even to lose height before landing). For this reason it is unwise to risk a long glide with little height margin to join a circling glider, especially if it is in the vicinity of a gliding club.

If, on the other hand, a glider or a gaggle of gliders is seen to be gaining height, then the risk of not finding lift is reduced but never eliminated. Thermalling gliders flown by known, competent pilots, whose thermal can be joined at the same height or higher, will offer the best chance of contacting a usable thermal.

Where competition gliders are found thermalling in a gaggle, you will usually (but not always) be able to assume that they are in the best lift available. The density of traffic in such gaggles may cause concern to the less experienced pilot – to the extent that you may prefer to find your own less crowded thermal.

Smoke

The presence of smoke can be an indication of anything from weak or unusable lift, to a strong, rapidly ascending thermal, depending on the source of the smoke.

The large, dark columns of smoke caused by stubble or heath fires often indicate the presence of phenomenal updraughts. These can give rapid rates of climb, depending on the ferocity of the fire below. The lift is often very turbulent and accompanied by neighbouring heavy sink. As visibility can be poor in such thermals, cloud flying techniques may have to be used. Depending on the care taken in burning the source land, little or no usable lift may be found, and so care should be taken not to arrive too low at the smoke.

When smoke is not part of a usable thermal, its behaviour is often a good indicator of where a thermal can be found. If a thermal starts to

Fig 3.10 Thermals indicated by smoke. Deflected smoke can often indicate air being entrained into a thermal.

ascend close to some smoke, the entrainment of the neighbouring air often changes the angle of ascent of the smoke. Observation of any column of smoke, from whatever source, may therefore reveal the presence of a nearby thermal.

Dust and straw

The inflow of air into the place where a thermal breaks away from the surface will cause a disturbance, which may lift up dust or straw. A thermal may even entrain such debris and carry it off in a swirling motion, thus creating a straw or dust devil. In some warm countries, such dust devils may persist for several minutes, giving a vivid picture of part of the thermal. Contacting a dust devil thermal can provide a glider with an excellent climb rate.

Crop movement

Occasionally, in calm or very light wind conditions, as a thermal leaves a field of crop, the surface of the crop can be seen rippling and 'snaking' in an erratic manner, indicating the source field of a thermal. (This should not be confused with the more general wind ripples one often sees affecting crops on windy days.)

It was a miserable British soaring day with almost total stratus cover at around 3,000 feet. While flying cross-country in a Twin Astir, we were attempting to reach a gliding site at which to land. As we started on the downwind leg of our landing pattern, we could see the club members clearing the single runway for our arrival.

Then, just to the left of our downwind track, I noticed that the crop in a field started moving. The movement looked like a thousand snakes writhing rapidly and randomly through the crop.

We overflew the field and found just enough lift to almost halt our rate of descent. After about three minutes of hard work to stay airborne, the snaking of the crop ceased and our variometers showed a smooth increase to indicate 2 knots up.

We had just observed the birth of a thermal!

Wind changes

The local wind changes caused by nearby thermals can often be seen affecting windsocks or any nearby smoke. The smoke or the windsock will point towards the ascending air.

Watching for and acting upon such relatively subtle signs can result in your placing the glider right in the path of a rising thermal. All you need to find a thermal is a sharp eye, some imagination and a little bit of luck.

Search patterns

Whether you are using visual clues or attempting to locate a thermal source by deduction, it is important that you plan your search methodically.

The advantage of approaching a cumulus cloud or a thermal source into wind or downwind has already been mentioned – but there are many other ways of increasing your success rate when thermal hunting.

When searching for a thermal, select a track, and a visual feature at which to aim (usually a cloud or a suspected thermal source), and then fly directly towards it. If the nose of the glider moves off this track, it will probably be due to rising air lifting a wing and thus causing a slight turn.

Any lift will tend to tip the glider away from the thermal. This tipping action may be very gentle, almost imperceptible, but can still be enough to alter the glider's course away from the thermal. If the glider is allowed to wander in this way, it will follow a flight path which will take it between most of the thermals in the sky.

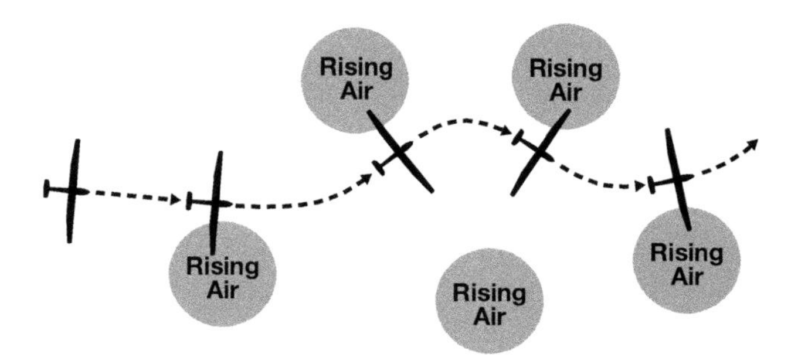

Fig 3.11 Heading deviations due to thermals. Thermals are anti-social and will tend to tip the glider, turning it away from the thermal.

Small, unintentional changes of track should be resisted. If the tipping sensation is pronounced, turning towards the rising wing may reveal a usable thermal.

The height available for thermal searching will usually be the determining factor in how far you can search before returning to the airfield or a preselected field. In order to use this height to best advantage, you should plan your search, so that if lift is not found on your outbound track, you can return to the field without retracing your exact track. Even a return track displaced by a short distance from the outbound route will take the glider through different air and double the chance of finding lift.

When thermal streeting is seen to be present from the alignment of clouds, or is suspected on a blue day, then the search pattern should be oriented relative to the wind. When cloud streeting exists, flying into wind under the clouds should reveal many thermals. When approaching a cloud street, do so at 90 degrees to the street in order to spend as little time as possible in the sinking air between the streets.

When no clouds are present to show where the streets are, heading directly into wind or downwind from a source will again give a better chance of encountering a thermal. If nothing but sink is encountered, you are probably in a SINK STREET. In this event, altering course by 30 to 45 degrees for a short period may well place you in lift. On encountering lift, you should return the glider to a heading which is directly into wind or downwind, to follow the lift street. If the sink is very severe then it will be necessary to alter course by an even greater angle (80 to 90 degrees) to get out of the sink as quickly as possible.

Whatever search pattern you adopt, fly it at a sensible airspeed. This will mean that in air that is giving little sink, the airspeed used will be lower than when flying through air which is sinking at a faster rate. This will ensure that you are not wasting height by flying unnecessarily fast, or alternatively not spending longer than necessary in descending air.

Techniques for using thermals

When you first start to thermal soar, you will be happy to spend hours climbing by circling in thermals, and quite rightly so. Most of your climbs will be achieved in this way and this early practice will be invaluable. However, there are other ways of using thermals without circling. Once you become more choosy about which thermals you want to use to climb, these other techniques will help you move about the sky with the minimum loss of height, while you look for a thermal which will give a good rate of climb.

If you fly around all the time at the airspeed at which the glider achieves its best glide angle, you will gain close to the maximum amount of height which can be achieved while flying through any thermals you encounter (assuming that you do not circle). However, your low airspeed through sinking air will cause you to lose much more height than is necessary, because of the time you will spend in sink. Flying faster will result in less time spent in any rising air (less height gain) but also less time in the sinking air (less height loss). What is needed is a compromise, and this is achieved by adjusting the airspeed to suit the vertical movements of the air.

As you fly into a thermal, you should reduce airspeed, and as you leave the lift, you should increase your airspeed. It is possible to gain some extra time (and height) in a thermal by doing a slight 'S' turn as you transit the thermal. (This subject will be covered in more detail in Chapters 12 and 13, on 'SPEED FLYING' and 'DOLPHIN FLYING'.)

Unfortunately, not all weather situations support these so-called 'dolphin flying' techniques, and so classic-style thermalling, by circling in lift, is still necessary.

Centring in thermals

Should you decide to climb by circling in a thermal, your main task is to position the glider in that part of the thermal which is ascending at the greatest rate, that is, the core. The act of manoeuvring the glider into the thermal's core is known as CENTRING, and due to the thermal's changing shape as it ascends, centring may be a continuous task throughout the whole climb.

If you ask ten top glider pilots how they go about centring in a thermal, you will probably get ten different answers. Most of them may use more than one technique, varying which is used depending on the size, roughness and strength of the thermal. Centring technique is therefore a highly personal skill. It will depend on the rate of roll of the glider being flown, the sensitivity of the instruments and, probably most of all, on how sensitive the pilot is to the gusts and lulls which accompany most thermals.

Some gliders buck and roll readily in thermals and almost tell the pilot where the lift is. Others are not as informative and seem to dampen out gusts, thus denying the pilot vital clues about the position of a thermal's core. Gliders which are sluggish in roll are a liability when thermalling, while those with a good rate of roll help the pilot duel with the roughest thermals.

The variometer will normally take second place to the pilot's natural sensitivity – indeed an over-sensitive or sluggish instrument can often lead to a confused pilot and a thermal being lost. Although many pilots become obsessed with what strength of lift is shown on the variometer, the rate and direction of movement of the variometer pointer is just as valuable an indication when centring in thermals. Because of the delay in indicating the actual rate of climb, the tendency of the variometer's pointer shows whether the situation is improving or getting worse. As it is *improvement* in the rate of climb which you are seeking, the variometer's trend is very important.

As far as the actual process of centring in a thermal is concerned, it would be wrong to say that there is one correct or infallible method. Many books, instructors and competent pilots will show you various methods of centring in thermals. You should try all of them and then use the one or more methods which suit you and your glider. Once you have established a successful method, never let yourself be forced into a method that does not suit you. The following method is one which many pilots use in a number of variations, and is one which can be developed as experience is gained, so that quicker centring can be achieved.

If the glider flies into air which is rising at a faster rate than that which it was in previously, there will be a momentary change in the angle of attack of the wing, and a resulting change in the forces it produces. The noticeable effect will be a slight surge in the airspeed and a physical push on the 'seat of the pants' of the pilot. A slight increase in air noise may even be perceived. How pronounced these sensations are will depend on the sharpness of the transition between the air in which the glider was flying and the rising air which it is entering. This in turn will be relative to the increase in lift.

What all this means is that if a surge is encountered, the glider is flying into air which is rising faster than the air which it was in. On feeling this surge, you should momentarily reduce the angle of bank of the wings (assuming you are already turning) to move the glider towards the lift. This reduction of the bank angle should only be for a very short period of time – perhaps two to three seconds. The wings should then be returned to the original degree of bank. This manoeuvring is continued as each gust is felt, until the glider is circling within the core of the thermal.

Let us look at the technique diagrammatically in order to visualise what we are trying to achieve. For simplicity, we will use a cross-section through an idealised thermal and assume that in its core, the glider will climb at 4 knots (400 ft/min), reducing as we go further out from the core, until eventually it will descend at 6 knots in the sink on the periphery of the thermal (figure 3.12).

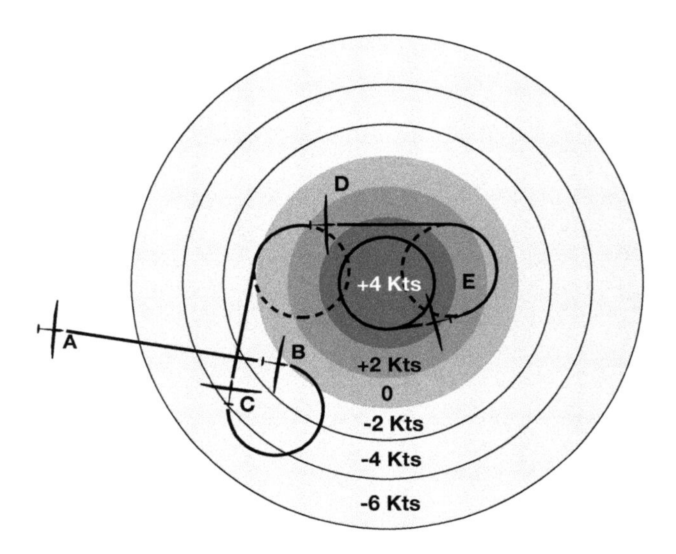

Fig 3.12 Thermal centring. Reducing the bank angle each time a gust is experienced will move the glider towards the thermal's centre.

At A in figure 3.12, the glider is cruising in sink. As a result of the descending air around the thermal, it is common to experience an increase in the rate of descent as a thermal is approached. In this instance you need to have the courage to press on, height permitting, or else you will spend your life turning away from thermals.

At B, the rate of descent will have reduced and the air will probably be somewhat more 'lively'. There may be a slight or even a pronounced tipping effect away from the thermal as the left wing is raised. If this is felt, you should turn towards the rising wing, in this case to the left. (Remember that any other gliders already in the thermal will determine the direction in which you must turn.)

Do not wait for the variometer to start indicating lift before turning, as even the best variometers require the glider to have gained some height before they can register that it is ascending. By the same token, do not turn immediately you sense lift, as you will be in danger of having most of your turn outside the lift and in sink.

Let us consider the worst case and assume that you get it wrong or traffic forces you to turn the wrong way – that is, to the right. In this situation, continue the turn.

In the vicinity of C, you will be moving from air which is descending at 4 knots to air which is descending at 2 knots. As you fly into this better air, you will feel a slight surge or gust. When you sense this, reduce the bank angle, or even level the wings completely. This will move your circle towards the better air. After a second or two, re-establish the original amount of bank.

A similar sensation requiring the same manoeuvre will occur at D, this time as you move into rising air.

Note that although the human body is very sensitive to the changing rates of ascent or descent of the air, it is not very well calibrated – and as a result, flying into air that is descending less will give the same sensations as flying from air that is descending, into air that is rising, or even into air that is rising faster. This is where the glider's variometer becomes useful, and must be used to confirm that the glider has indeed been manoeuvred into rising air each time a gust was felt (E). Using this technique the glider will eventually be placed in a circle in the core of the thermal.

As the glider climbs in the thermal, any gusts experienced should be acted upon in the same way, thus continually re-centring the glider in the thermal's core.

If you miss a gust, or are not sure, note the direction in which you were heading when you thought you felt it, and continue the turn, anticipating the onset of the gust and your action the next time the glider nears that part of the circle.

The duration of levelling of the wings (or reduction of bank) depends on the strength of the gust felt. If you are positive about where the lift is, act positively. If in doubt, act cautiously. The longer you delay re-establishing the turn, the further the turn will be moved in the direction of straightening. Moving the turn when you are uncertain of exactly where the better lift is located is a quick way to lose a thermal.

If you feel the outer wing of your turn being pushed up, you might correctly assume that the better lift is on that side. Unless you are absolutely certain that this is the case, do not reverse the turn. Such a manoeuvre will take several seconds to perform and, if you were wrong about the position of the lift, could result in your losing track of the thermal's position completely. At any rate, you will have to start re-centring all over again.

Every gust, lull, push on a wing and variometer indication will give you a clue to the thermal's shape and the position of the core. All of this allows you to build up a mental map so that if you do lose the core (or even the whole thermal) then there is a better chance of being able to find it again. It is essential that you gain this mental mapping ability. If you fly with an instructor who is also a competent soaring pilot, you will find that he or she will often be able to talk you into the core of a thermal (or even relocate a thermal after you lose it) despite the fact that you are doing the flying. As this cannot be a result of feedback from the glider's controls, it can only be because the instructor has been building up a mental map. However, even the best soaring instructor can lose track of the whereabouts of a thermal's core if the student makes enough erratic manoeuvres!

If you do lose the lift after having been established in a thermal, reduce the bank slightly. This will result in a larger radius of turn and more air being covered. Providing you have not been overdramatic with your previous manoeuvres, you should re-encounter the lift. You can then begin careful re-centring.

If there is any appreciable wind, the chances are that you have lost the lift by descending out of the downwind side of the thermal. This is because in windy conditions the thermal's ascent will be sloping and the glider will be descending relative to the air in the thermal. On windy days it will usually pay to look for the lost thermal by straightening up when the glider is heading into wind and then re-centring once the thermal has been re-discovered.

An alternative centring technique

Having stated that there are many ways of centring in a thermal, it is only fair that I offer you at least one other technique. Like the gust method described above, it does not place its total dependence on variometer readings but, as with the gust method, uses the variometer for confirmation that each manoeuvre is placing the glider in better lift.

Let us assume that you have found the thermal and are circling. Half of your circle is in lift and half is in sink (position A in figure 3.13). As you leave the lift and go towards the sink, you will get the opposite of the push up 'from the seat' that you got when you entered lift. Most glider pilots will recognise this sinking feeling – a feeling of very slightly reduced g.

On feeling this sensation, you should tighten the turn for slightly under

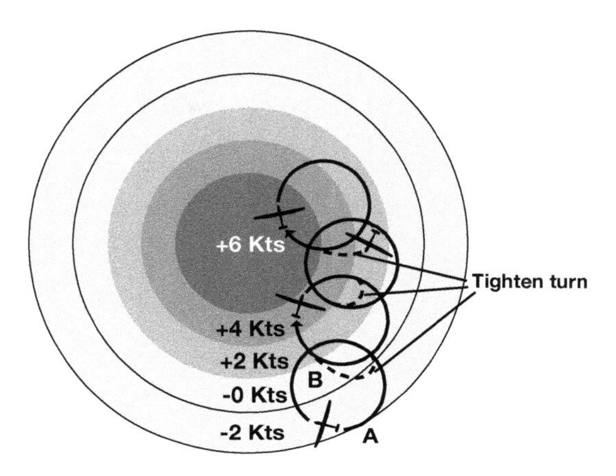

Fig 3.13 Alternative centring technique. Tightening the turn for about 160 to 170 degrees of turn when sink is felt will move the glider quickly away from the sink and towards the thermal core.

half a turn – about 160 to 170 degrees (position B). After this angle of turn, return to your original bank angle. Your circle should now be in better lift. Repeat this manoeuvre each time you sense you are moving into air that is not rising as quickly. These manoeuvres should move you towards the thermal's core until ultimately you are circling in the core.

The advantage of this technique is that by tightening the turn when entering air that is sinking (or not rising as fast as that just flown through) the glider will spend less time in the poorer air.

Once you are established in the best lift, use as steep an angle of bank as is necessary to keep you in the thermal's core.

There are many other techniques for centring in thermals, but I have given you these two for two reasons. Firstly, they can be combined – so that once you are certain where the thermal is, you can turn quickly away from sink and straighten up, to speedily reach the lift. Secondly, they do not rely too much on a variometer which, if badly set up (as is sadly and unnecessarily the case in many club gliders), can ruin your mental map of the air.

Whichever centring technique you decide to use, your aim should be to perfect it so that you can get the glider into the centre of the thermal within one or two turns. To be successful at centring requires practice and constant mental mapping of the air, noting where the surges and gusts occur and comparing these with the readings on the variometer. With a bit of practice you will learn to use all of these inputs to centre quickly in the best lift.

Vertical position within the thermal

The fact that a thermal bubble ascends with a vortex motion means that at different levels within the thermal bubble, the airflow will have different vertical and horizontal velocities. For instance, near the bottom of the bubble the air will be flowing more horizontally than vertically as the thermal draws in the neighbouring and descending air. The direction of this inflow will become increasingly vertical as it approaches the thermal's core, reaching its maximum vertical velocity in the centre of the bubble. As the air approaches the top of the core, its vertical trajectory will again become more and more horizontal as it starts to flow outward before starting its descent down the outside of the bubble.

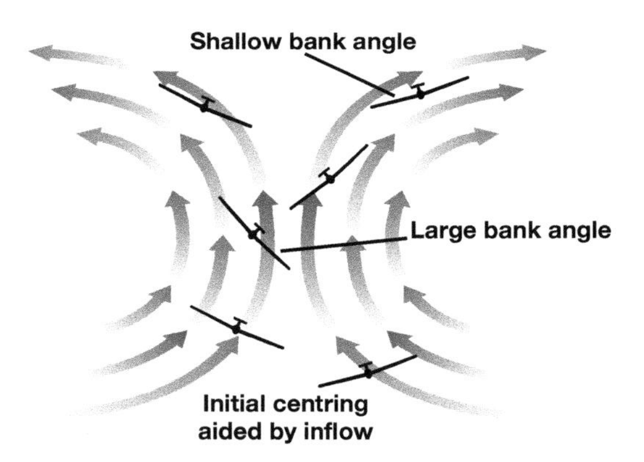

Fig 3.14 Vertical position in the thermal. The amount of centring and the angle of bank necessary will depend on the vertical position within the thermal bubble.

In practical terms, what these inflows and outflows mean is that, depending on how far up the bubble you are when you hit the thermal, there may be a tendency for the glider to be either 'sucked into' or 'pushed out of' the thermal. If the glider enters the lower parts of the thermal bubble, despite the lower climb rate achieved, the inflow will tend to 'self-centre' the glider. As the glider reaches the upper part of the bubble, the climb rate will again be less than it was in the core as a result of the increasingly horizontal outflow of air, which will tend to ease the glider out of the thermal.

These effects will require different actions on the part of the pilot. At the bottom of the bubble, where the inflow of air is helping you to centre, dramatic centring manoeuvres may not be necessary. As the rate of climb increases and the glider is assumed to be entering the core area, then more positive centring manoeuvres can be made. Near the top of the bubble, the rate of climb will start to decrease, and at this point the thermal will be more difficult to use. A reduced bank angle will contribute to the glider's rate of climb. Better still would be to go and find another thermal.

One of the great fallacies which student glider pilots are often told by at least one of their instructors is that thermals 'do not just peter out'. From the above description, you can assume that they do. However, do not use this to make excuses for losing a thermal. Losing a thermal when it is powering the glider upwards means that you have lost it when you were near or in the core. Losing it as the lift gradually decays probably means that you have indeed topped out of the useful part of the bubble, or the thermal has reached the top of its ascent.

Common faults while centring

Thermals vary in size and strength. Every glider has its own performance characteristics once it is established in a turn. It is therefore necessary to match these two variables to get the optimum rate of climb. Failure to do so, along with other faults, will degrade your rates of climb or may even lose you the thermal altogether.

Insufficient angle of bank

The first criterion is to place the glider in a turn, the whole of which lies within the best area of lift. The greater the angle of bank used, the smaller the radius of the glider's turn. Therefore, by using a reasonable angle of bank, you stand a better chance of keeping the glider in the area of best lift.

If you are taking more than 20 seconds to turn through 360 degrees then the chances are that your rate of turn and the glider's turning circle is too large. As a result, part of the time the glider will be turning in air that is ascending at a lesser rate, or even in air that is sinking. If this is the case then you need to use more bank.

Too much bank

Conversely, too much bank, while keeping the glider in a smaller turn, will have a performance penalty due to what is known as the glider's TURNING or CIRCLING POLAR.

When turning, the lift force produced by the wing must be increased in order to turn the glider as well as support the glider's weight. Increasing the lift force will also increase the amount of drag that the glider produces. The effect of this is to decrease the glider's performance while turning, resulting in a greater still air rate of descent. Therefore, the greater the angle of bank used, the greater will be the glider's rate of descent through the air. Table 1 shows the minimum rate of descent in still air of a typical medium-performance glider when different bank angles are used.

TABLE 1

BANK ANGLE (deg.)	MINIMUM RATE OF DESCENT IN STILL AIR (knots)
0	1.38
30	1.71
45	2.36
60	3.94
70	6.89

If the thermal is weak and too much bank is used, then it is possible to cancel out any rate of climb which could be achieved, because of the increased rate of descent incurred as a result of having too much bank.

The optimum angle of bank will depend on the size, strength and degree of roughness of the thermal. In a strong, narrow thermal, angles of bank in the region of 45 degrees or more may be appropriate; whereas in a large, weak thermal, a smaller amount of bank will often give better results. Around 30 degrees of bank is usually a good starting angle until you have assessed conditions. If the thermal core is large and strong then you may be able to maximise the climb rate by using a smaller angle of bank.

The more bank you use, the higher the airspeed must be. Therefore, to reduce the glider's turning circle, you can either fly at a slow airspeed with a shallower angle of bank, or alternatively use a large bank angle and a higher airspeed. Which combination of airspeed and bank angle you use will depend on the glider's handling and performance as well as the nature of the thermal.

Unnecessary variation of the bank angle

Changing the angle of bank, even for a short period of time, will move the circle that the glider describes. The gusts associated with some thermals will, unless countered, cause the bank angle to change. Add to this any unnecessary variation of the amount of bank by the pilot, and it can be very easy to lose track of where the lift is in relation to the glider. To avoid losing the best lift or perhaps even the whole thermal, only make considered control inputs and avoid unnecessary changes to the angle of bank.

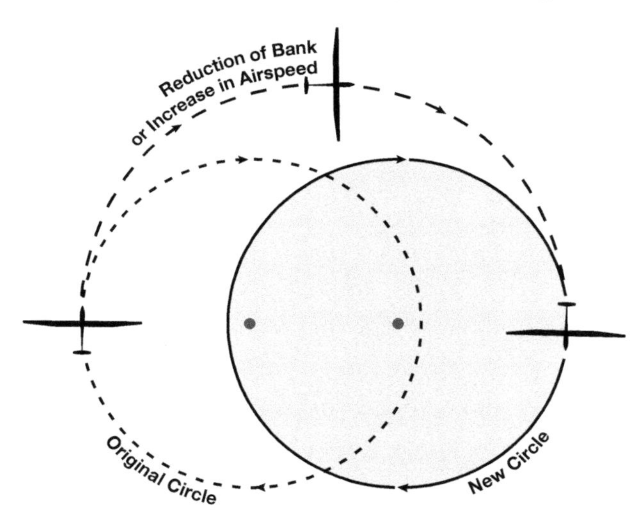

Fig 3.15 The need for accurate flying when thermalling. Any reduction in bank angle or increase in airspeed will increase the glider's turning circle.

Unintentional airspeed changes

Just as altering the angle of bank will result in a different radius of turn, so too will changing the airspeed. Flying the glider faster will result in a larger radius of turn for the same angle of bank than will flying at a lower airspeed. This means that if the airspeed is allowed to increase unintentionally, the glider's turn will be elongated. Once the airspeed is returned to its original value, the glider will return to its original radius of turn, but the centre of the turn will have moved. Therefore, poor airspeed (or attitude) control can result in the glider leaving the thermal and possibly cause the pilot to lose track of where the lift is to be found. To avoid this occurrence, regularly monitor attitude and airspeed.

Unnecessary control inputs

Every time you move the control column or rudder pedals, you disturb the airflow past the controls and create extra drag. This increased drag will reduce the glider's climb rate. It is especially important when circling in weak thermals (typical of those you get late in the day or under overcast skies) that only necessary control inputs are made – otherwise what may have been a marginal climb rate in a weak thermal may become a slow descent.

In competitions, or anywhere where gaggles of gliders are found thermalling together, it is amazing to watch the amount of rudder waggling going on in a glider in front of you, despite no obvious positioning manoeuvres! Such pedalling around the sky has to be detrimental to the efficiency of both the pilot and the glider. If you see another pilot doing this, use it as a reminder to ask yourself if you are also guilty of such over-controlling.

Lookout suffering due to obsession with the variometer

Climbing quickly to the top of a thermal would be a total waste of time, if it were followed by a slow descent by parachute or an even faster descent without one!

Mid-air collision is one of the greatest risks in gliding. Gliders tend to congregate in thermals. Because of centring manoeuvres, differing airspeeds, turn rates and climb performance, there will always be the risk of collision. Despite rules which are designed to reduce this risk, the only answer is to maintain a good lookout.

Looking out, using a regular scan, will also help the accuracy of your flying and, as a result, your climb rate. A good scan will not only take in the sky and other gliders, but also the glider's nose attitude and wing tip, which in turn indicate the consistency of the airspeed and angle of bank. The variometer and airspeed indicator can be included in the scan, although airflow noise and an audio system fitted to the variometer can be used to free some of your attention from these internal 'distractions'.

Airmanship and rules when thermal soaring

* When joining another glider in a thermal, you are required to circle in the same direction as the glider already in the thermal.
* The first glider established in a thermal dictates the direction of turn.
* When joining a thermal you should enter the thermal in such a

way that your entering will not force other pilots already in the thermal to alter their turns to avoid you.

* Never fly for any length of time in another glider's blind spot. (Typically, this will be close behind and/or below the other glider.)
* Should you lose sight of another glider known to be in the thermal, avoid sudden changes of track. If you suspect that this other glider is too close, you should consider leaving the thermal. (In this situation avoid manoeuvring rapidly – unless necessary.)
* On leaving a thermal, do so in a way that reduces the risk of straightening up in front of another glider.
* Do not fly in the company of other gliders near the base of cloud, where the visibility is often poor.

Chapter 4

Hill Lift

What causes hill lift

When the wind meets any form of barrier, such as a hill or mountain, it has to go either around it or over it. If the obstruction is a long hill (or a line of hills), then most of the air mass will be deflected upwards and will pass over the top of the hill. This upward deflection will provide a large area of rising air in which it is possible to soar a glider for as long as the wind continues to blow against the hillside. Such HILL LIFT may take the glider to altitudes of 1,000 feet or more above the top of the hill being soared.

After the air flows over the crest of the hill, it will start to descend and become turbulent. This can produce heavy sink, and can be a danger to a glider which is allowed to drift downwind of the area of hill lift. This downdraught is often known as CURL OVER, or the CLUTCHING HAND, which, in strong wind conditions, accurately describes the disastrous effect it can have on any glider caught in it. Depending on the shape of the hill and the wind conditions, the transition from what may be a narrow band of hill lift into the curl over area may be quite sudden.

When thermals are present, this curl over may be more severe. As thermals form, the wind will normally increase and become gustier. The stronger the thermic activity, the more turbulent will be the approach through the curl over at a hill-top site.

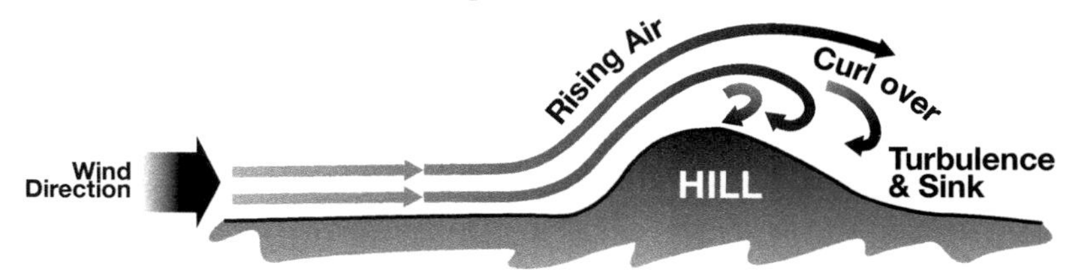

Fig 4.1 Hill lift. As the wind blows against a hillside, the air is forced upwards. As the air flows over the crest, it becomes turbulent and can cause severe downdraughts.

The rate of climb and the height which a glider can achieve in hill lift depends on many factors.

Wind strength and direction

Generally, the stronger the wind, the stronger the hill lift and the higher a glider will be able to climb in front of any particular hill. The direction of the wind is critical, as apart from requiring the wind to be blowing against the hillside, a wind blowing at right angles to the face of the hill will give better lift. If the angle between the wind direction and the hillside is less than 90 degrees the amount of hill lift will be reduced. This is because the air mass will be deflected along the face of the hill as well as over it. In the extreme case of the wind meeting the face of the hill at a shallow angle, then little or no hill lift may be produced.

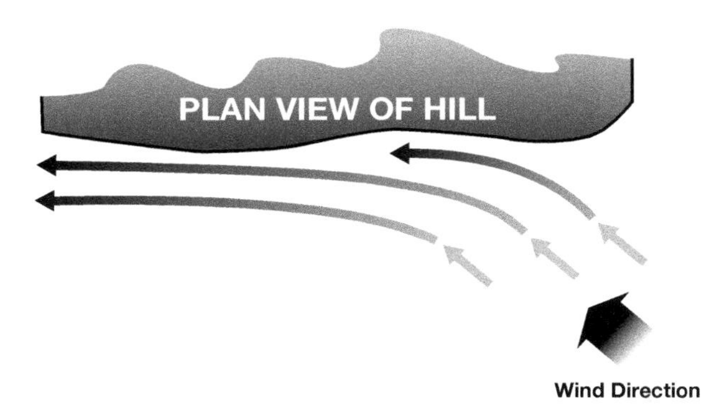

Fig 4.2 A wind at an angle to the hill. A wind blowing at an angle to a ridge is unlikely to produce as much lift as one blowing perpendicular to it.

Fortunately, the sides of hills and mountains are seldom straight. There are often spurs and outcrops on the face of a hill or mountain, and these may provide a surface perpendicular to the wind to give usable hill lift. Any bowls or gullies which face into wind will also produce good lift.

In mountainous areas, where there are many ridges, slopes and valleys, it can often be difficult to assess the wind direction and, in turn, which slopes are likely to be producing hill lift. One problem is that local winds tend to flow along valleys, rather than in the same direction as the prevailing wind on the day. This situation is further complicated by the fact that there are often many minor valleys branching out from the main valley. Add to this the fact that the strength and direction of flow of these valley

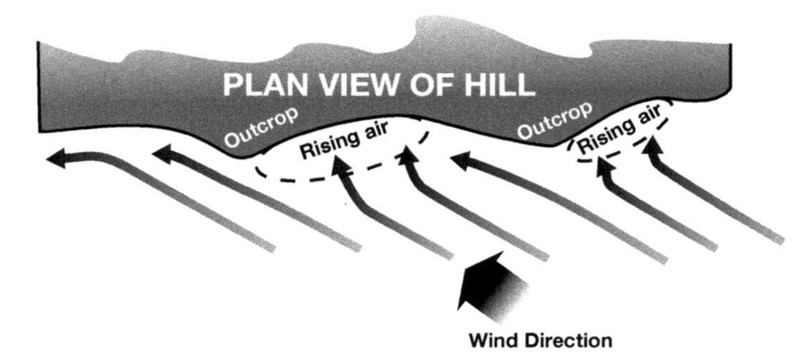

Fig 4.3 Outcrops producing hill lift. Spurs and outcrops on a hill may produce better lift than the main hill face if the wind is at an angle to the main face.

winds will alter as the day goes on, and you have all the elements which make hill, and especially mountain soaring a fascinating experience.

Hill size and slope

First impressions might suggest that the bigger the hill, the better the lift. The size of the hill will influence the potential height that can be gained in hill lift. One would quite rightly expect that a glider using the lift produced by a mountain which is 3,000 feet in height would gain more height above the mountain than one soaring a hill slope 400 feet high. However, more important than the size of the hill is the shape of the windward face of the hill.

For good hill lift, the ideal hillside is one which is steep enough to deflect the air smoothly upwards, but not so steep that the deflected air becomes turbulent.

If the hillside is too steep, as in a cliff face or a vertical ridge, the air mass will need to be deflected through a large angle. This will result in the

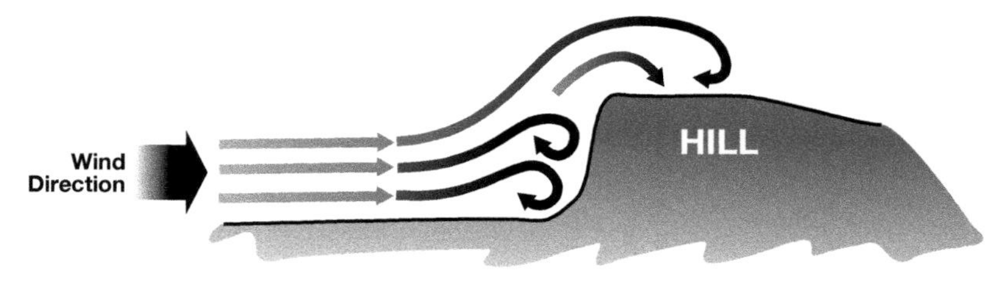

Fig 4.4 Sink near a steep hill face. If the face of the hill is steep, then the wind may cause eddies and downdraughts to form instead of usable lift.

air becoming turbulent, causing eddies and downdraughts as well as hill lift. These effects may well be so pronounced as to make what hill lift exists difficult to use safely, and give only marginal gains of height.

In fact, such eddies reduce the chance of finding lift to such an extent that a glider approaching the face of a steep slope low down may fail to find usable lift, even although gliders are seen soaring the ridge or hill face higher up.

Surface of the hillside

Just as the surface of the glider's wing needs to be smooth in order to achieve laminar airflow over it for optimum results, the surface of the hill-side also needs to be of a nature which will present little drag to the air-flow, if this is to be kept laminar. Rough terrain on the lower slopes, trees or rocky outcrops on the hillside will disrupt the air mass as it progresses up the slope, leading to turbulence and reducing the hill lift. If this is the case, then better lift may be found further out from the hill face than would be the case if the slope's surface were less rough.

Any knolls or steps in the hill face will cause the airflow to break away and will cause curl over effects which will either disrupt the hill lift or make it difficult to use.

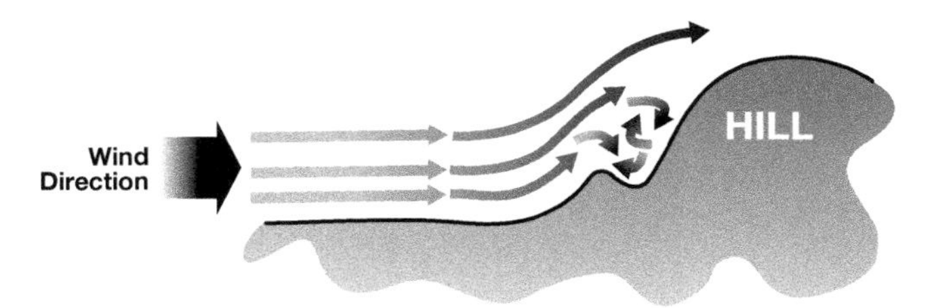

Fig 4.5 Irregularities on a hill face. Knolls or steps in a hill face can cause turbulence and break up the lift.

Upwind terrain

The terrain upwind of the intended soaring slope can also have a dramatic influence on whether or not the hillside will produce any hill lift. If there are other hills or mountains immediately upwind of the hillside, these can disrupt the airflow and shield the hillside from the wind. In addition, such hills may make the air very turbulent, degrading soaring conditions and making flying on the ridge quite unpleasant.

Hills or mountains further upwind, as well as having the above effects, may create LEE WAVES (see Chapter 6). If the descending part of one of these waves occurs in the vicinity of the intended soaring slope, it may be powerful enough to cancel out the hill lift and even cause large amounts of sink where you might expect to find lift.

Where a club operates within reach of a hill, mountain or ridge, local knowledge will be of benefit in anticipating what conditions are to be found when embarking on a hill soaring flight. When approaching a slope with the intention of soaring along it, some consideration of the above points may avoid the disappointment of not finding lift, or prepare you for what might be a rough ride. In this respect, a cautious approach is a safer approach; the more ill-founded your suspicions, the better the surprise as your glider soars up the hillside.

Stability of the air mass

If the air mass is very cold and stable, there will be a tendency for the wind to attempt to flow around the sides of the ridge, despite the general wind direction being towards the hill or ridge. This may result in the wind being deflected along the face of the hill or ridge, instead of over it. If there is a strong inversion below the height of the hills, there may be a reasonable breeze blowing across the top of the hills, but no hill lift on the lower slopes. This effect may be pronounced on cold winter mornings, changing only as the day progresses and the valley air warms up, resulting in usable hill lift.

If, on the other hand, the air is unstable enough for thermals to form, then as a thermal drifts across the hill or leaves the hill top, the hill lift will temporarily be enhanced. Once the thermal has passed, the down-draughts from the thermal may disrupt the hill lift to the extent that, where weak hill lift was previously found, only sink will exist. Once the thermal has moved far enough away, normal hill lift will be restored.

Anabatic lift

In addition to, and sometimes instead of, the hill lift created by the effect of the general wind, the glider may be able to soar on a local wind, known as an ANABATIC WIND. When the sun heats the surface of a hillside, the air close to the hillside will also be heated. Being lighter, this air will start to move up the hillside, creating what is called an ANABATIC FLOW and giving ANABATIC LIFT. This flow can be strong enough to enable a glider to gain height.

The strength of the lift will depend greatly on the amount of solar heating, which in turn will depend on the size and surface of the slope and its orientation relative to the sun. Anabatic lift will be poor or non-existent early in the morning when the sun is low, gradually improving during the day and reducing in strength again towards evening. The anabatic flow will be upset and cease over any part of a slope which is in shadow for any length of time.

The narrow band of lift will tend to be very close to the face of the slope. If the slope is very steep or very rough, the rising airflow may become disrupted and broken. A smoother, gentler slope will produce a better anabatic flow.

As air ascending a slope anabatically will be heated as it creeps up the face of the slope, it may acquire a high enough temperature by the time it reaches the top of the slope, to form a usable thermal. At the same time, the valley air, which may lack this extra heating, may not become unstable enough to form thermals.

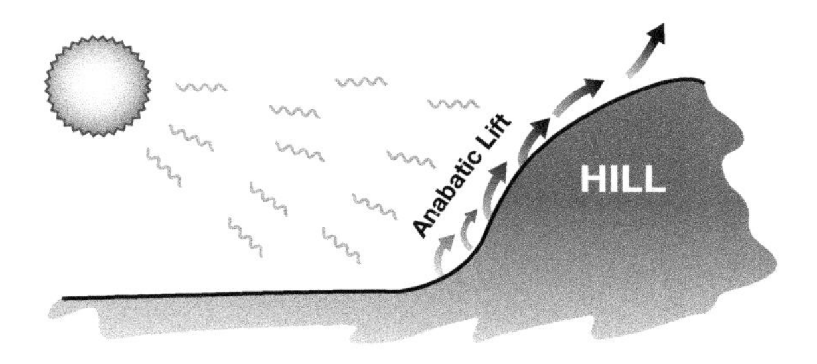

Fig 4.6 Anabatic lift. Solar heating of a hill or mountain slope can create anabatic lift.

Even when soaring in pure, wind-driven hill lift, the effect of any anabatic wind should not be disregarded. Often this will manifest itself by the best 'hill lift', being at the sunny end of a curved ridge!

It had been a hot, anticyclonic day. It was late afternoon and I had done about 18 instructional winch launches in a Bocian, none of them reaching more than 700 feet above ground level. On the ground, there was no wind.

The next launch took us to a similar height, but instead of the expected slow descent after releasing the winch cable, we had a hint of lift. We started to circle and climbed very slowly (a quarter of a knot – sometimes less), and as we did, it became apparent that we were drifting very slowly towards the hill which reaches 1,300 feet above the airfield and lies just half a mile to the south-east. This hill marks the edge of the Cairngorm Mountains, which, a short distance further east, rise to nearly 4,000 feet above sea-level.

The lift steadied at just under 1 knot, and after a while it became apparent that, if we continued our present angle of climb, we would cross the ridge top with only 100 feet or so to spare. Would there be curl over in such a light wind or any sink? As the area behind the hill was totally unlandable, and as the 'escape route' around the valleys behind the hill involved more track miles than we could cover safely, I had to decide whether to let the climb take us behind the hill or whether to throw away the thermal.

The fact that there was no noticeable hill lift as we passed the face of the hill made me decide to continue the climb, in the theory that no hill lift meant no curl over (whether that theory is reasonable, I do not know). We continued circling, ready to throw ourselves back over the front of the ridge at the first sign of sink. Gladly, no sink appeared. In fact, once above the plateau behind the hill, the rate of climb increased to 2 knots in air which continued to be beautifully smooth.

We gained enough height in this lift to go exploring the northerly faces of the higher mountain slopes, to find that flying close to the slopes produced steady lift of about the same strength. Then to our delight, when we did a close inspection of the south facing slopes, these gave similar consistently smooth lift. The air on every slope we ran along seemed to be rising.

After an hour of fascinating soaring, we landed back on the airfield – to the stares of envious course students whose best flight so far that week had been a 4 minute hop.

Even an instructor having to bash circuits in hot unsoarable conditions can have his enthusiasm topped up by a soaring flight above mountains where the whole sky is rising gently.

Chapter 5

Hill Soaring

Using hill lift to maintain or gain height is called HILL SOARING and was the first type of soaring to be achieved. These days, hill soaring is used as a means of both local soaring and cross-country flying, or when no other types of lift are available.

Soaring in hill lift obviously has advantages when attempting duration flights, such as those required for the FAI Silver Badge. (Duration records were discontinued many years ago, as the ability to squat in hill lift for as long as the wind continued to blow in a favourable direction led to fatigued pilots. Skill did not play a significant part in such records and boredom must have contributed to pilot fatigue.)

When cross-country flying, the possession of hill soaring skills can be useful if a long line of hills lies roughly on the glider's track. The hill lift from such hills can be used to increase the glider's average cross-country speed, and the utilisation of such a ridge has contributed to the success of non-stop glider flights over distances in excess of 1,000 miles on the Appalachian Mountains in the USA.

On a more modest scale, even smaller ridges, no greater than 700 to 800 feet high, have allowed pilots flying along the South Downs in Southern England to complete 175 kilometres of a 500 kilometre task before the first thermals of the day had formed.

Even lesser hills may save a glider from a field landing when the thermals temporarily 'let it down', thus providing an airborne 'parking place' until a usable thermal can be contacted and cross-country flight resumed.

Never underestimate the assistance that a ridge (even a slight rise in the terrain) may give you when you are final gliding. It may just make the difference that gets you home – providing, of course, that you are on the upwind side of it.

Add to the above the fact that hills and mountains are excellent thermal sources, and you will see that the potential uses of hill soaring extend much further than beating aimlessly up and down a local ridge.

On a long, marginal final glide from the east back to Husbands Bosworth airfield, I found myself flying along the line of a very small ridge (not much more than 200 feet high) which runs east–west just to the north-east of the airfield. The wind was a moderate northerly, and the reduced rate of descent offered by the air rising up the ridge made the difference between a nail-biting end to the final glide, which could possibly have ended in a field in the valley, and what turned out to be a more comfortable approach.

On the other hand, I should have read the warning signs during a marginal final glide into Dunstable airfield when the finish line was reporting an easterly wind of 5 knots. Dunstable airfield lies on the side of a hill which rises on its eastern boundary. The upper wind, undoubtedly being stronger, caused sinking air, which resulted in my ignominious arrival in a wheat field just short of the airfield boundary.

General technique

Given that a hill, ridge or mountain is producing hill lift, all that you need do is fly the glider on the windward side of the hill, and in a position relative to the hill where the best lift is to be found. This position will depend on the wind strength and the glider's height relative to the top of the hill.

In light winds the glider may well have to be flown very close to the hillside, in order to make the best use of the hill lift. In stronger winds, it may be possible to climb while flying the glider at a more comfortable distance from the hillside. In any event, the glider will still be relatively close to the hillside (usually never more than a few wingspans away), and with the ground this close, a safe margin of airspeed is essential.

As a general rule, when the glider is within approximately 500 feet of the hill, or if conditions are turbulent, the airspeeds used when hill soaring should be similar to those used on an approach to land. By maintaining a safe airspeed, the glider's controls will be more responsive and the danger of stalling or spinning while close to the hill will be reduced. This extra airspeed will also give the glider some reserve of energy if a sudden downdraught is encountered.

As height is gained, the area of best lift will be found further out from the hill's face. The glider should therefore be flown progressively further out from the hill as height is gained.

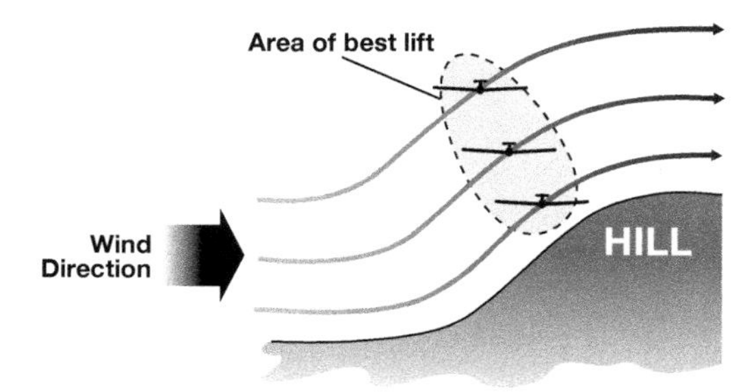

Fig 5.1 Area of best hill lift. The area of best hill lift will be further out from the hill's face as height is gained.

With this increase in the distance from the hill, the glider's airspeed can be reduced to one which will better suit its climbing characteristics.

Once established in the hill lift, it is necessary to track along the windward face of the hill to stay in the best lift. Depending on the wind strength, this may involve flying the glider with its nose pointing out from the hillside to avoid being drifted either towards it or, if the glider is above the hill, behind the face of the hill.

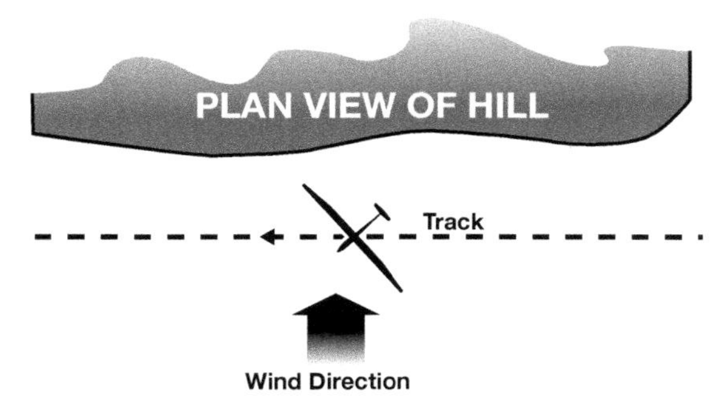

Fig 5.2 Tracking in hill lift. It may be necessary to 'track' the glider along the hill's face to prevent the wind drifting the glider towards the hill.

The degree to which it will be necessary to track in this way will depend on the wind strength. In very strong winds, it may be possible to remain in the hill lift with the glider pointing directly into wind and more or less stationary relative to the hill.

As the effect of the wind will be to drift the glider towards or behind the hill, the turn made at the end of each beat along the hill should be made away from the hill. Such turns can be put to good advantage by making them in an area of good lift, thus increasing the time spent in such an area, rather than waiting until the glider moves out of the lift at the end of the ridge before turning. The timing of such turns must take other hill soaring traffic into account.

It is important that any turns are accurate as far as airspeed and control co-ordination are concerned. Inaccurate flying will increase the glider's rate of descent through the air, which, if the hill lift is weak, may make the difference between staying airborne or having to land. When close to the hill face, accurate flying and airspeed control is also essential for safety reasons.

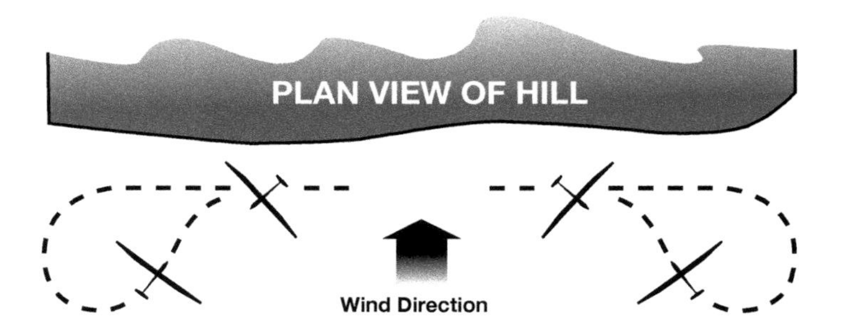

Fig 5.3 Turning at the end of each beat in hill lift. The glider is turned away from the hill at the end of each beat.

Approaching the hill

The initial approach to the hill should be carried out with caution and with some extra airspeed in hand. As well as allowing for any turbulence close to the hill, this extra airspeed will also allow for the fact that the glider is entering a vertically moving air mass which, at the moment of entering it, may cause an increase in the glider's angle of attack. This is more likely if the rising air forms a narrow band with a sudden interface with the general air mass.

Despite the fact that the wind appears to be blowing in a favourable direction for the production of hill lift, you can never be sure that lift is present, unless other gliders are hill soaring at a similar height. It is therefore important to set yourself a minimum height, and a position at which to abandon your attempts to hill soar if lift is not encountered. Normally this will be determined by the height needed to return to the gliding site or a suitable field in which to land.

If approaching the hill head-on, the turn along the hill's face should be started early, in order to avoid flying into the hill because of the tail wind and the elongated turn that this will cause. The stronger the wind, the earlier this turn must be initiated. A safer tactic is to plan your entry into the hill lift from a more oblique angle, thus giving more time to turn and also to assess the conditions near the hill.

Landing out near a hill

Should an attempt to hill soar fail, or the hill lift become unusable, it may be necessary to land. If the hill on which you are flying is not within range of an airfield, a landing will have to be made in a field.

There are many factors to be taken into account when considering the suitability of a field for a landing. Apart from the field's size and surface, slope will usually be a major problem when the field is near a hill. A landing into wind is normally preferred to a downwind landing. This will mean that, on the face of it, the ideal landing direction will be away from the hill. However, as the terrain at the base of a hill will often slope downwards for a long way out from the hill, there is every chance that many fields will be unsuitable because they slope downwards in the direction of an into-wind landing. NEVER ATTEMPT A DOWNHILL LANDING, as even a slight downward slope will make it difficult to place the glider on the ground. Even if this is accomplished, the ground run will be lengthened and it will be difficult to stop the glider before the boundary fence is reached.

The answer is to select a field well away from the hill and to abandon any attempt to hill soar early enough to ensure that you can reach this field and still be able to fly a circuit.

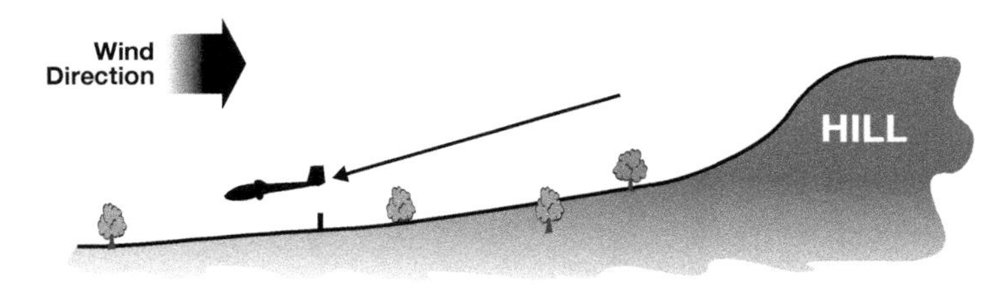

Fig 5.4 Field landing near a hill. Should a field landing be necessary, watch out for slope in the fields anywhere near the hill.

If this is not possible, and a sloping field is all that is available, then an uphill, downwind landing is safer than attempting to land into wind and downhill. In moderate or strong winds, the only safe solution is not to get yourself into such a predicament.

Incidentally, it is easy to believe that if a hill has stopped producing usable lift, the wind must therefore have decreased in strength or changed direction. This is not necessarily the case, as other influences, such as wave interference, may have caused a temporary cessation of the hill lift. Thus, it is unwise to jump to conclusions regarding the wind strength or direction in such instances. Doing so may wrongly affect your choice of landing field. The only safe answer is to look for wind indicators near where you want to land.

Mountain soaring

It would be easy to think that mountain soaring is simply hill soaring on a larger scale. While to some extent this is true, there are some startling differences which add to the enjoyment and challenge of mountain soaring.

If a hill soaring flight begins from the top of the hill, then the glider may never be below the crest of the hill. If the glider is launched by winch from a site near the hill's base, it may arrive at the hill near the top of the hill's face or be above the crest after only a few beats in the hill lift.

When mountain soaring, a pilot can expect to arrive at the face of the mountain (even after an aerotow launch) with most of the mountain still above the glider. This can make a pilot who is new to soaring in mountains feel very small and very nervous indeed. From a winch launch, the pilot may arrive at the mountain at a height where, if lift is not contacted immediately, a return to the airfield will have to be made.

From this lowly start, the glider can be worked up the slope using normal hill soaring techniques. One problem is that the wind at this low level may not be the same, in direction or strength, as that on the upper slopes.

Winds at lower levels in valleys do not follow the rules for the general airflow, but instead tend to flow along the valley sides. This means that some or maybe all of the slopes will not produce hill lift at their lower levels. It is often necessary to use a narrow band of lift on some knoll or outcrop protruding from the main ridge in order to gain enough height to move on to and soar the middle and upper slopes.

Often the shoulder of a hill, situated where the main valley meets a secondary valley, will be the starting point of the climb when mountain soaring. The beats in the lift on this shoulder may be no more than 'S' turns, and gaining height initially may be painfully slow. Once the jump is made on to the larger slopes above, the beats will usually be longer and the lift much better and easier to use.

Once the glider is high enough, it may be possible to go behind the forward ridges and soar higher ridges and bowls to gain height. Although, on the face of it, this breaks the basic rule of hill soaring, which is never to go downwind of a ridge, such tactics are often possible when mountain soaring because of the relative heights of the various ridges in the area and the vagaries of mountain winds.

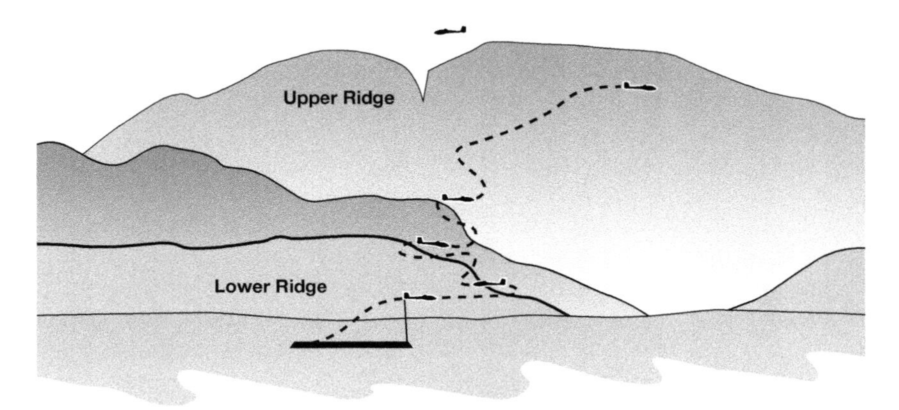

Fig 5.5 Mountain soaring. Mountain soaring may require starting on lower ridges to gain enough height to safely exploit upper slopes.

Exploring mountains by slope soaring can be very exhilarating – and as the lift may consist of a narrow band close to the mountain face, you may only be a couple of wingspans away from the rock face. The impression of speed which this gives is more obvious than when soaring above a more lowly ridge, thus adding to the thrill.

It is unlikely that all your mountain soaring will be carried out in hill lift alone. Anabatic lift, thermals and mountain lee waves may also be encountered, adding to the interest and excitement of such soaring.

Each group of mountains has its own peculiarities and offers soaring of different types as well as different hazards. By far the best way to be introduced to soaring a particular area of mountains is by flying with an instructor or pilot who has local experience. Such an introduction is invaluable if you are to enjoy soaring mountains in safety.

Hazards when hill or mountain soaring

Curl over

The curl over zone downwind of the hill lift can cause a straying glider to descend rapidly, perhaps even forcing it on to the hilltop. Care must be taken to keep the glider out of this zone and to avoid being drifted back over the edge of the hill, where much turbulence and sink may be encountered. Generally, the stronger the wind, and the narrower and more vertical the summit of the ridge, the more severe the curl over will be.

If conditions exist for lee wave propagation, and the wavelength of the waves matches the ridge width, the flow of air over the hill may plunge down the lee slope, producing much stronger winds in this region than in the free air over the ridge top. (The streamlines of airflow are concentrated down the lee slope instead of being uniformly spaced.) In such circumstances, the wind down the lee slope may reach twice the speed of the wind over the crest of the ridge, making the sink in this area even more ferocious.

Not only the main hill top can cause curl over. Any ridges, steps or small peaks on the face of the hill or mountain will cause breakaway of the airflow, and cause an area of potential curl over and sink. Even with the more gentle lift from anabatic flow, such prominences can cause a localised area of sink where lift may have been expected.

Orographic cloud

The air rising up the face of a hillside will cool as it rises. As air cools it is no longer capable of holding as much water vapour. If the air is very moist, the cooling effect of its ascent will result in cloud forming close to the hillside. This OROGRAPHIC CLOUD can form quickly (especially if a moist air mass arrives), often shrouding the hill in fog. If this occurs while a glider is hill soaring, then there is a great danger of the glider suddenly finding itself in cloud in close proximity to the hill.

It is difficult to predict this event, and the only safeguard is always to know which heading to adopt in order to fly directly away from the hill. Turning on to this heading will result in the glider quickly reaching clear air as soon as it exits the area of lift, and therefore the orographic cloud.

If soaring in mountains, cumulus cloud may also form well below the mountain tops. Obviously, it would be extremely foolish to penetrate any clouds when below or close to the height of the highest mountain tops in the area. Picturing the outcome of discovering that the cloud being penetrated is full of mountain does not need much imagination.

Outcrops and trees

The surface of a hill or mountain is rarely smooth. When seen from close up, most hill or mountain faces have outcrops of rock, or sometimes stacks, sticking up from the slope or face. Care must be taken not to fly into or touch a wing on these carbuncles. Solitary trees often grow from a slope, again offering another trap for the unwary. The only way to be safe is to concentrate on the slope ahead, especially if you are close to the hill or mountain face, always being ready to move rapidly away from the slope should such a hazard be seen. Special care should be taken when flying around a corner where such objects could await unseen.

Cables and wires

In some areas, such as the European Alps, various cables and wires are strung from the mountainsides towards the valley. Wires may be for electricity or communications, whereas cables may be to support cable cars or transport materials such as timber. Spotting such obstructions may be difficult because of their thinness and the huge span between the supports.

Navigation

Often, when viewed from above, one mountain valley can look very much like another. It is easy to follow the wrong valley, only to find that by the time the glider exits the valley, you are several miles from where you thought you would be. This is a particular danger if there are several almost parallel valleys leading away from an upper slope. In an extreme case, the result of such a mistake could be a descent into an area of unlandable terrain.

Lack of landing fields

Often in hilly or mountainous countryside, there is a distinct lack of reasonable fields in which to land. There may even be no fields at all in the vicinity. Fortunately, if the glider is high enough, it should be able to fly a considerable distance away from the high ground in order to reach an area with suitable fields. The most important decision that has to be made is at what height any attempt at slope soaring should be abandoned and a course set for an 'escape' to a landable area.

It may even be possible to have the location of a number of preselected fields marked on your map, along with distance rings giving heights required to reach each of them. This technique is especially useful if flying cross-country in mountainous terrain. Local knowledge will be essential and invaluable for soaring in such an area and should be sought by any visiting pilot.

Hill soaring on cross-country flights

Any attempt to fly cross-country using solely hill lift has to be meticulously planned. Unless you are familiar with the route to be flown, some time will have to be spent poring over accurate maps to assess the practicality of the task. Much will depend on the continuity of the ridges and their orientation relative to the expected wind. Such ridge running will utilise basic hill soaring techniques, ideally without the need to waste time beating back and forth over any one section of the ridge.

Every so often it will be necessary to cross a valley or a gap in a line of hills. At these places, careful consideration will have to be given to the height required for the glide to the next likely hill producing lift, especially if the height of the line of hills being soared is not great. Before jumping such a gap, it is wise to assess the fields ahead, in case a possible outlanding becomes a reality.

If crossing from one ridge to another involves a jump upwind, then, if possible, it is best to avoid the lee side of the upwind ridge. Failure to do so will risk encountering descending air, which will reduce your chances of making it across the gap. Curl over and turbulence are inherent dangers in such a situation.

Attempting any large task by using hill lift alone is not common, as normally you would hope to top up height with thermals, or contact a wave system. Both these possibilities offer less exciting flying but are more flexible in terms of the glider's track. Entering wave lift from hill lift will be dealt with in Chapter 7 WAVE SOARING. Contacting and using thermals when hill soaring requires quite different techniques from normal 'flatland' thermalling.

When hill soaring, different parts of the hill or ridge will give different strengths of lift. If hill lift increases greatly and suddenly, then you have probably come across a thermal. This is more likely if the sun is shining on the land below the part of the hill which you are soaring, or if a cumulus cloud is passing over the face of the hill. It is still possible, and indeed essential, to search for likely thermal sources while hill soaring, using all of the techniques described previously.

The need for a different thermalling technique comes from the fact that, in all probability, there is a hill either beside you or slightly below you. If any attempt is made to thermal by circling, the glider will either collide with the face of the hill or be drifted into the curl over area.

The answer is to start a turn away from the hill as soon as you pass through the best lift. As your heading begins to come round to bring you back to the hill, reverse the direction of the turn. Repeating this 'S' turn manoeuvre will keep you in the area of best lift without getting too close to the hill. Remember to maintain a safe airspeed and start your turns in good time to allow for the glider's rate of roll. After gaining enough height to be clear of the hill and the curl over band, you can continue climbing using normal circling techniques.

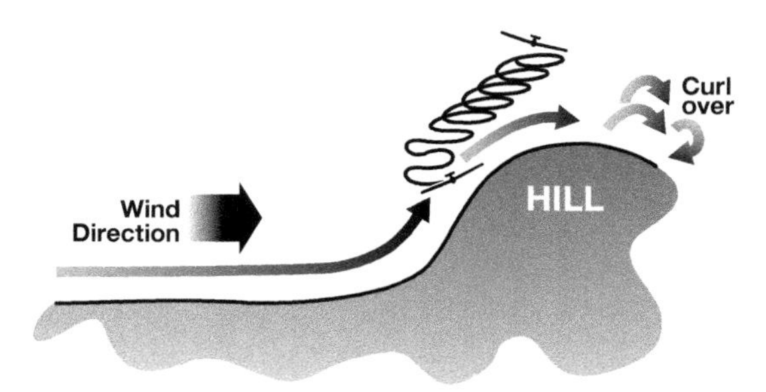

Fig 5.6 'S' turning in thermals. Thermals encountered near the hill face will require the flying of 'S' turns in the lift until clear of the hill.

Occasionally during a thermal cross-country flight, when for some reason thermals have temporarily ceased (such as cloud over-development and spreadout), you may find that using hill lift, even on an isolated hill, may keep you airborne until thermals recommence. Although frustrating, the fact that you are still in the air rather than on the ground at least retains some hope of completing the flight. It is at times like these that you will be glad, firstly that you can hill soar and secondly that you have practised thermalling away from a hill.

It is very easy, especially if your own airfield has a soarable hill or ridge available, to simply beat back and forth in hill lift with very little effort or enterprise. Do not just hill squat at the ceiling of the hill lift. Watch for changes in the air which may indicate the presence of thermals or wave. When height is adequate, you can explore the air away from the ridge, and practise handling skills, field selection or crop identification, safe in the knowledge that you can scamper back to the hill lift to top up your height. Hill soaring is a useful skill, providing you make proper use of the flying time it provides.

Airmanship and rules while hill soaring

As always, the need to maintain a good lookout is essential. When a hill which is producing lift is within range of a gliding site, there is every chance that gliders will congregate in the area in front of the hill. Such traffic density increases the risk of a mid-air collision. Not only will gliders be coming from a head-on direction, but they will also be joining the hill lift from other directions.

As the maximum height possible in the hill lift is achieved, your glider and the others will all find themselves more or less at the same height. Safety depends on everyone searching the sky for other gliders.

Despite your maintaining a good lookout, the following factors increase the risk of a mid-air collision.

Flying into a low or setting sun will make oncoming gliders difficult to see. Hazy conditions or a dirty canopy (or both) aggravate this situation. Never fly with a dirty canopy; your life could depend on its clarity! In winter, when a low sun-angle is common, misting of the canopy is also common. Watch out for these situations, and if you are concerned about your chances of seeing other traffic, move away from the hill lift, or plan your beats along the hill so that when you are in the thick of the traffic you are travelling down-sun. Never assume that the other pilots will see you.

In order to reduce the risk of a collision when hill soaring, the following rules apply regarding your behaviour with respect to other gliders.

* When two aircraft are approaching head-on at a similar altitude while hill soaring, the glider with the hill on its left shall alter course to the right. (This is common sense as the other glider is in a poor position to obey the normal convention because the hill or the curl over area is on its right.)

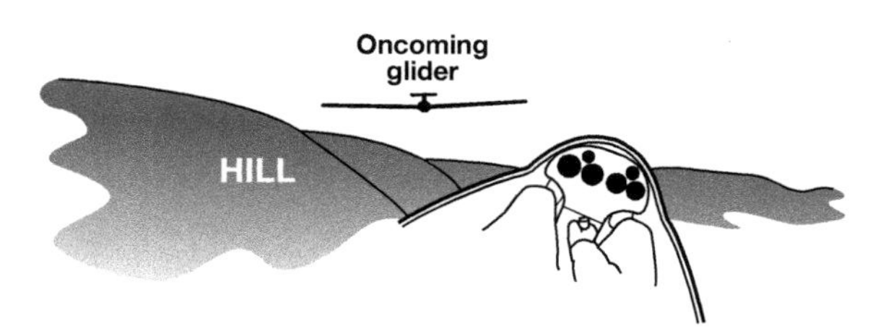

Fig 5.7 Approaching another glider head-on while hill soaring. When two gliders are approaching head-on, the one with the hill on its left shall alter course to the right.

* In the United Kingdom and in the USA, when overtaking another glider while hill soaring, you should overtake between the other glider and the hill. (In practice there is often not enough space between the hill and the other glider for this manoeuvre, in which case it is better to end your beat of the hill early, turning outward and then back along the ridge.) In other countries, hill soaring gliders overtake on the side away from the hill. It is therefore essential that you know the rules for the country in which you are flying.

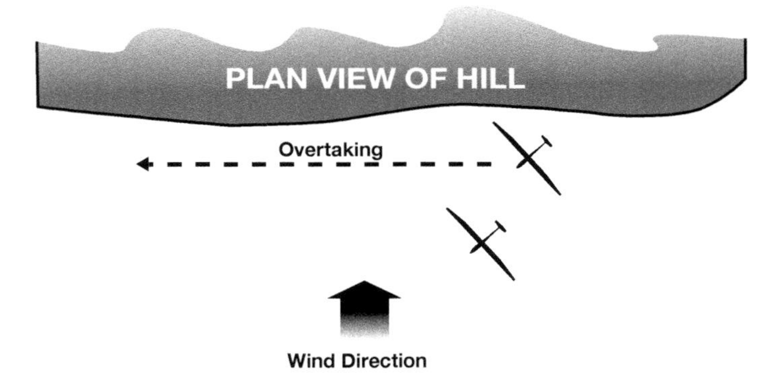

Fig 5.8 Overtaking while hill soaring. When overtaking another glider (in the UK and USA), overtake between the other glider and the hill.

Chapter 6

Lee Waves

Providing atmospheric conditions are suitable, when air is caused to flow over a barrier, it starts to undulate up and down in a wave motion downwind (or to the lee) of the obstruction. Commonly, the barrier which starts off this LEE WAVE effect is a line of hills or mountains, but on some occasions the triggering mechanism for wave may be solely atmospheric with no ground feature directly involved. In this chapter, we will look firstly at the more common topographically triggered waves, and then at those waves which are caused by purely atmospheric influences.

The formation and characteristics of mountain lee waves

The best place to visualise how a lee wave is formed is a shallow stream. Look for an area where the water flows over a completely submerged rock, causing a series of waves to form in the water downstream of it. These are lee waves. Often, despite the fact that the water is flowing quite fast, the waves will remain more or less in the same position; in this case, they are known as STANDING WAVES.

If you imagine that the water was the earth's lower atmosphere, and that the rock was a hill or mountain, then you might be able to visualise a similar effect in the air.

As the wind blows against an obstruction such as a line of hills, initially the air will be deflected upwards. (We have already seen this effect giving hill lift.) If certain conditions exist, the air will flow up over the hill and down the hill's lee side. It will then rebound upwards, forming the start of a wave system.

The initial wave is called the PRIMARY WAVE. The air will rise until it reaches the crest of this wave and then descend again. This wave effect often continues with decreasing intensity through many cycles and for many miles downwind of the original line of hills. The effect may also be magnified vertically, creating waves, the top of which may reach altitudes of 80,000 feet or more. If a number of subsequent waves are present downwind, then this is known as a WAVE TRAIN.

The expression used to describe the vertical depth of a wave is known as the WAVE AMPLITUDE. This is measured as half of the vertical distance between the crest and the trough of a wave.

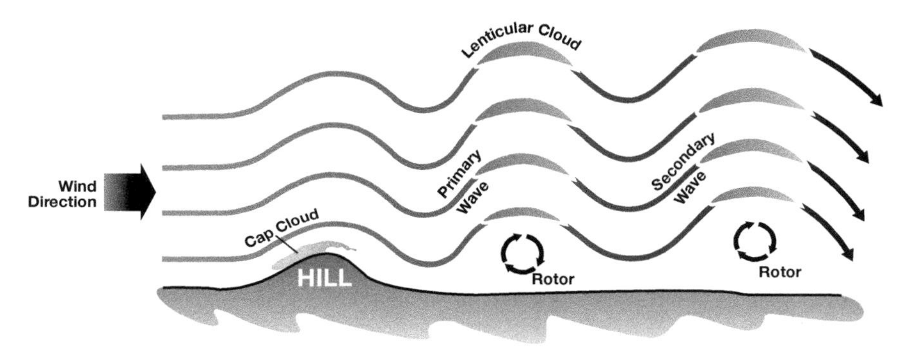

Fig 6.1 Mountain lee waves. When the wind blows over a range of hills or mountains, it often forms a series of waves downwind of the high ground.

Lee waves can be very powerful, with the air rising and descending at thousands of feet per minute when strong winds blow across large mountain ranges. Rates of ascent in excess of 25 knots have been recorded. Unfortunately, the downdraughts are every bit as strong and as plentiful as the updraughts. In fact, the effect can be so great that warnings of the presence of 'mountain lee waves' are frequently issued to airline crews.

Factors affecting the formation of mountain lee waves

The formation of lee waves needs more than just a line of hills or mountains and some wind. On many windy days in mountainous terrain, no soarable waves may be encountered. This is probably because one or more of the following conditions have not been met.

Wind

The wind strength and direction is critical to the formation and characteristics of the waves produced.

To stand a chance of triggering wave, the wind at lower levels (up to around 3,000 feet) must be within 30 degrees of the perpendicular from the line of hills or mountains.

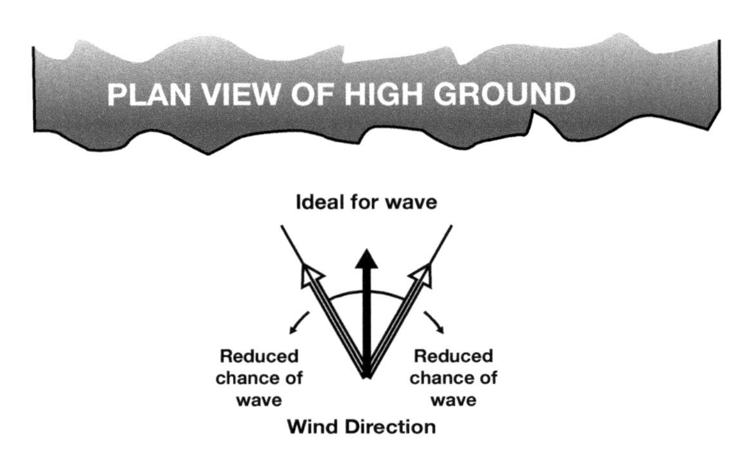

Fig 6.2 Wind direction for wave. To stand a chance of triggering wave, the wind must be within 30 degrees of the perpendicular to the high ground.

The surface wind must be of a reasonable strength (usually 15 knots or greater). As a general rule, the larger the mountains, the stronger the wind will have to be. The strength of the wind should increase as altitude increases, without changing its direction with height. (The requirement of increasing wind speed with height for wave was noticeably absent in October 1995, when the British height record was increased with a flight to 37,730 feet. One reporting station showed the wind as almost constant at 65 knots at all levels between 5,000 feet and 45,000 feet. This seems to be an exceptional case.)

The wind strength will affect the distance between the waves – that is, the WAVELENGTH. The stronger the wind, the further the air will travel during one up-and-down oscillation, and the greater will be the wavelength. The wind speed will not directly affect the wave amplitude.

Note, however, that when the wind strength increases greatly with height, the waves tend to be flattened, thus reducing their amplitude. Strong high-level winds, such as a jet stream, will limit the height of any waves. Should the wind decrease with height, the waves may become steeper. If this decrease in wind strength is sudden, the waves may become so steep that they break up into turbulent air.

The waves formed will be aligned parallel to the high ground causing the wave, which means that they need not be at exactly 90 degrees to the wind direction.

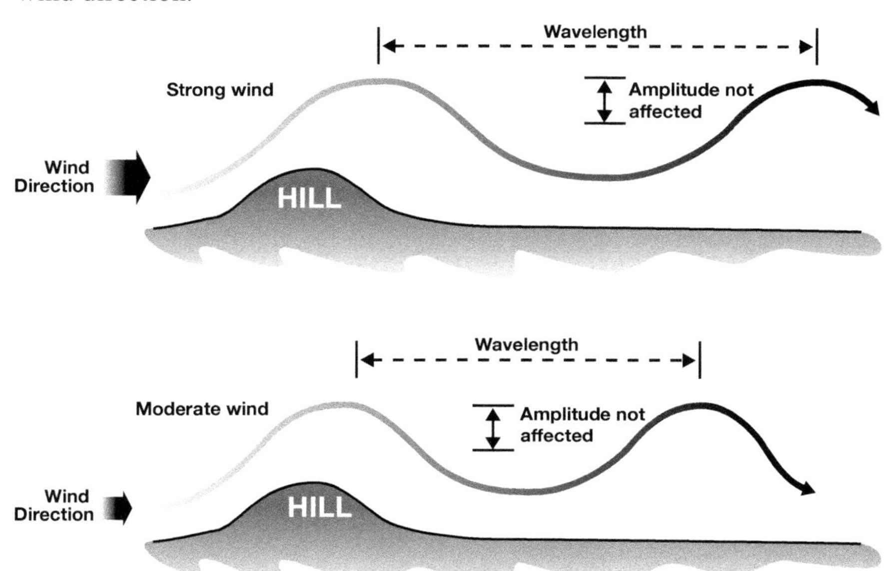

Fig 6.3 The wind strength affects the wavelength of the wave system.

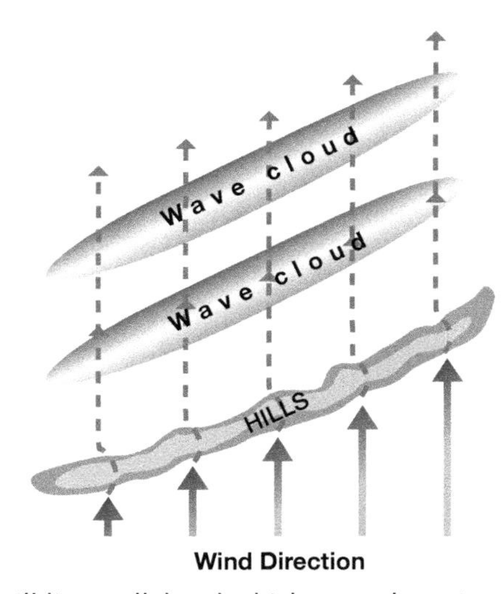

Fig 6.4 Lee waves will lie parallel to the high ground causing them.

Shape and size of the hill or mountain

Certain hills or mountains are capable of producing lee waves which reach many times their own height. For example, the British height record mentioned previously was set in an area where the maximum height of the mountains is barely 4,000 feet.

If a hill or ridge does not present a reasonable length of face to the wind, then it is likely that much of the air will flow around the edges of the high ground rather than over it. This will reduce the initial upward deflection of the air and the likelihood of wave.

As well as the length of a hill or mountain, its shape will determine the existence as well as the characteristics of any wave formed. Too steep a windward face may lead to a turbulent flow and thus no laminar flow down the lee slope to initiate the wave effect.

Probably more important is the shape of the lee slope. For good wave, the summit should be rounded and the downwind side of the hill should be smooth but with a definite gradient. Too steep a gradient, as with a cliff, will lead to breakaway of the airflow and reduce the likelihood of regular wave propagation. Such a slope may well produce wave, but vortices created behind the hill may cause fluctuations in the airflow, which will cause the position of the wave to vary. This will contribute to the bafflement of a pilot attempting to establish a pattern in his or her variometer readings.

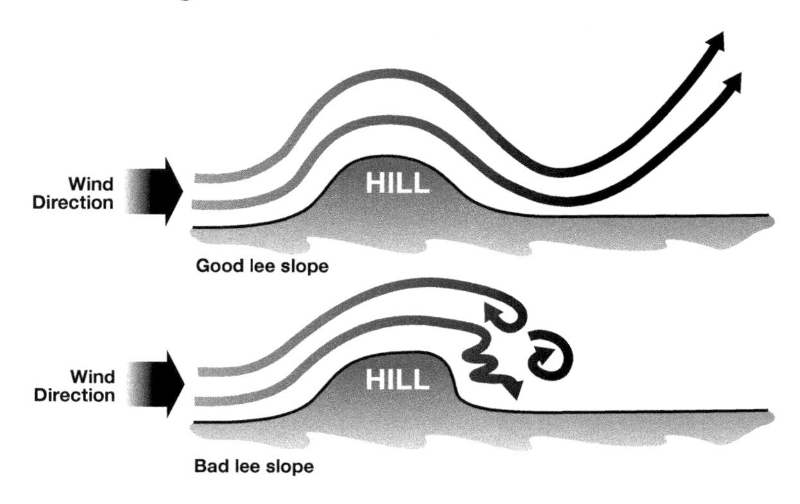

Fig 6.5 The shape of the lee slope is important. Too steep a lee slope will disrupt the wave pattern.

The size and width of a ridge affects the wave amplitude. If the lee slope is very shallow, the wave produced will also be shallow with low wave amplitude. A well-defined hill, with moderately steep slopes, will have the potential to produce a wave with greater amplitude. The largest amplitude will occur when the wavelength of the system (dictated by wind speed) is the same as the width of the hill or mountain. Therefore, in a strong wind, a narrow, steep hill will not produce waves as deep as it would in moderate winds. In a strong wind, a wider area of high ground would produce waves with higher amplitude.

Where a wave flow encounters further hills or mountains downwind of the primary obstruction, the wave amplitude may be modified. If the high ground downwind is situated such that it fits the wavelength of the wave system, the amplitude will be enhanced downwind of the secondary line of hills. Should the wavelength be out of phase with the terrain, the wave amplitude will be reduced, and the system may even break down, resulting in no further waves downwind.

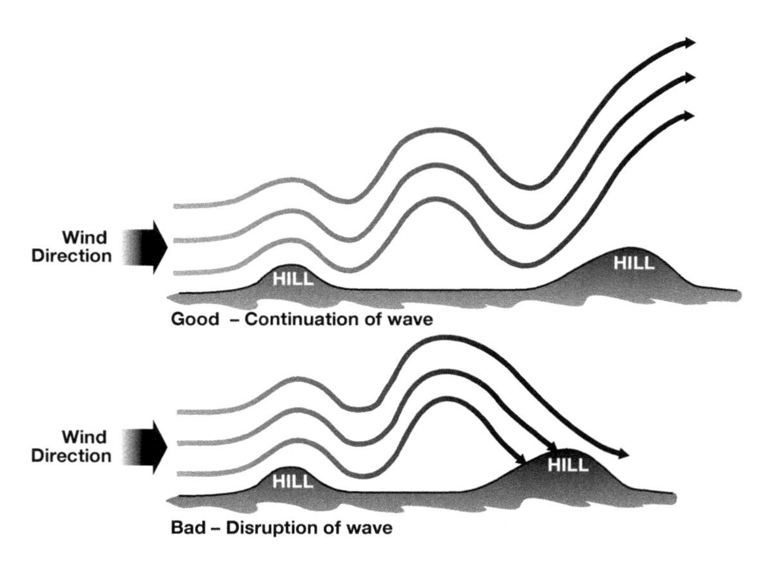

Fig 6.6 Effect of the downwind terrain on the wave. Terrain downwind of the high ground causing the wave will affect the wave amplitude or can even disrupt the wave system.

In the text above, I have used expressions such as 'hill', 'mountain' and 'ridge'. It should be appreciated that, for the purpose of describing lee wave, these expressions are interchangeable. In addition, when discussing a mountain which triggers a wave system, the descriptions are to a large extent theoretical as, with a few exceptions, the glider pilot is often unaware of the initial triggering feature. A pilot may be using a wave downwind of a 'known source', but this mountain may be merely a 'booster' feature in phase with a wave system which was triggered many miles upwind by another mountain.

Atmospheric stability

Wind and hills are only two of the ingredients necessary for wave propagation. Probably the criterion on which the glider pilot can find it most difficult to gain information is the degree of stability of the atmosphere. The required atmospheric stability for the existence of mountain lee waves exhibits a peculiar variation in the degree of atmospheric stability with altitude.

If stable air is disturbed (as when the wind blows over a ridge) it will tend to return to its original level. As it does so, it will tend to overshoot this level, causing a vertical oscillation or wave. The more stable the air mass, the stronger the restoring force and the shorter the period of oscillation. If the air is unstable it will not display this tendency to return to equilibrium, and therefore no oscillation will result.

To get a series of waves, the energy of the wave has to be trapped within a layer of the atmosphere, known as a WAVE DUCT. A wave duct normally consists of a layer of stable air above an inversion starting at about the height of the high ground. Above this layer of stable air is usually a deep layer of less stable air.

The air below the inversion is often unstable enough for convection to take place. Although this unstable lower layer is not essential for the formation of waves, its presence tends to increase the wavelength of the wave system. However, if this unstable layer becomes too deep, then it may have the effect of killing off the wave system.

An atmosphere displaying the correct layers of stability causes the descending lee-side air to rebound to form a wave system. Ideally, the stable layer should be several thousand feet deep and the unstable layer above this should continue for tens of thousands of feet.

Occasionally, atmospheric conditions are not conducive to allowing the air mass to continue to wave for repeated cycles downwind of the triggering feature, but instead only allow the air to rebound once, forming a single lee wave. This type of wave may still allow soaring to considerable heights immediately to the lee of the high ground.

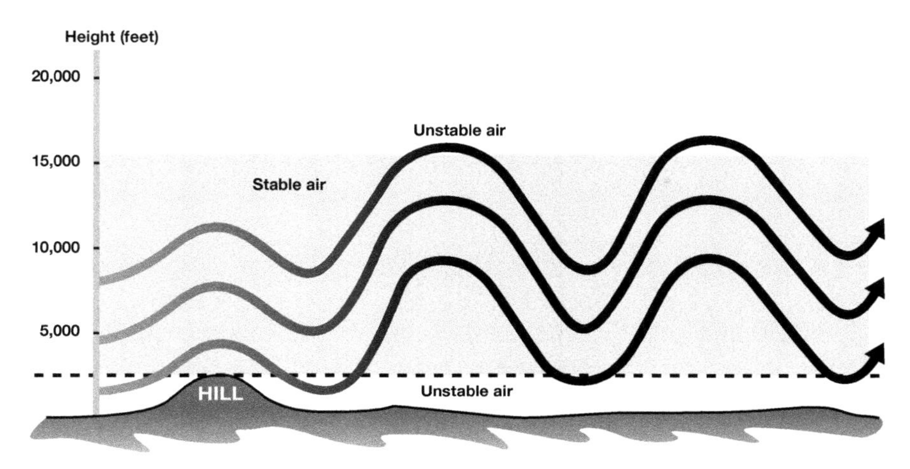

Fig 6.7 Atmospheric stability for lee waves. The correct variation of stability with altitude is important for the formation of lee waves.

If there is a deep layer of cold, stable air at low level on the downwind side of the high ground, this will prevent the air from descending the lee slope, resulting in no lee waves.

The conditions conducive to wave mean that thermal streeting can occur beneath lee waves. If this is the case, it is not unreasonable to expect lee waves above the thermal streets. If cumulus clouds mark the

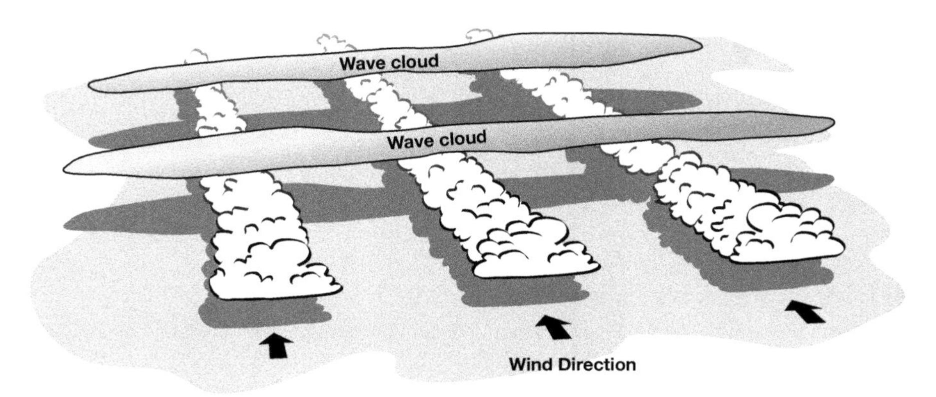

Fig 6.8 Waves above cloud streets. Lee waves can exist above cloud streets. Often they will affect the strength of thermals.

thermal streets, certain clouds may be more developed (and their thermals stronger) due to their being in phase with the lee waves above. Wave clouds may even be seen above the cumulus streets.

Synoptic situation that favours the formation of lee waves

The forecasting of wave has progressed enormously since glider pilots first started exploiting wave lift. However, such are the local variables that local knowledge still adds to the accuracy of predictions at most wave sites.

On the large scale, the probability of wave can be assessed from synoptic charts. This is done by looking for the large scale air mass features which are likely to give the atmospheric conditions mentioned previously – that is, suitable winds and the desired stability.

These conditions often exist when the soaring area lies under a warm sector (the area between warm and cold fronts). Anticyclones can also give the required stable layer, but unfortunately without the moderate or strong winds to cause the wave. However, if such a high-pressure area is moving away, or its centre is sufficiently far away, so that approaching depressions are causing the winds to increase, then there is a better chance of wave.

Features associated with mountain lee waves

Apart from the large areas of rising and sinking air, there are several other features which are associated with lee waves.

Laminar airflow

For reasonable lee waves to exist, the air will ideally adopt a smooth or laminar flow. At lower levels the air may suffer from mixing due to turbulence or thermals. However, once clear of this COBBLESTONE EFFECT, flying conditions are normally smooth.

Foehn gap and lee downdraught

Where air is forced to ascend the face of a hill or mountain, its temperature will decrease. If the air temperature reduces to its dew point, moisture will condense out in the form of cloud. If rain falls from this cloud, the air which flows down the lee slope will contain less moisture, and thus have a lower dew point than it had for its ascent of the windward side.

This descending air will be heated during its descent and will exceed its new dew point temperature at a greater height than that at which condensation occurred during ascent. This will cause any cloud to evaporate. The result is often a cloudless gap immediately downwind of the high ground, known as a FOEHN GAP. Where cloud does reform, it may have a higher base than the cloud on the upwind side of the high ground as a result of its having a lower dew point and therefore a higher condensation level.

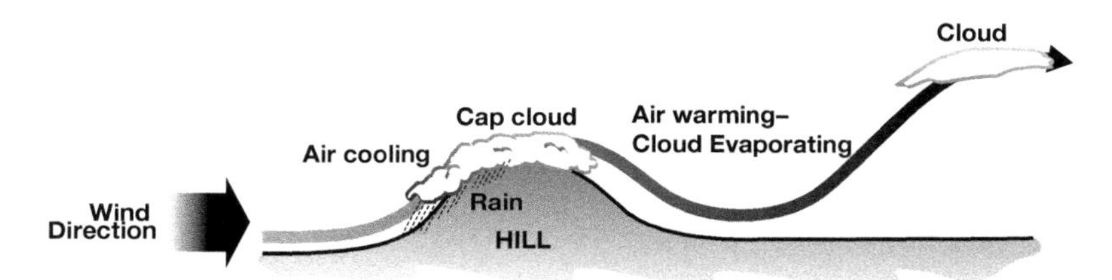

Fig 6.9 Foehn gap. Air made drier during ascent of high ground may cause a cloudless gap on the lee of high ground.

A foehn gap can also occur if the low level air on the upwind side of the high ground is cold enough and stable enough to resist being forced over the high ground. In this case drier, upper air may be deflected downwards, evaporating any low cloud to the lee of the high ground.

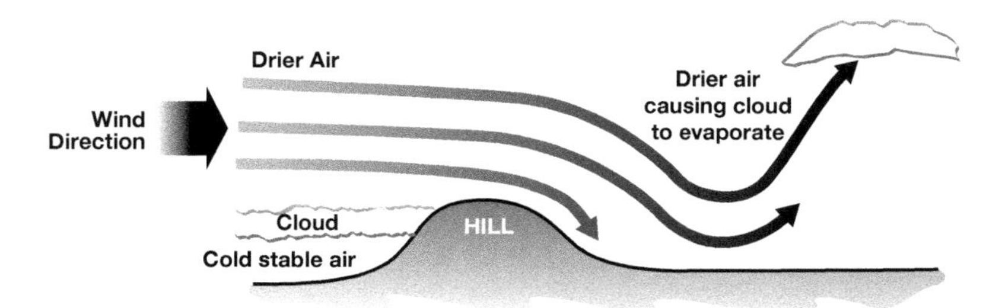

Fig 6.10 Drier upper air causing a foehn gap. If drier air from upper levels descends the lee slope, a foehn gap may occur.

If the wind is moderate or strong, the downdraught on the lee side of high ground can be severe, and the air in this region very turbulent.

Orographic cloud

Although an orographic cloud is not a result of wave, the presence of such a cloud may indicate that wave is present. This is especially the case if the airflow through the cloud appears laminar, and the cloud is seen to initially descend the lee side of the hill before evaporating.

Lenticular clouds

A classic wave system is accompanied by classic clouds, called LENTICULAR CLOUDS. These clouds, although they get their name from their more common 'lens-shaped' cross-section, can come in a variety of shapes. They are normally long, thin clouds orientated parallel to the ridge of high ground which is causing the wave system, and therefore generally lie close to, but not necessarily at, 90 degrees to the wind direction. Two features distinguish them from cumulus clouds. Firstly, they tend to have a very smooth profile due to the laminar characteristics of the airflow causing them; and secondly, they are usually stationary in the sky despite the strong wind which may be blowing.

Their formation is totally due to the wave system. As the air rises up the front of the wave, it cools. As it cools it is no longer capable of retaining all of its water content as vapour. On reaching its dew point, this water vapour will condense out as cloud – that is, a lenticular cloud will form. As the air descends again on the down side of the wave, the air will be warmed and the water droplets of the cloud will be evaporated back into invisible water vapour. Therefore, in classic wave conditions, the clouds will be present only as a strip at the position where each wave reaches the condensation level. The fact that the wave system is standing still relative to the high ground which is causing the wave means that the clouds also remain stationary and, rather conveniently, signpost the position of the rising air.

It is often possible to see the streamlines of the air as the moisture condenses out as cloud, and even where it ascends the cloud's leading edge. The steepness of these streamlines at the leading edge of the cloud gives some indication of the position and strength of wave lift. The thickness of the cloud may simply be an indication of a large amount of moisture at that level, and not necessarily of strong lift.

If conditions are favourable, then similar wave systems may exist above the lower one. These systems may also be marked by lenticular clouds at higher levels, depending on the moisture content of the air at these levels.

Often the clouds caused by a wave system are less than classically lenticular. Sometimes they will look more like smooth, isolated towering cumulus, or often stacks of plates piled one above the other. Bars of low wave cloud can often look like a long cumulus street when viewed from the ground below. The one thing that they will all have in common is the fact that they tend to maintain their position – for long periods at least.

Rotor and rotor cloud

Beneath the crest of each wave, especially beneath the primary and those waves closer to the high ground, the airflow is often greatly disrupted, causing severe turbulence. This area of turbulent air is called a ROTOR. Depending on the humidity of the air mass at this level, the rotor may be marked by a ragged cloud, whose shape is constantly and rapidly changing. This cloud is known as a ROTOR CLOUD or ROLL CLOUD. Rotor, if present, will normally be found level with, or just below, the tops of the source hills or mountains – and should be avoided at all costs.

Where rotor extends down to the ground, it can cause gusts, severe turbulence, wind shear and a complete reversal of the wind direction. Glider pilots, from sites where lee waves are common, are all too aware of the damage rotor can cause to trailers and club buildings, and know when to suspend launching until there is a favourable change in wind direction.

Cloudless wave days

If the air mass does not contain enough moisture, then no condensation will occur, even although a strong wave system may exist. This will mean that the presence of lee waves will not be indicated by lenticular clouds, and that the position of any rotor will also be unmarked by cloud.

Other types of waves

Wake waves

Isolated peaks (as opposed to lines of hills) can also cause waves to form, although these waves tend to align themselves in a more downwind direction similar to the wake of a ship. They may interfere with other waves, causing localised variations in the rate of ascent of the air.

Waves over cumulus

Not all wave is triggered by terrain features. Clouds can also provide enough of a barrier to start the air mass into a wave motion.

When a cumulus cloud forms, it moves downwind at the speed of the wind at the height where the thermal feeding it became an entity in its own right. If the wind speed increases with height then the faster-moving upper air, on encountering this slower-moving convective barrier, will be deflected upwards to flow over the cloud, thus potentially initiating a wave motion.

Unfortunately, unlike a hill or mountain, a cloud will only act as a wave trigger as long as the thermal continues to feed the cloud. After the thermal ceases, the cloud will start to break down and drift downwind. This will herald the end of this temporary wave.

Fig 6.11 Convective barriers. Cumulus clouds may form enough of a barrier to deflect the air upwards, causing a temporary wave.

When thermal streets form, whether or not they are indicated by cloud streets, each thermal will have an effect on the inversion above it. The inversion above the streets will be pushed up slightly, giving it an undulating nature. If the upper wind is at a large angle to the wind direction below the level of the inversion, then the undulations of the inversion may be enough to cause waves to form. A wave system may then exist above the streets.

Whether or not these waves can be used efficiently when cross-country flying, they are certainly worth exploring while local soaring if their presence is suspected. If cumulus clouds have smooth, curved tops, or if thermal climbs are of variable strengths with heavy sink in between, then think 'wave'.

Travelling waves

Not all waves are stationary. In some countries large, travelling waves occur either as solitary waves or as multiple waves. They are thought to be triggered by large atmospheric phenomena, such as the colliding of

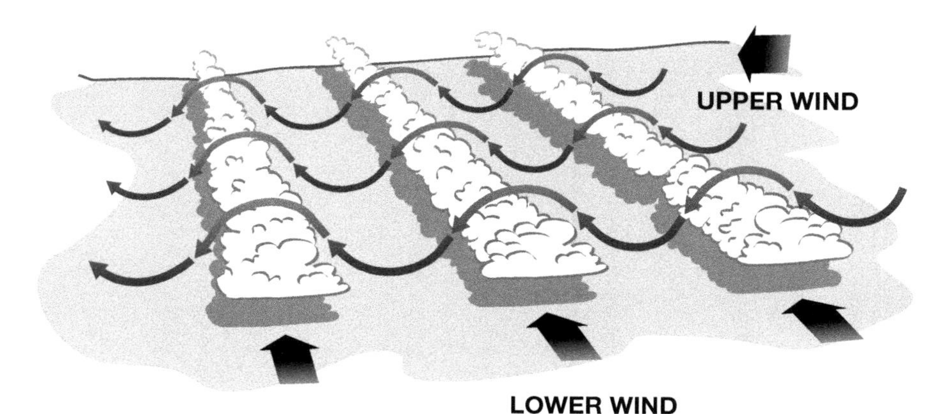

Fig 6.12 Waves parallel to thermal streets. If the upper wind is at a large angle to thermal streets, wave may occur above the thermal streets.

opposing sea breeze fronts or the downdraughts from large cumulonimbus clouds. They may travel for hundreds of miles and be eminently soarable – but to remain in lift, a glider pilot would have to go cross-country with the travelling wave.

Chapter 7

Wave Soaring

Is there wave present?

The first problem that you face when you turn up at the gliding site is deciding whether or not there is wave present. If there are classic lenticular clouds all over the sky, or reports from other pilots already airborne, then it may be fairly obvious that wave exists. On the other hand, some days will not offer such an easy answer to the question, 'Is it waving?'

The weather forecast or advice from local pundits might be the first reason to believe that wave soaring is likely. Tow-plane pilots will often have more up-to-date information, not only on whether there is wave, but also on where it has been contacted and how high it will be necessary to be towed to contact the lift.

If information is not available from these sources, then it will be necessary to look for more subtle signs to aid your deduction.

There may be clouds around, but although they may not be classic, their nature may give some hints that wave is responsible for their formation. Firstly, look at whether or not any clouds remain stationary over a period of time. Secondly, see whether they have smooth sides or tops. If the answer to either or both of these observations is 'yes', then wave is probably present.

Other signs which suggest that wave flow may exist may be observed near the high ground likely to be triggering or enhancing the wave. An orographic (cap) cloud, which initially descends the lee slope but evaporates shortly after beginning the descent, may indicate the start of a wave motion.

Smoke from heath fires or chimneys on the lee slope which descends the slope (and, in some instances, may even be seen ascending again in a gentle curve some distance downwind) shows the existence and even the position of a wave (possibly even the primary wave).

Erratic and unpredictable variations in the surface wind strength, causing periods of calm when the wind has generally been moderate or strong, are signs of position changes of the wave crests aloft.

On cloudy days, the only indication of wave activity may be regular, if narrow, clear slots in an otherwise overcast sky.

On days when thermal soaring is possible, the presence of lee waves may make expected rates of climb unpredictable and areas of sink unexpectedly severe. If cumulus clouds are present, then their spacing may be non-uniform, with large areas persistently devoid of cloud, especially near the lee of high ground. On the other hand, the wave influence may be strong enough to determine the pattern of the cumulus. In this situation, 'streets' of cumulus may form *across* the direction of the wind instead of in line with the wind. Some cumulus clouds may display smooth tops or have little wave clouds above them (known as pileus cloud).

It was a beautiful day. The wind was light. Visibility was excellent and the sky clear of cloud – except for an orographic cloud shrouding the front and top of the nearby hill, which was commonly used for hill soaring. Because of this cloud, no one was prepared to take a launch.

On the basis that the presence of orographic cloud suggests some uplift of air at the hill face (although perhaps not enough to keep a glider airborne) a fellow club member with an Olympia 463, and myself flying a Pirat, took a winch launch and flew towards the hill. Several 'experienced' instructors muttered disapprovingly – it was not the 'done thing' to attempt to soar a hill shrouded in cloud.

The first thing that I noticed as I turned to parallel the edge of the cloud which was clinging to the hill was that, despite the light wind on the airfield, the wind at 1,500 feet was very much stronger. In fact, it was impossible to stay in front of the hill unless the glider's nose was pointed well towards the wind.

The next surprise was how well the 'hill lift' was working, both in height and in strength. It was soon apparent that the lift in front of the hill was usable well beyond the normal heights expected from ordinary hill lift on this particular hill. We were in wave.

Soon we were joined by other gliders and enjoyed one of those special days when what seemed like half of the club fleet was strung out in a stationary line at around 10,000 feet before we all went off individually to explore the wave.

Finding the wave lift

Many of the features mentioned above can also be spotted from the air, and therefore it is essential that you continually assess the chance and likely position of any wave while airborne.

The rising air associated with a wave system is the air which is moving up the front of the wave. Therefore, the glider must be positioned in this area if it is to gain height.

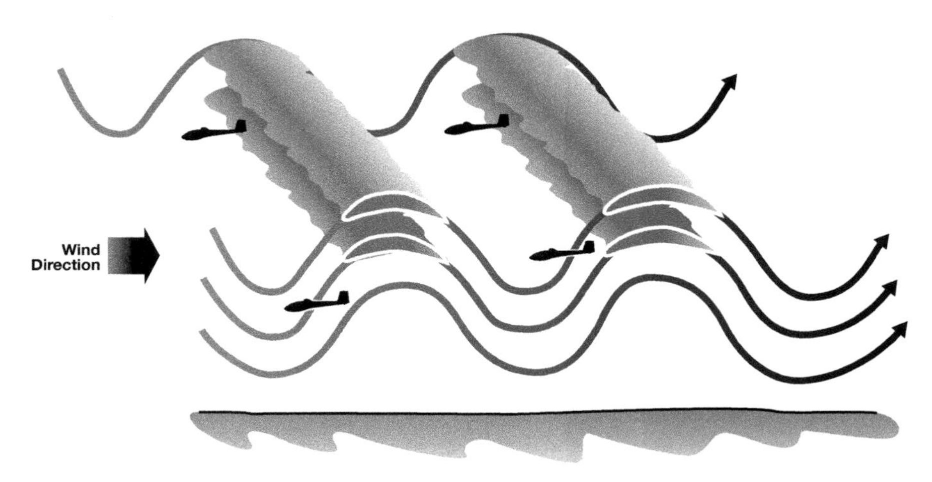

Fig 7.1 Position of wave lift. To climb, the glider must be positioned in the up-going part of the wave.

If the sky is cloudless, then finding this area of lift can be difficult, although local knowledge at such times is invaluable and well worth seeking before take-off.

Finding the wave lift becomes somewhat easier if there are lenticular clouds indicating the position of the crests of the wave. In this instance, you should fly the glider into a position upwind of the associated lenticular cloud. Typically, the lift will be found about a quarter, to one third, of the distance between one cloud and the next cloud upwind. This relationship will vary, and as height is gained, the lift will be closer to the cloud.

Lenticular clouds that have very little vertical depth may be disappointing as wave markers. The airflow producing these, although ascending, may not be rising at as great a rate as the air elsewhere.

At low levels, the wave may manifest itself only by the presence of small wisps of cumulus-like cloud which form at the downwind end of a cloudless gap. These increase in size as they move downwind and join the main wave updraught.

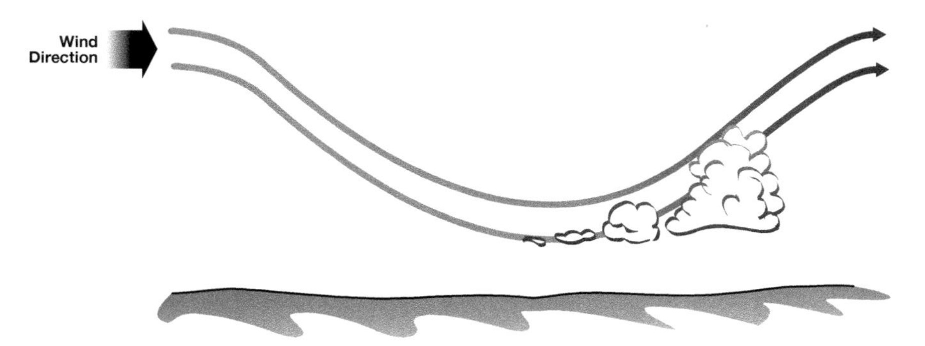

Fig 7.2 Fractocumulus indicating position of wave lift. Small wisps of cumulus-like cloud drifting and congregating at the downwind edge of a cloudless gap may indicate the position of wave lift.

Whether or not lenticular clouds are present, entry into the wave lift will be marked not only by an indication of lift on the variometer, but also by the smoothness of the air in which the glider is flying. This air may be so smooth that often you will not need to make any control inputs for minutes on end. Indeed, the first time you experience good wave lift, you can easily be forgiven for distrusting the variometer and altimeter readings, both of which will be showing climb rates normally associated with large, turbulent thermals or strong hill lift.

Methods of entering the wave lift

The easiest method of entering wave lift is to be aerotowed into it. This is common at many wave sites, and local tow-plane pilots usually know where the wave lift can be contacted in most wind conditions. This may mean towing the glider a considerable distance from the airfield and/or to a height greater than normal launch heights. The technique most likely to succeed is for the glider to be aerotowed directly into wind towards the high ground causing the wave. By doing so, the glider should be towed through successive areas of lift and sink until it is high enough and in lift good enough to release from aerotow.

While on tow, it is essential to monitor the variometer, especially if the air becomes smooth. Do not be tempted to release the towrope too early, as in wave conditions this is a sure way to end up back on the ground prematurely. Instead, wait until the variometer shows a large improvement over the normal rate of climb on aerotow before releasing. Do not be fooled by a relative, but still mediocre, increase in the rate of climb after being towed through an area of strong sink. If in doubt, do not release the towrope. Instead, take the aerotow higher and further into wind.

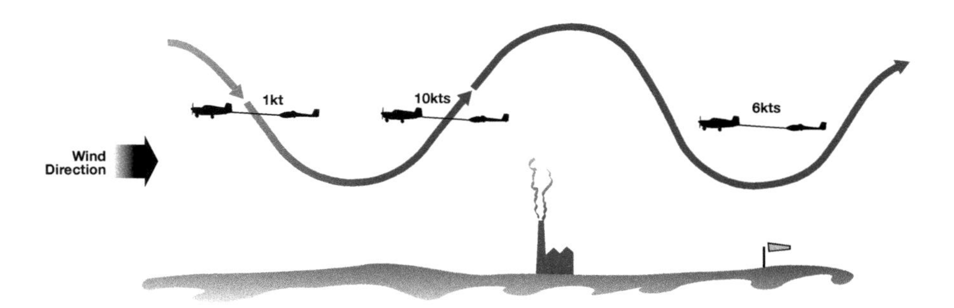

Fig 7.3 Variometer readings on aerotow. Learn to recognise what is a normal rate of climb while on aerotow.

It is possible to contact wave lift directly from the top of a wire launch. However, contacting wave lift directly from a wire launch is much less likely than from an aerotow and will depend greatly on the wave system being low enough, and in a position within reach with the height available.

Occasionally, a glider is able to gain enough height in hill lift to reach the wave lift. This may involve gliding out from the hill lift in the hope of contacting the wave lift. If such an attempt to reach the wave fails, then the glider can be flown back to the safety of the hill lift to regain height. As any change in the wind conditions may result in a change in the position of the standing waves, the attempt to reach the wave lift should be repeated regularly.

At some hill soaring sites, the up-going part of the wave may coincide with the hill lift. If this is the case, not only will the rate of climb which the glider can achieve in the hill lift be increased, but it will also be possible for the glider to climb directly from hill lift into the wave system. It should be appreciated that any shift in the position of the wave system may result in the opposite effect – that is, the hill lift may be cancelled out completely by the down-going part of the wave. Therefore, caution should be exercised on any occasion when the glider is scampering back to the hill, hoping for hill lift to save it from an early landing.

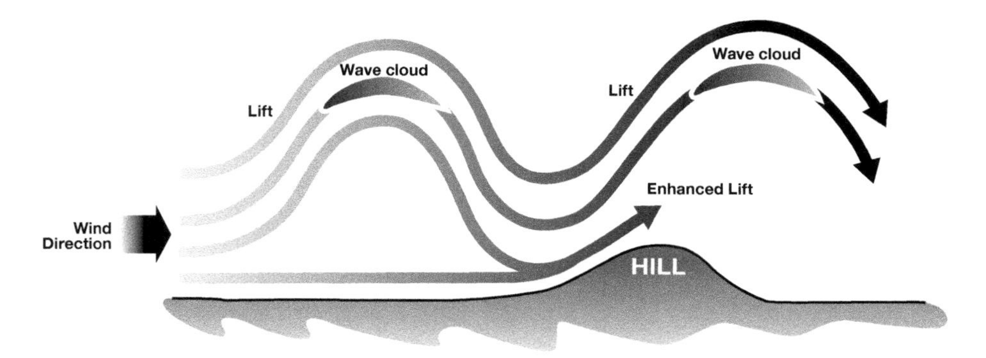

Fig 7.4 Wave affecting hill lift. When the up-going part of the wave coincides with the hill lift, a much-increased rate of climb will be achieved in the hill lift. Hill lift may be cancelled out completely if the wave is out of phase with it.

The first sign that wave is being contacted may be the smoothness of the air, accompanied by an equally smooth increase in the variometer reading. On the other hand, it may be the entry of the glider into rough, choppy air. This 'cobblestone' effect is often experienced (along with broken, intermittent lift) just before the glider climbs into the smooth airflow typical of wave. While encountering this turbulence, you should attempt to gain and jealously guard every foot of height, as only by doing so do you stand a chance of contacting decent wave lift.

When thermal soaring is being carried out, mountain lee waves, if present, may interfere with the distribution and the strength of the thermals. This wave interference may cause those thermals which coincide with the rising part of the wave to give a higher rate of climb; whereas the rates of climb in those thermals in the descending part of the wave will be reduced. Some of these thermals may be very distorted, making climbing difficult. If a thermal giving an enhanced rate of climb can be used, it may be possible to climb into the wave system and enjoy some exceptional flying on what started out as a thermalling flight.

On some occasions, it may even be possible to leave the thermal when cloud base is reached, fly into wind and climb up the front of the cumulus cloud in wave lift. Increasing airspeed as much as possible while still in

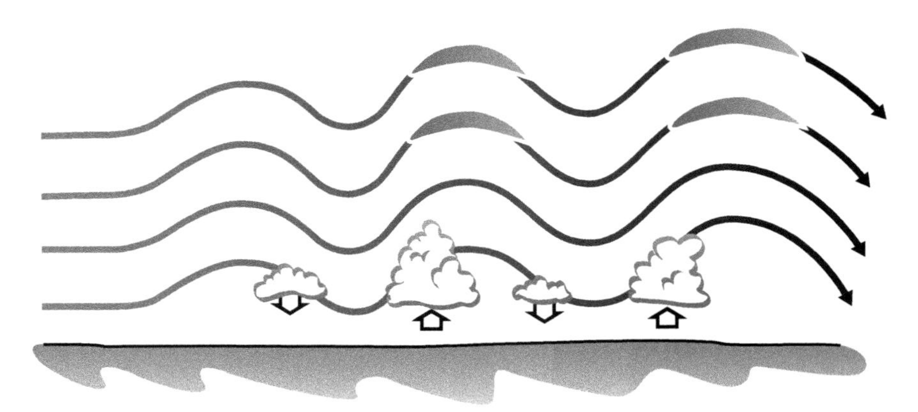

Fig 7.5 Wave interfering with thermals. Wave lift will often affect thermal distribution, enhancing some thermals and suppressing others.

the lift of the thermal, and then converting this excess airspeed to height as you clear the edge of the cloud, may increase your chances of contacting the wave lift. In this way a pilot may gain the extra height required for a Silver badge without entering cloud. On other occasions, the wave may only be reached by taking a cloud climb.

Fig 7.6 Contacting wave from thermals. Waves may occasionally be contacted by climbing to cloud base in a thermal and then flying into wind.

Whichever method is used to enter the wave lift, if the glider encounters the wave system near its lower levels, the lift may be broken and difficult to use. If this is the case, flying the glider at the airspeed which gives the minimum sink rate might help gain enough height to place the glider into the better lift which may exist slightly higher up. Once established in better lift, you can afford to be more adventurous in trying to map how far the area of lift extends, and the position where the best rate of climb can be found.

Wave soaring technique

Once you have contacted consistent wave lift, the way in which you use it will depend, to a large extent, on what your goals are for the flight. The techniques which follow are written for the pilot who wishes to use wave lift to gain the maximum height, possibly to achieve a badge requirement or a record. Some of the techniques will also apply to the pilot who wishes to utilise the wave as a means of flying cross-country. However, as the cross-country pilot may not be as concerned about gaining or conserving height as the pilot striving for altitude, cross-country flying in wave will be dealt with in Chapter 15.

The normal technique for using wave lift is to fly the glider in beats, back and forth in the area of lift, just as one would when soaring in hill lift. If there are lenticular clouds present, you will be able to refer to the cloud to maintain the glider's position in the area of lift, as well as using it to mark the extent of the lift zone. When the glider is well below the level of the lenticular cloud, the best lift may be well forward of the cloud. As you gain altitude, the lift will be closer to the front of the cloud. At this point, the analogy with hill soaring will be an accurate one. Once above cloud level, better lift may be found further upwind of the attendant cloud. If the cloud has a near vertical windward face, then strong lift may be found in a narrow band in clear air close to the upwind side of the cloud.

The length of each beat in lift will vary depending on conditions. In some cases, the area of lift may be very long, allowing you the chance to explore the lift while flying each beat. On other occasions, the lift area may be quite small and require you to fly small beats, or even to soar in it by doing a series of 'S' turns until something better can be found. During these beats in lift, the maximum rate of climb will be achieved by flying the glider at an airspeed close to that giving the minimum sink rate in still air.

Occasionally an area of lift may be encountered which takes the form of a narrow core of turbulent, rapidly rising air. If such a core is encountered, it is often best to exploit it using a thermalling technique, with large angles of bank. Higher up, such narrow lift may be smooth, and if lift remains strong, any tendency to drift may be negligible. This is often the case if the glider is close to a cloud which has a steep vertical face.

The wind will have an effect on the glider's track during its beats in the lift. In light winds, the effect may be small and can be countered by tracking the glider with its nose pointing slightly into wind. The amount by which this tracking will be necessary depends on the wind strength. As wind strength will increase with altitude, it is not uncommon to find that it is necessary to point the glider directly into wind, simply to remain in the lift, and to avoid drifting back into the sink behind the lift. Even by flying the glider into wind in this way, there will still be occasions when the only way to remain in the lift will be to increase airspeed to counter what may be a very strong wind at flying levels.

Having established the glider in lift, any subsequent loss of the area of lift will probably be caused by one of the following events.

1. You may have flown the glider beyond the end limits of the area of lift, in which case, turn the glider through about 160 degrees (turning into wind – never downwind) and fly back the way you came. This should bring you back to the area of lift.
2. The glider may have drifted back behind the area of lift. It is quite common to drift back behind the lift area, so firstly you should fly the glider forward into wind to see if the lift is re-encountered in this area. This can be done either by turning more into wind; or if the glider is already heading directly into wind, by increasing the airspeed. If the lift is rediscovered, the original track or airspeed can be reselected.
3. The glider may have been flown to a position upwind of the area of lift. If you suspect that this is the case then turning across the direction of the wind may be enough to drift the glider back into the lift. In strong winds, this should be enough to displace the glider sufficiently downwind. On regaining the lift, turn back into wind, or track as necessary. On no account should you turn directly downwind if the wind is strong, as this will increase your groundspeed so much that you may quickly fly through the lift and lose it altogether.

In all of these situations, resuming a track or heading close to the original track may lead to the lift being lost in the same way again. Annoying as this may seem, at least you now know how to regain the lift, whereas adopting a different soaring track or airspeed may result in losing the lift in some other way, necessitating a completely new search. Monitoring your position relative to the cloud or a prominent ground feature may help in deciding how the lift was lost, but the greater your altitude, the less accurate a ground reference will be.

Over or close to mountains, the area of wave lift will often lean upwind as it goes higher. This means that the top of a wave climb may be several miles upwind of the place where the climb started. Further downwind of the mountains there may be less tilt and the area of wave lift may be almost vertical. In this latter case, you may find that you are over exactly the same place at the top of the climb as you were when you contacted the lift at the start of the climb.

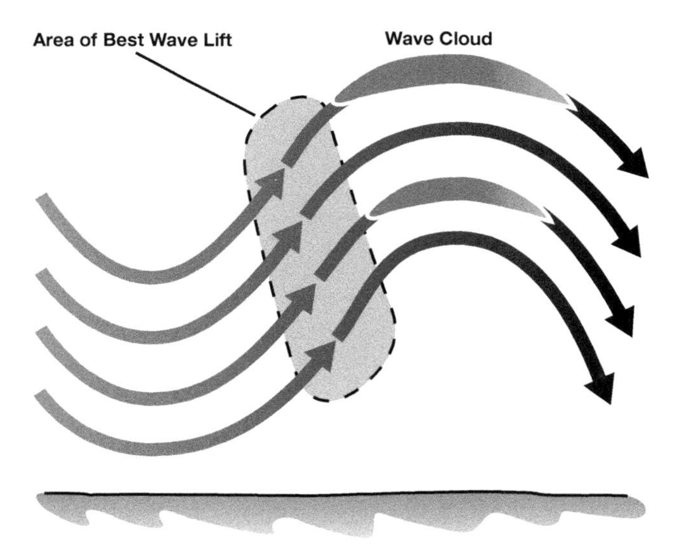

Fig 7.7 Wave lift position sloping with altitude. As altitude is gained the position of the best lift will often be further upwind.

Possibly, through no fault of yours, you may lose the lift because the wave system may have changed, and the lift ceased or gone somewhere else. If the wave system changes, the position of the lift, sink and lenticular clouds will change.

In fact, the system may break down completely, making any search for the new position of the lift fruitless. All of this may occur almost instantly. All it takes is a change of wind direction or strength, or perhaps the influence of a large shower many miles upwind, and you may find yourself sitting in an area of sink or even in cloud.

Climbing at a steady 4 knots past 5,000 feet, the Bocian was stationary with the mountains below and a lenticular cloud behind us. We were enjoying the view which the fantastic visibility gave us.

Then, as I looked up, what had been a sunny day appeared to be growing overcast. In the next instant, we were engulfed in cloud. The variometer needle was now showing the maximum down indication of 10 knots – maybe the actual descent rate was more.

We increased airspeed, thinking that this would soon bring us out of the upwind edge of the cloud and into more favourable air. For what seemed a long time, we stayed in cloud with the altimeter unwinding. We broke out into the sunshine over the valley at 2,000 feet. In the area we had left, the mountains rose to 3,500 feet!

The good news is that such changes in the wave system can provide better conditions as well as worse. There is nothing more pleasing than to be searching for the next area of wave lift and to see a lenticular cloud forming beside your wing tip, as if by magic!

Your rate of climb may even have reduced to zero because you have reached the top of the particular wave system. If this is thought to be the case, the lift will probably have diminished at a very gradual rate. If it is more altitude you are after, you will have to make contact with another, higher wave or wave system. As Murphy's Law would have it, this 'topping out' of the lift always seems to happen just below that vital badge height. (I am convinced, albeit unscientifically, that if the heights for badge awards were made lower, the height at which wave systems 'top out' would also be lower!) It is at this point that your greed for height must be discarded. If you are to contact a higher wave system, you will probably have to penetrate into wind. To do this you will have to sacrifice some of your height. Hard as this may seem, if you stay where you are you will not gain any more height, and will still be short of that badge or record. So you may as well go for it, and hopefully, the few thousand feet you have to throw away will be rewarded by a higher climb further upwind.

In making this assessment of where to fly to reach greater heights, look

Having been soaring at 17,000 feet, contact with the wave lift was lost quite suddenly and severe sink was encountered. It was not long before I was down to 10,000 feet, still descending rapidly and (being over the mountains) several valleys away from a safe landing area.

Despite frantic searching for better air, the high rate of descent continued, as did my attempt to fly away from the mountain plateau. I could not figure out what had happened to the wave!

At 7,000 feet there appeared in front of the glider what looked like grey smoke, as if someone had lit a bonfire and the smoke was streaming in the wind – but at 7,000 feet? Maybe I should have remained on oxygen longer?

As I watched this phenomenon, the smoke thickened rapidly and I realised that what I was seeing was a lenticular cloud forming just in front of the glider.

A quick change of track upwind of the 'smoke' and I was in strong lift with what had become a well-formed lenticular cloud just off my left wing. Soon I was at 23,000 feet.

around and search for other lenticular clouds above your level. If the sky is overcast, this may make it difficult to see the outlines of lenticular clouds above. There is always the possibility that the overcast is being caused by a lenticular cloud only a couple of thousand feet above you. If this is the case, a move upwind may be all that is required to reach the lift which is causing the cloud. In contrast, turning downwind when at the top of a wave usually guarantees a greater loss of height than moving upwind.

Hazards associated with wave flying

The following is a description of the hazards which can be present when wave flying. The fact that few of these lead to problems is a tribute to the likelihood that pilots embarking on wave soaring flights are generally aware of the potential problems, and take precautions to prevent them occurring. The following paragraphs are aimed at improving your awareness, and thus helping you to prepare yourself and equipment for wave flying. The knowledge that you are well prepared for this form of soaring will give you confidence and add to the comfort and enjoyment of one of the most fascinating forms of flying.

Sudden sustained height loss

'What goes up must come down.' When applied to wave flying, this expression is certainly true on a large scale. It is very easy to find oneself in the down side of a wave, either accidentally or when moving between areas of lift. When the still air sink rate of a glider is added to the rate of descent of the air mass, wave downdraughts can appear very severe and seem very wide. Gliding back to base or to a safe landing area will require very large safety margins in terms of height for the distance to be covered.

> *Most of the elements of the winch launch were normal: good initial acceleration and good airspeed during the launch. However, we were not climbing as well as we should. This fact was evident both by the low variometer reading and by the view from the cockpit.*
>
> *The top of the launch was at a much lower height than would be expected in what were favourable wind conditions. There followed a very short, exciting circuit to land, most of which was carried out in horrendous sink.*
>
> *We had launched in the descending part of a wave.*

Complacency

Once established in lift, wave soaring can be remarkably easy. The air is smooth, the lift can be strong and it is easy to become obsessed with the quest for altitude. All of this can lead to a false sense of security. The fact that the glider may be operating at altitudes and in wind strengths which are not normally associated with other forms of lift, means that safety during wave soaring flights depends on the pilot remaining vigilant.

Strong or variable winds

Many good wave soaring days are accompanied by strong winds, even at ground level. Extra care must be exercised during take-off and landing, as severe turbulence is a common feature of airfields situated near high ground. When landing, especially after a long flight, you may find that conditions have changed somewhat since your launch. This could mean that the wind strength or direction may have changed, possibly resulting in an increase in turbulence on the approach. It is always a good policy to keep in radio contact with the base airfield, so that you can be informed if any deterioration in conditions has occurred. In turn, many gliding sites at which wave soaring takes place will broadcast warnings of such deteriorations and may in bad cases advise an immediate landing (or even a landing away from base) if local rotor is becoming a problem.

At other times, the wind may have dropped to a light breeze over the airfield, as a result of this area being under the crest of a wave. This can be embarrassing as you 'eat up' runway, having started the approach with extra airspeed to allow for a strong headwind or wind gradient.

The strong winds will make it easy to lose track of one's position unless sight of the ground is maintained. Even when the ground is in sight, one part of a mountain plateau can look very much like another from 20,000 feet, if even a fraction of the ground is obscured by lenticular or orographic cloud. Maintain your orientation by keeping known ground features in view. If your glider is equipped with a satellite navigation system (generally known as a Global Positioning System, or GPS), this can be used to constantly monitor your position, despite the problem of partial cloud cover.

Turbulence

The slight buffeting type of turbulence, described previously as 'cobblestone effect', is more of an indication that the air is up to something, rather than a nuisance. On the other hand, the turbulence associated with mountains and wave can be frightful. It is essential before taking off on a wave flight, that there are no loose objects in the cockpit, that heavy items (such as oxygen equipment) are firmly fixed, and that your harness is tight.

The fact that wind, mountains and wave flying go together means that there is often some degree of turbulence when near the ground. Take-offs and landings may take place in extreme turbulence due to curl over from nearby hills, or because of rotor. When the wind is strong, there is a good chance that approaches will be made through a steep wind gradient. This will require extra airspeed from the beginning of the approach if you are not to be caught out.

If attempting to contact wave lift when at lower altitudes, perhaps when moving directly from hill lift, you may have to penetrate part of the rotor before reaching the wave lift. The air in this area is likely to be very turbulent.

Turbulence is often encountered at lower altitudes. Special care should be taken when descending at high airspeed, as what may have been smooth air at altitude may suddenly become very turbulent air lower down. It is wise to slow down to normal cruising airspeeds as the glider descends towards the height of the local mountains, or to 4,000 feet above airfield level.

Upper air turbulence can also be severe. There is an account of an RAF Canberra aircraft being inverted by wave turbulence while flying at 40,000 feet over the mountains of Scotland.

Cloud cover

Unlike other types of soaring, wave soaring can result in the glider climbing quickly above cloud, without the need to deliberately enter cloud. The danger lies in the fact that what may have been scattered clouds below the glider, when altitude was first gained, can quickly become a thick layer of cloud, between the glider and the ground. This may happen for several reasons.

As the gaps in cloud may have been caused by the presence of the wave system, any change in the conditions supporting the wave system may cause the system to collapse and the cloud gaps to fill in. If fronts are approaching, the moisture content of the air mass may increase, causing layer cloud to form at a level below the glider.

The first signs that the cloud gaps are disappearing is often the appearance of streamers of cloud forming across the gaps between the lenticular clouds. Alternatively, the lenticular clouds may simply lose their structure and merge, obscuring the ground from view.

Even if your glider is equipped with cloud flying instruments, you will be in the unenviable position of having to descend through cloud, which may extend below the heights of the high ground in the area. Whether or not your glider is equipped with navigation aids, descending into mountain-filled clouds is an unhealthy pastime. The only safe answer to this problem is not to get into this predicament in the first place. With this in mind, never climb through small cloud gaps. Always watch out for the areas of cloud below extending to join up and fill the cloud gaps. Start your descent early and do not underestimate how long it will take to get rid of your height. Losing 20,000 feet, for instance, can take several minutes, even when diving with full airbrake.

Should you be unfortunate enough to get caught out above total cloud cover, then you should consider flying a heading which will take you towards flatter terrain where a descent through cloud will be less dangerous. This contingency should be assessed before take-off, along with the heading required and timings to cover the necessary distance at the likely airspeed. Such calculations should take the likely wind conditions at flying levels into account.

If you have GPS on board, this can be used to navigate to a position where a descent can be made, knowing that at this location there is no high ground below the likely base of the clouds. If GPS is to be used to break cloud in this way then it is best to follow a pre-planned flight pattern which prevents you straying too far from the 'safe' descent position. This, of course, assumes that the wave system has not collapsed, and that you can manage to remain airborne to reach such a descent point.

These 'cloud break' techniques are only for use in an emergency, and as such, you should not climb above cloud knowing that such a technique will be your only way of descending.

Lack of oxygen

From 10,000 feet upwards, the amount of oxygen in the atmosphere reduces to a level where a pilot will suffer a rapid decline in mental ability, unless breathing air supplemented by an oxygen system carried in the glider. The onset of HYPOXIA, as the effect of insufficient oxygen is known, can be quite insidious. In fact, one of the main symptoms of its onset is a feeling of well-being which may well mask other symptoms, such as clumsiness, lack of co-ordination, narrowing field of vision and tiredness. Depending on the time spent at altitude, unconsciousness may occur.

Above 20,000 feet, consciousness will be lost within ten to fifteen minutes unless supplementary oxygen is used. This period will reduce rapidly to between two and a half to six minutes at 25,000 feet. At 30,000 feet, it will be reduced to between one and a half to three minutes.

The height at which a pilot will suffer from hypoxia also depends on the fitness of the pilot. An unfit pilot can expect to show symptoms of hypoxia at a lower altitude than one who is in good physical condition. Cold temperatures add to the problem.

Any increase in physical activity will increase the effect of insufficient oxygen. While it is hard to imagine the pilot of a glider exercising in an energetic manner while in flight, actions such as struggling in a cramped cockpit to reach stowed objects may be enough to lead to exhaustion or light-headedness if insufficient oxygen is available.

Because of these dangers, the British Gliding Association recommends that an oxygen system should be carried if you intend flying above 12,000 feet. It should be used at all times when the glider is above 10,000 feet. In the event of its suspected malfunction, or if you feel unwell, you should descend below 10,000 feet as quickly as possible.

The oxygen system used should be suitable for the intended flight, and be turned on at the main valve on the cylinder BEFORE take-off, unless this valve can be reached easily in flight.

CONSTANT FLOW oxygen systems are used by many glider pilots. As the name suggests, these systems deliver a steady flow of oxygen (either two or four litres per minute depending on the regulator setting). They are simple, relatively cheap and reliable. However, as they supply oxygen continuously, whether the pilot is inhaling or not, they are wasteful in their use of oxygen.

DILUTER DEMAND oxygen systems automatically supply the required percentage of oxygen to the inhaled air depending on atmospheric pressure. Unlike constant flow systems, they supply oxygen only when the pilot inhales, making them more economical in their use of oxygen.

ELECTRONIC OXYGEN DELIVERY SYSTEMS are also available. These are lightweight and economical in their use of oxygen and supply the required amount of oxygen to the inhaled air as dictated by atmospheric pressure.

Above 40,000 feet, the atmospheric pressure is not sufficient to force enough oxygen into the body tissues, and therefore even the diluter demand system is insufficient. A PRESSURE BREATHING system is required which delivers oxygen under pressure to the lungs. Using such equipment requires special training.

The capacity of an oxygen cylinder and the pressure of the oxygen within it will determine the length of time it can deliver oxygen. This should be for at least the time that the pilot intends to be above 10,000 feet, plus a safety margin. (There are few things more frustrating than having to curtail a good wave climb because your oxygen supply is becoming depleted.) With this in mind, small, portable oxygen bottles are useless except as a back up. A badly fitting mask will waste oxygen and will render even the best system useless.

The system components must be compatible. For example, if the system is a demand system, the mask must be a demand mask. Consult the system's agent or manufacturer if you have any doubts about the system's operation or limitations.

The oxygen cylinder should have been tested, and tubing and connectors must not be perished. (Leaving equipment such as masks and rubber tubing exposed to direct sunlight for long periods accelerates their ageing.) Oxygen equipment should be kept away from oil or grease. These substances are explosive when they come into contact with oxygen at high pressure. All of this means that it is best to get an expert to assemble and fit your oxygen system. At any rate, the glider will require secure fittings to take a heavy item such as an oxygen cylinder.

During flight when oxygen is being used, monitor the contents gauge regularly. Do not attempt to stretch the endurance of the system by using a lower flow rate than is required, and never hesitate to increase the rate if you feel you need it. When assessing the amount of oxygen remaining, remember to take into account the amount you will require during the descent.

Also check regularly that the oxygen system is functioning. With a constant flow system, the re-breathing bag should slowly inflate when you are not inhaling or exhaling. Demand oxygen systems normally have an indicator to show you that oxygen is flowing through the regulator. The most likely cause of a malfunction (assuming that you have not run out of oxygen or inadvertently turned it off) is either a tube becoming crushed or disconnected, or a valve in the mask freezing. The latter is always a possibility because of the combination of the low ambient temperatures at altitude and the moisture caused by exhalation. Flexing the mask occasionally is usually enough to break up any ice formation in the mask valves before it becomes a problem.

Finally, monitor both yourself and your partner if flying in a two-seat glider. Look for any of the symptoms mentioned above. If in doubt, increase the oxygen flow rate, try to breathe normally and descend.

Decompression sickness

When you open a bottle of lemonade, the reduction of pressure on the liquid causes bubbles to form. The human body also contains gases, which at normal (around sea-level) atmospheric pressure are dissolved in fluids within the body's tissues.

As the atmospheric pressure on the pilot's body reduces during ascent, these gases can form bubbles which increase in size as altitude increases. If these bubbles obstruct blood vessels or form in joints, the pilot will suffer discomfort or pain – and in an extreme case, loss of consciousness and death.

Decompression sickness (DCS), as this condition is known, is well understood in the world of sub-aqua and deep sea diving, where it is commonly referred to as the BENDS. Although decompression sickness can occur at altitudes from 18,000 feet upwards, it is more likely at altitudes above 22,000 feet. Above 25,000 feet, the risk of suffering from DCS increases greatly. However, cases of DCS affecting glider pilots are unknown, although some experts believe this is a result of lack of reporting.

The main symptom to look out for is a stiffness or pain in a joint or joints. Should such a symptom be experienced, start using oxygen (or increase the oxygen flow rate if oxygen is already being used), descend immediately and seek medical advice. A doctor should be consulted even if the symptoms disappear after landing. Symptoms of decompression sickness may not occur until some time after an altitude flight. Further altitude flying should not be undertaken for a period of at least forty-eight hours.

Avoid flying within 12 hours of sub-aqua diving to a depth of 10 metres, as this increases the risks of suffering decompression sickness. This interval should be increased to 24 hours if a dive has been made to a depth greater than 10 metres.

Cold

The outside air temperatures encountered when altitude flying may be very low; possibly lower than -50°C at higher altitudes. Unless both the pilot and the glider are prepared for this extreme, then wave flying can be an unpleasant and dangerous experience.

In the cramped space of the average glider cockpit, it is extremely unlikely that you will be able to exercise enough to keep warm. The only answer is to dress warmly for the flight. Several layers of light comfortable clothing will insulate you better than just one or two heavy garments. Avoid working up a sweat before flying, as any moisture will reduce your clothing's insulation properties and lead to discomfort. Never launch with wet clothing or footwear. Warm boots, such as 'moon boots' over a couple of pairs of socks, are ideal. Ski gloves over thinner cotton gloves will protect your hands, while Balaclava-style headgear will prevent heat loss and draughts affecting your neck. Many pilots wear several layers over their upper body and feet but forget the legs and lower abdomen. These can chill just as much as the feet, and so 'long johns', and two pairs of trousers or some form of 'cover-all' may save you discomfort. Whatever clothes you wear, make sure that they are comfortable, do not hinder mobility and do not leave areas of skin exposed.

The glider should be sealed so that cold draughts do not affect the pilot. Care should be taken that such sealing still allows ventilation air both in and out, otherwise canopy misting becomes a problem. Ideally, ventilation systems should be directed at the canopy and not at the pilot's face or feet.

Icing

The low air temperatures involved in altitude flying create the risk of airframe icing. Water in the atmosphere can remain in a liquid state even at temperatures well below 0°C. On making contact with an object which has a temperature below 0°C, the water will immediately freeze and form ice on the surface of the object. A cold glider is a perfect object on which ice will form. Such ice accretion can have numerous adverse affects on the glider, making it wise to stay clear of any clouds when the glider has been flying at levels where the outside air temperature is below 0°C. Lenticular clouds can cause airframe icing to a far greater extent than layer-type clouds. However, even flying in clear air, your glider may still pick up ice if its surface is cold enough and the air it flies through is moist enough. A good example of this effect is when, in apparently dry, but cold conditions, you bring the glider out of its trailer and suddenly what was a dry glider is covered in hoar frost.

The changes that ice accretion makes to the profile of the wing and fuselage will reduce the glider's performance and possibly its handling. Even before take-off, ice or hoar frost may form on a cold glider if the air is moist enough. Any such ice must be removed and the wings dried BEFORE flight. NEVER launch with contaminated wings.

Ice accretion may cause controls to become stiff or even jam, especially airbrakes. As flying in the smooth air of a wave system often requires only a few small control inputs, you would be wise to move the controls regularly to reduce the chances of them icing up. The sealing surfaces of airbrakes, if wet, will provide a large surface along which water can freeze, thus preventing their opening. This is particularly the case with trailing edge airbrakes with their larger sealing surfaces. Occasionally opening the airbrakes just enough to break any ice may keep them free for later use. However, take care not to let ice build up in the airbrakes while they are unlocked, as this may prevent their subsequent closure. For this reason, do not unlock the airbrakes for such ice test purposes when close to or at the same level as cloud. After altitude flying, such a function test of the airbrakes on the downwind leg of the circuit to land is a good idea that may save some embarrassment on the approach. The grease used to lubricate the glider's control systems (including undercarriage linkages) should have a specification which will not freeze, even at low temperatures.

Ice may build on the canopy, obscuring vision. This icing may not be limited to the outside of the canopy. The pilot's breath may moisten the air enough to cause ice to form on the inside surface of the canopy.

Wiping condensation from the inside of the canopy usually smears the canopy, making visibility worse. Using cockpit ventilation to introduce colder but possibly drier air into the cockpit may be an uncomfortable but more efficient method of demisting the canopy and thus reducing the chance of icing. Do not use de-icing products on the canopy unless their use is specifically approved by the glider's manufacturer.

If visibility is significantly reduced, the direct vision (DV) window should be opened fully (despite any icy blast) in order to navigate, and perhaps even for landing.

Instruments may misread or fail if ice builds up on sensing probes or static vents. It is wise to have a changeover switch which will allow an alternative static source to be used for essential instruments. (Such alternative static sources almost always cause indication errors which are mentioned in the glider's flight manual, and are best assessed before you go wave flying.)

Although icing may occur at altitude, ice may not melt off the airframe and canopy as readily as it formed. It may remain on the glider during the approach and the landing. This will increase the glider's stalling speed, requiring extra airspeed on the approach. Add to this the reduced forward visibility, and icing becomes a serious hazard.

Onset of darkness

The sunset, when seen from altitude, occurs later than it does when observed from the ground below. This means that it is possible to be soaring in daylight at altitude when it is already dark on the ground. Add to this the time it takes to descend from altitude, and it becomes quite conceivable that, unless you are careful, you may find yourself searching for and attempting a landing at an unlit airfield in total darkness. To avoid such excitement, make a note of the local sunset times before take-off and make sure that you can get back to the airfield and land well before the onset of darkness. As a general rule, if you allow one minute to lose each 1,000 feet of your altitude, plus ten minutes to fly the circuit and land, then you should not get caught out. This assumes that you are close enough to your airfield to allow a descent with the airbrakes open.

The effect of altitude on limiting airspeeds

When an airspeed indicator is calibrated, it is assumed that the density and temperature of the air never varies from those of a so-called Standard Atmosphere. This means that, at low altitudes, the airspeed indicator will give a fairly accurate indication of the actual airspeed. As altitude

The Twin Astir in which we were flying had a retractable main wheel, which, unlike most gliders, retracted by swinging up sideways into the fuselage rather than hinging backwards in line with the fuselage axis.

We were joining an intentionally high circuit pattern after a wave flight which had taken us to 17,000 feet, when the pilot in the front seat attempted to lower the undercarriage for landing. The undercarriage lever would only go halfway towards the 'DOWN' position.

Thinking that it might simply be a matter of more force required on the lever, I offered to assist by pushing on the lever in the rear cockpit. The lever did not move in the slightest. I tried to retract the undercarriage to have another go but it would not come up either. Even with both of us heaving and hauling in unison, nothing moved.

We could only assume that either the linkages had contracted with the low temperatures at altitude, causing some kind of over-centre lock, or that water or mud in the wheel bay had frozen, jamming the levers.

A radio call to the airfield made sure that the best grass area was cleared for our landing, but with a wheel which was locked at a sideways angle of 45 degrees to the vertical, I could imagine the state of the mechanism after the landing.

We circled and flew up and down the downwind leg while we heaved and pushed on that wretched lever. We applied positive g to the aircraft – but to no avail. A landing was imminent.

On final approach, I gave the lever one last push – still solid. Then I decided to try to retract the undercarriage again. If this worked, at least the damage would be minimised. To my relief the undercarriage came up. It was free. Now I was faced with the decision of whether to attempt to lower it again and risk it jamming halfway again.

At about 100 feet, I pushed the lever firmly forward. The 'clunk' confirmed that we would be landing on a wheel rather than on the glider's belly.

On subsequent flights, the undercarriage was lowered while we were a lot higher.

increases, the density of the air reduces, resulting in the airspeed indicator showing less than the actual airspeed. This difference will increase as the glider's altitude increases. Therefore, the TRUE AIRSPEED (the actual speed at which the glider is moving through the air) will be greater than the INDICATED AIRSPEED.

This difference does not affect the glider's indicated stalling speed. If the glider's indicated stalling speed in a particular phase of flight at lower altitudes is, say, 36 knots, then at altitude it will still stall at an indicated airspeed of 36 knots. Best speeds for cruising will also be as indicated on the airspeed indicator, although in fact the true airspeed being achieved will be greater.

The problem occurs when we deal with the maximum airspeed allowed (Vne). At speeds above this airspeed, there is a danger of control flutter. As control flutter depends on true airspeed and not indicated airspeed, if the glider is flown at indicated airspeeds close to Vne at altitude, then it may well be flying outside its safe limits. For this reason manufacturers often state lower limiting indicated airspeeds for different altitudes. These will be given in the glider's flight manual and must be observed.

Table 2 is typical of the restrictions which appear in a glider's flight manual and shows the maximum indicated airspeed which must not be exceeded at various altitudes while flying that type of glider.

TABLE 2

ALTITUDE (feet)	INDICATED AIRSPEED (knots)
Sea-level	135
5,000	135
10,000	135
15,000	132
20,000	123
25,000	113
30,000	103
35,000	93
40,000	83

Chapter 8

Sea Breeze Fronts

We have already seen how temperature affects the density of the air and creates thermals. Although it may be hard to believe, two adjacent air masses with different densities will not readily mix. When an air mass meets a less dense air mass, it forces its way under the less dense air mass, causing the less dense air to ascend. This is the mechanism which we see on a grand scale when weather fronts are formed. However, such situations also occur on much smaller scales.

When sea air encroaches over warmer land, the cooler sea air, being denser than the warmer air over the land, will undercut the land air, forcing it upwards. The interface of these two air masses is known as a SEA BREEZE FRONT.

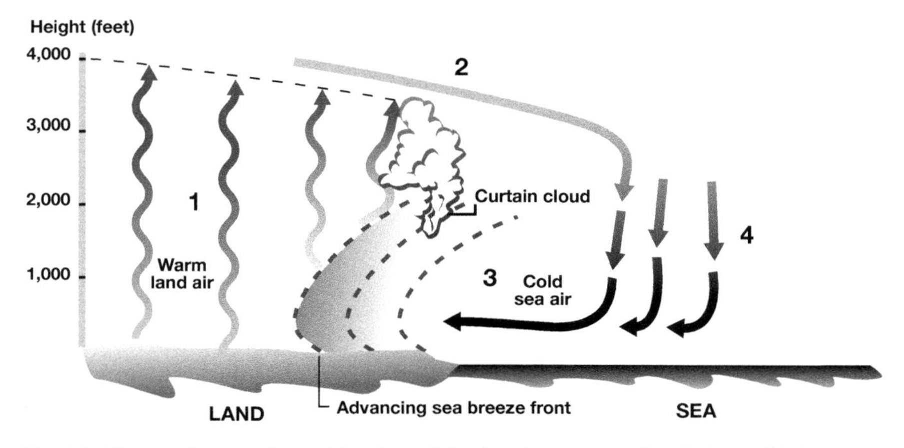

Fig 8.1 The sea breeze front. Heating of the land creates a circulation which causes ingress of sea air, resulting in convergence of two air masses.

Unfortunately, the mechanism which causes the formation of a sea breeze front is not simple, and whether such a front will be soarable in a glider depends on the conditions on the day.

Formation of a sea breeze and sea breeze front

When the sun heats the land, the air above the land warms up and expands (stage 1 in figure 8.1). The rising air causes a slight increase in atmospheric pressure at around 4,000 feet over the land. This results in a difference between the air pressure at this level above the land, and that above the sea.

This pressure gradient causes a flow of upper air seaward which reduces the surface pressure over the land (stage 2). Air at lower levels above the sea begins to flow inland towards this area of reduced pressure (stage 3). The air that moved out from the higher level above the land will descend to replace the sea air that flowed towards the land (stage 4). Thus, a sea breeze circulation has begun.

This colder, denser sea air, on meeting the warmer, less dense land air, forces the land air upwards, forming a sea breeze front. A sea breeze will only start to form when the land warms enough relative to the sea to trigger the necessary circulation. This may mean that often a sea breeze will not start until close to midday, beginning near the coast and gradually moving inland as surface heating increases throughout the day.

Sea breeze fronts have been experienced well inland, late in the afternoon and evening. A sea breeze front may even continue moving inland during the hours of darkness, only halting when the land cools. On occasions, sea breeze fronts have been observed almost 100 miles from the coast in temperate latitudes, although more commonly they will be encountered within 25 to 30 miles of the coast. In countries situated closer to the equator, where the inland progression of a sea breeze is not deflected as much by the Coriolis effect (caused by the earth's rotation), they may penetrate inland by as much as 200 miles or more.

As the sea breeze requires a temperature differential between the land and the sea, the strongest sea breezes will be likely when the land heats up well and the sea remains relatively cold, as in spring and summer.

Whether or not a soarable sea breeze front is formed will depend greatly on the direction of the prevailing wind *before* the sea breeze sets in. If the general wind is blowing towards the land, the temperature of the land near the coast may not rise high enough to warm the air sufficiently. If this is the case, the required upper level pressure gradient will not exist to create the necessary circulation for a sea breeze front.

A light, off-shore wind or a wind blowing along the line of the coast will favour the formation of a sea breeze front, creating enough of a conflict between the two air masses to produce a zone of usable lift at the front. If the offshore wind is strong (15 knots or more), then it may be too strong to allow a sea breeze front to progress inland.

The speed at which a sea breeze front will move inland will depend on several factors.

The instability of the air mass over the land will determine the depth of thermic activity. The stronger the thermals and the higher they ascend, then the stronger the sea breeze and the further the sea breeze front will push inland. The exception to this rule is when the air is so unstable that cumulonimbus clouds form, as they will disrupt the necessary circulation for the sea breeze and halt its progress. Cumulonimbus clouds often form if potentially very unstable air is undercut by sea air. In this situation, the sea breeze front may be marked by a line of cumulonimbus clouds.

The speed at which a sea breeze front moves inland during the day is typically between 5 and 15 knots, although in some circumstances it may even be faster than this. However, its progress is not necessarily constant, pulsing inland in some instances and becoming stationary in others.

The low-level flow, made up of sea air, is relatively shallow, typically extending no more than 1,000 feet up from the surface. This air will be relatively cold and stable. This means that any line of hills or mountains which it encounters will act as a barrier, delaying or even halting its progress. Eventually the sea air will tend to escape through any valleys or around the edges of a line of hills, causing any sea breeze front formed to alter shape. (A visual analogy would be the sea engulfing a child's sand castle. As the tide pushes in, the seawater firstly creeps around the edges and invades through any gaps in the sand walls.) There may even be further convergence zones where these deflected air masses meet behind the hills, depending on their relative temperatures after following their differing land tracks.

Where a broad peninsula of land heats up sufficiently, this can create sea breezes from both coasts. The result may be a sea breeze front down the centre of the peninsula. Should the prevailing wind be blowing across the peninsula, the sea breeze front may be displaced towards the downwind coast. In extreme cases, it may even lie some distance out to sea.

Characteristics of a sea breeze front

As seen from the ground

When a sea breeze front passes a point on the ground, the temperature of the air will drop, perhaps noticeable as a distinct chill to the observer wearing light summer clothes. The humidity rises sharply due to the ingress of moist sea air.

A thin line of haze or cumulus-type cloud may be seen just before the sea breeze front arrives, but after its passage, thermic activity (as indicated by cumulus cloud) will decrease or possibly cease.

The arrival of humid sea air often causes visibility to decrease. However, if the day has been hot and hazy with shallow cumulus clouds inland, the arrival of the sea air may actually improve the visibility. The main point to note is the change in visibility.

As the sea breeze front arrives, there is likely to be a change in wind speed and direction. After a while the wind will gradually veer if you are in the northern hemisphere, while in the southern hemisphere it will gradually back. The wind speed is likely to increase.

Occasionally, depending on the observer's distance from the coast, the new air mass will even smell of sea air.

As seen from the air

The first sign you may see while thermalling which shows that sea air is encroaching is a line towards the coast beyond which the sky is blue and devoid of cumulus clouds. The line where the cumulus clouds end and blue conditions take over is where the sea breeze front is to be found.

If no cumulus clouds are present inland, then the only visible indication of the encroaching sea air may be a change in the visibility towards the coast. If there is a definite boundary where the visibility changes, then it is along this that the sea breeze front will be found.

If you are close to the front, you may see a line of 'curtain' clouds. These veil-like clouds, which are often ragged in appearance, are caused by moisture condensing out of the air as it is lifted up the interface between the two air masses. Above this line of cloud there may be larger cumulus clouds which have been created by thermals triggered by the undercutting cold air.

The visibility on the seaward side of this line may be quite poor, depending on the moisture content of the sea air and the amount of smog from coastal towns, which is trapped in this stable air mass. However, on other occasions the sea air may actually improve visibility, so again it is the sudden change in visibility which heralds the arrival of a different air mass.

Soaring a sea breeze front

Having identified the presence of a sea breeze front, you are faced with two choices. You can 'make a run for it', heading inland in the hope that you can get away from the stable sea air. Alternatively, you can explore it and enjoy some fascinating soaring, the like of which you may not often experience.

The area of ascending air will be found along the line of the front. If the front is marked by curtain clouds, then the lift will be to the landward side of this cloud. The lift band will be narrow (perhaps only a few wingspans wide at most) with areas of sinking air on either side of the lift.

You may be able to soar in the lift by running along the line of the front. However, if there is no curtain cloud or marked change in visibility, it may be difficult to follow the line of the front accurately. Remember that the front might not necessarily be straight. More than likely, it will have dents and bulges where terrain has influenced its landward progress.

Soaring along a sea breeze front can be quite a nail-biting experience, often operating at times at heights of less than 1,000 feet, until contact with a better area of lift is made. This will require you either to be close to a usable airfield or to have fields selected as you soar.

The rates of climb along the front can be variable, with stretches of weak to moderate lift (less than 1 knot to 2 knots) and areas of strong thermic type lift (6 knots or more).

If you encounter an area of strong lift, you should use it by circling, assuming of course that it is wide enough. Such lift is likely to be thermic and is probably due to the front triggering a thermal in its path. This may coincide with a part of the front where large cumulus clouds loom overhead. By using thermals in this way, you might be able to get up above the level of the lower front and soar along the face of the curtain cloud.

Once at the level of the curtain cloud, you can soar along it as if it were a vertical hill face. This region can be quite gloomy, with the sun obscured by both the curtain cloud and the cloud above. It is not uncommon to encounter countless swifts and swallows, which dart about close to the cloud as they catch insects carried up in the rising air. On occasions, such are the densities of these small birds that they show up on radar as a line marking the front.

Through breaks in the curtain cloud you will be able to see the stable, blue sky on the seaward side of the front. It is possible that there may be some weak lift on the seaward side of the front, but this will probably be too weak and at too low a level to exploit. Such air is better thought of as being slightly turbulent rather than thermic. Should you stray through the front and into the sea air, you will probably encounter sinking air and perhaps some light turbulence. As a general rule, if you lose the lift and find yourself in a noticeably different air mass, you should turn inland immediately in order to reach the land side of the front (and hopefully lift) before you lose too much height.

The soaring conditions on the 245 kilometre competition flight out of Lasham had been good, with excellent visibility and well-spaced cumulus marking strong thermals. It was on the last part of the last leg that conditions took a turn for the worse.

Seventeen miles north of the airfield the good conditions changed abruptly. Visibility suddenly dropped from more than 50 kilometres to less than two. The finish line radio was broadcasting drizzle in a southerly wind. I could hear experienced local pilots complaining that they could not find the airfield. The sea air had arrived with a vengeance.

I had no choice other than to climb as high as possible on the edge of the gloom and start a final glide.

In these conditions, with little height to spare (and before GPS was available) the final glide was uncomfortable to say the least. Four miles out, with a forward visibility of no more than three fields ahead, I decided to land in a field rather than risk the LS3A, which a friend had kindly loaned me for the championships.

The push-pins on the land-out board showed that lots of gliders had only just failed to reach the airfield. Later reports mentioned that a couple of pilots had actually flown past it, missing it by less than a mile!

The effect of a sea breeze front on a gliding operation

If you are in the UK the sea is always relatively close. Many gliding sites are within a few miles of the coast and suffer the effects of sea air. If a gliding site is on or very near the coast, then as soon as thermals start, there is a danger of a sea breeze setting in.

Assuming that gliding operations have begun, the increase of wind that often accompanies the start of thermals may make it necessary to change the runway in use. As the sea breeze front arrives, a further runway change (perhaps back to the original direction) may be necessary because of the shift it causes in the wind direction.

After the passage of the sea breeze front, usable thermals will normally cease. Some pilots may have been lucky enough to soar the front; others may have made use of the pre-frontal thermals. Pilots who have headed off cross-country may have been fortunate enough to escape and enjoy good thermal conditions well inland, but they may find it difficult or impossible to return to the airfield later in the day once the sea air has extended inland.

At gliding sites well inland (15 miles or more) the sea breeze front may not appear until late afternoon. This can produce some interesting local soaring at the end of a cross-country flight and is certainly worth another launch if you have already landed.

Pseudo sea breeze fronts

So-called pseudo sea breeze fronts can also appear anywhere where there is differential heating. These features are not dependent on sea air, but are set up because a cloud layer obscures the sun's heating from one area, while an adjacent area is clear and benefiting from solar heating. This leads to more than simply a local area of differential heating where thermals will form, as a complete sea breeze front type of circulation can be initiated with the front moving forward into the previously sunny region.

In general, sea breeze front soaring does not appear to get as much attention as it did in the past. This is certainly the case in the UK. This may be because, with higher performance gliders, we are all too busy flying fast, thermal cross-country flights which tend to shy away from sea air.

Even on these flights, a sea breeze front orientated along the line of the homeward track can increase your average speed or maybe even save the day and get you home. If anything, the increase in glider performance should mean that we are better equipped than ever to explore and exploit sea breeze fronts.

In August 1961, John Williamson (British Champion, and later British Team member and National Coach) completed a 275 nautical mile flight in an Olympia 419 from Upavon in Wiltshire, England to Ayton, north of the Scottish border. Apart from the distance, the flight was exceptional in that the thermal soaring conditions on the day were not all that great. In fact, had it not been for his correctly identifying a sea breeze front, he would probably have landed near Darlington, after flying 190 nautical miles. Carefully following the front, by reading its position from such signs as the visibility and wind direction and using the lift it triggered, made it possible to fly the last 85 nautical miles of this Diamond Distance flight.

Section 2

Cross-Country Soaring

Chapter 9

Task Selection

The first skill which any pilot needs in order to set a task that will make the best use of a day is a reasonable understanding of weather patterns and forecasts. Unless you are a professional meteorologist, such knowledge will be based on self-study, and on previous observations of what makes a good or bad soaring day. Much of the basic understanding can be gained from textbooks on general meteorology written for the lay person and for general aviation. Once this is gained, books more dedicated to soaring meteorology can be studied. There are many books available which deal with general meteorology. As far as meteorology specific to soaring is concerned, books by Tom Bradbury and C.E. Wallington are highly recommended. After that, only weather observation and experience will help your judgement.

Even once you are capable of judging the potential of a soaring day, it still remains necessary to gather information and make observations to allow you to decide which is the best task for the day. Typically, this should be done by weather watching, not only on the morning of the intended flight but for several days ahead of it. Only by being aware of the general synoptic situation can you anticipate how the day may develop. Note how the high pressure areas are moving. Are they building or collapsing? Are the forecast fronts moving at the rate you expected? Is the wind changing towards the direction you had hoped, and is its strength becoming acceptable?

Make up a list of the ideal conditions for your area (wind direction and maximum strength, minimum temperature, dew point depression, position of high pressure areas, distance of the fronts and period since their passage). Track the weather for the days leading up to your arrival at the airfield and see if these 'ideals' are being met.

Only by such dedication to weather watching can you arrive at the gliding site with at least some idea of the quality of the soaring conditions which may follow. Once you have obtained more detailed local weather information (such as a specific gliding forecast, or perhaps opinions from trusted pundits) you can start to select your task.

The duration of the soaring day

The first deduction you will have to make is how long soaring conditions are likely to last. The time when thermals are likely to start and give lift strong enough to use will normally be given in the local soaring weather forecast – if one is available. This will be calculated on the basis of the surface temperature rising high enough to break down any temperature inversion. If this trigger temperature is given, you can monitor whether the temperature is rising as forecast, and therefore whether thermals are likely to start near the promised time. The forecast will also give the time at which usable thermals are expected to cease. This allows you to estimate the maximum period during which soaring will be possible.

Let us suppose that thermals are forecast, and look likely to begin at around 10.00 hours and cease at around 18.00 hours. On the face of it, this will give you a soaring day which is eight hours long. However, unless the desired task is to be a distance attempt, it is unlikely that you will want to head off cross-country in the weak conditions that are normal in the first hour of the soaring day; or still be trying to get back to base as the thermals die at the end of the day. Therefore, it is more common to be cautious when task setting, and not plan on being able to use the whole forecast soaring period. In fact, if your intended task is to be a speed task, you may plan to use only the few hours in the middle of the day when thermals are likely to be at their strongest. This will offer the highest average speeds. For this exercise, however, let us assume that we will plan on getting airborne at about 11.00 hours and being back at base by 17.00 hours. Therefore, you could set a task which is six hours long.

The average cross-country speed achievable

The next thing you have to look at is the expected quality of thermals during the task period. For the best part of the day (from midday until late afternoon in Europe), you can expect the best thermals, and consequently the highest cross-country speeds. The periods before and after this time can be expected to offer less strong thermals and therefore lower average cross-country speeds. The early period will also consist of thermals which have less vertical depth.

It is therefore a mistake to look at the forecast and use the best thermal strengths expected for the day for your estimate of average speed possible. Use an average of the thermal strengths for the whole day, allowing

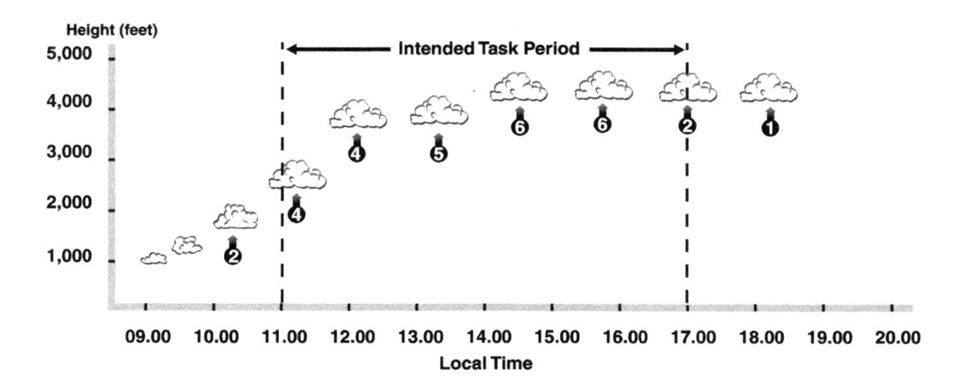

Fig 9.1 Task size achievable. The length of soaring day determines the task which can be achieved. As the day progresses, both the height to which thermals will rise and thermal strength normally increase. Allowance must be made for the weaker conditions which are typical early in the day.

for the fact that thermals, although weak to begin with, can improve quite rapidly. Often the timing of the strengthening of thermals is given on the soaring forecast. (Caution: the thermal strengths given on soaring forecasts tend to be the expected rate of ascent of the air, not of a glider. So, for planning purposes, reduce the thermal strengths given by about 1½ knots.)

Now you come to the first real problem. How fast can you fly? To work out a reasonable task you need to estimate what average speed you can achieve over the whole of the task period. Here I can only give a few rough guidelines, as there are many variables, such as pilot skill and experience, type of glider, and whether your task will take you over any bad areas for soaring or areas where your climbs will be limited by the base of controlled airspace.

In the early stages of cross-country soaring, pilots tend to overestimate their capabilities as far as average speeds are concerned. In the UK, it is not uncommon for pilots in an ASK 6 or similar to achieve about 50 kilometres per hour on their first successful attempt at a 300 kilometre flight. In modern gliders, experienced pilots will be doing this distance in less than 4 hours. In some hotter countries, such as Australia, South Africa and parts of the USA, average speeds will be much higher than this.

Therefore, it is impossible to generalise and say what an individual's average speed is likely to be. The best way to tackle this problem is to look at your previous performances. Only you will have an idea of what sort of average speeds you are capable of achieving in various conditions. Once you know this, you can apply it to the task-setting equation, and by simple arithmetic come up with a viable task distance for the day.

For instance, in the forecast conditions, if you think you can achieve 60 kph from 11.00 hours to 13.00 hours (120 kilometres distance) and 80 kph for the rest of the task period (320 kilometres in 4 hours) then you will be achieving almost 74 kph over the whole task period. On this basis, the maximum task it is sensible to set would be 440 kilometres.

In reality, the final glide will help keep up the average speed on the last leg. Other factors such as a launch later than expected, or a later start of convection, may require a change of task with a reduced task length.

Weather variables, other than the lowering of the sun as evening approaches, also affect the length and quality of the soaring day. The approach of weather fronts often brings with it high stratus cloud, which may reduce the strength of thermals or stop usable thermals from forming. If such a front is forecast or expected, a reduction in your task size will probably be necessary – either because average cross-country speeds will be reduced, or in order to get home before soaring conditions finally collapse. Unfortunately, areas of upper cloud are not always forecast, as they may be the result of a decaying front which has little general meteorological significance. Watch for such developments when task setting, and plan accordingly. Just because something is not forecast does not mean that it does not exist!

The nature of the task

Having decided the size of the task possible on the day, you can then decide on the type of task you will set.

Given a day with light variable winds, and subject to airspace or coastal considerations, you can set whatever shape of task you wish without having to worry about the effect of the wind. This will allow you to set a task that keeps you over good soaring country. It can be a triangle, a quadrilateral or a goal-and-return – anything you like. If you suspect soaring conditions may not be as good as forecast, you can plan a task which brings you past your base airfield on an intermediate leg. If the extent of the soaring area is small due to coastal, airspace or weather influences, you can set a smaller task around which you fly twice.

Winds of up to 15 knots at flying levels do not generally influence the tasks possible unless the wind is coming in off the sea. If this is the case, a large area near the coast will have to be avoided when task setting, as the cooling effect of the sea air will adversely affect the quality of thermals – and possibly their existence.

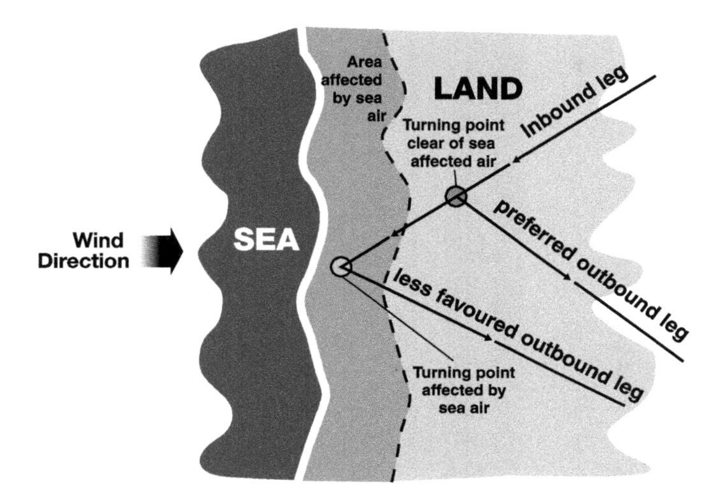

Fig 9.2 Effect of sea air. An onshore wind can result in lack of thermals downwind of a coast. It is unwise to declare a turning point in such an area.

Once the wind at cloud base exceeds 15 knots, thermal streeting is more likely to occur. It is therefore helpful to set a goal-and-return task, a narrow triangle, or a quadrilateral which has legs that are orientated more or less parallel to the wind direction, in order to make the best use of any thermal streeting which may occur.

When the wind is strong enough to have an effect on your progress, a task that has a tail wind on the first leg (while thermals are still weak) will help you achieve a reasonable average cross-country speed during this part of the flight. Hopefully, this will lead to your flying any into-wind leg(s) during the part of the day when thermals are at their strongest. This will make penetration into wind easier. If the task is well planned, the last leg and the final glide will have the benefit of a tail wind component, thus making life easier if this leg has to be flown when the thermals are again becoming weak.

One very useful aid to calculating the minimum thermal strength required to penetrate into a head wind is the JSW Final Glide Calculator (shown at figure 18.5). By setting the calculator so that the 'Airspeed'

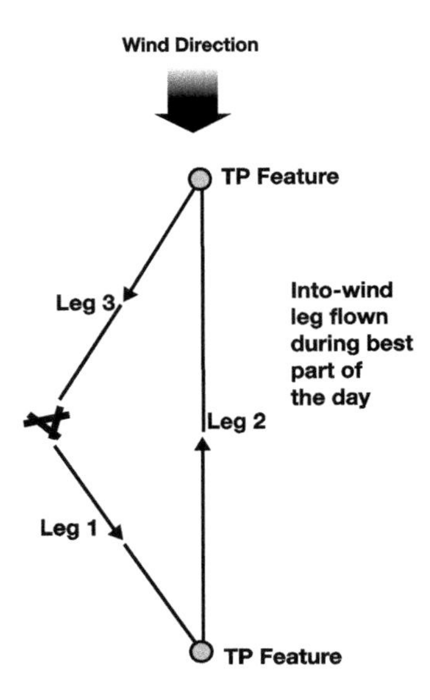

Fig 9.3 Making use of the wind when thermals are weak. Ideally, a large task will be orientated so that the first and last legs (when thermals are weak) are downwind.

scale is set to cross the 'Speed to Fly for Maximum Distance' curve at the appropriate head wind, you will be able to read off the minimum airspeed necessary to achieve progress into wind (where the 'Speed to Fly for Maximum Distance' curve meets the 'Airspeed' scale). At the top of the calculator (on the 'Speed-to-Fly' scale), you can read off the average rate of climb that would allow that airspeed to be used. If thermals weaker than this are forecast, then a task leg into wind using classic climb and glide thermal cross-country technique will not be possible. Dolphin flying (and by inference, some form of lift streeting) will be necessary if an into-wind leg is to be achieved. (*Note: a fuller description of the JSW Final Glide Calculator and its use is given in Chapter 18.*)

Although even moderate winds do not cause too many problems for modern gliders as long as thermals are strong, if the wind is very strong you may have to consider a task that is essentially downwind, either in a straight line or consisting of a dog-leg. Although these options are fairly unpopular in this day of closed-circuit tasks, if it is distance which you require for a badge claim or a record, then they may be worth considering, despite the inconvenience of a long retrieve.

Other factors which determine the nature of a task

Sea coasts

As mentioned earlier, tasks taking the glider close to a sea coast may create difficulties which can be avoided with a little consideration. Generally, if the prevailing wind is blowing onshore, then the area near the coast may suffer from poor soaring conditions. On the other hand, if a strong offshore wind is blowing then conditions may remain soarable up to the coast and even to some distance over the sea! It may even be intended to make use of the lift of any sea breeze front effect to extend your flight, but the determination of such effects in advance is difficult and injects an element of chance into task setting, to say the least.

Large estuaries can cause problems by allowing sea air to penetrate a long way inland, giving rise to the same problems as occur at coastal areas. In general, tasks should be set so that if an estuary (or a river approaching its estuary) has to be crossed, it is crossed as far inland as possible.

Poor thermal producing areas

Even areas remote from any coast can be renowned for producing poor soaring conditions. Large river valleys, irrigation areas and wetlands are all worth avoiding, either totally or early in the day when the ground has not had time to heat up sufficiently to produce reasonable thermals. The other problem of setting a task where you will arrive at such an area early in the day is that convection depth may not yet be great enough to give you the height you need to glide across the bad area to reach the better soaring conditions beyond.

High ground

Lines of hills or mountains often offer much better soaring conditions, with stronger thermals and higher cloud bases than low-lying areas. When task setting, look at the topography of the area over which you intend flying. Setting your task so that your track keeps you over high ground rather than valleys may just give you that extra speed and make achieving the task easier.

Controlled airspace

In the UK and many other countries, controlled airspace is the bugbear of most task setters. In England in particular, it is almost impossible to set a sizeable task that is not under some sort of controlled airspace, thus limiting climb heights. If the day promises to be one with very high cloud bases, then it is worth setting a task which takes you to an area where climbs are not limited by controlled airspace – assuming that this is practical, bearing in mind the foregoing considerations.

The other problem with controlled or restricted airspace is that it often forms a block across part of the route you would like to use. If this is the case, you will have to plan your task to fly past it (perhaps by changing your turning point) or around it (by using a dog-leg). Incorporating a dog-leg into your track can be wasteful of speed if the total distance covered has to be measured as a straight line. On the other hand, if it is possible to introduce a turning point (in this case often known as a control point) then the extra distance covered can be included as part of the task distance and will not affect your average speed. As some of the rules for claiming badges and awards now allow up to three turning points, setting and declaring a control point in this way can be useful rather than wasteful.

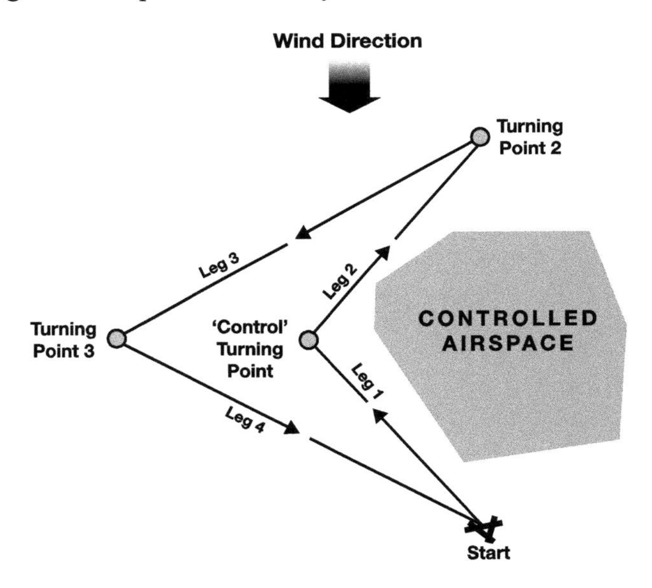

Fig 9.4 Use of a 'control' point. Often it is beneficial to declare a turning point which keeps you clear of controlled airspace while allowing you to claim the whole track distance.

Where it is possible to avoid controlled airspace by changing a turning point, this is best done with as large a margin as possible. To plan a task that runs along the edge of a control zone is to deny yourself the flexibility to divert from track to either side to find lift. (You must have noticed how the best cumulus clouds always appear to be in controlled airspace!) Another trap to be avoided, if possible, is to have a leg of a task which is just upwind of an area of controlled airspace when there is any appreciable wind. If you do find yourself low in such a position, not only are you not as free to look for thermals to the downwind side of your track, but also you have the added problem that, if forced to climb in any weak thermals, you will be drifted towards the controlled airspace.

Poor landing areas

The height you are likely to achieve in thermals will determine whether you can safely plan a task that takes you across areas in which you are unlikely to achieve a safe landing. For instance, large areas of forest, large lakes, marshlands, estuaries, moorlands, mountains or even cities are areas you might not wish to cross with a cloud base of 2,500 feet or less. So, try to avoid planning a leg which could present such problems. If the thermals are likely to be plentiful and go high, then crossing such 'badlands' may not be a problem.

Task setting for wave-soaring flights

If the flight is to be flown in wave lift, then many of the previous considerations will not apply. For instance, as wave lift is not reliant on surface heating, providing you can be aerotowed directly into the wave, or enter it from hill lift, you can start your task as soon as there is enough daylight to launch, and finish it as long as you can still see well enough to land safely.

Orientation of the task relative to the wind will be different from tasks relying on thermals. The task direction has to be considered in terms of wave alignment more or less at 90 degrees to the wind direction, rather than thermal streeting lying parallel to it.

The heights at which wave flying takes place reduce the need to plan the task to avoid unlandable areas. In fact, wave lift and areas of mountainous and unlandable terrain often go hand in hand.

Badge requirements

If the intended flight is an attempt to gain a badge or record, then the requirements to qualify for the award will normally dictate the shape of the task and the number of turning points that can be set. Make sure that you know the rules for the award for which you are attempting to qualify, or you could unwittingly have planned a failure even before you launch.

Task setting is a useful winter occupation

One very useful occupation when the winter weather sets in, or when your glider is receiving its annual inspection, is to plan tasks for the coming season. Preparing many tasks of all sizes and in various directions can make task setting on the day much easier. By using one of the many task-setting computer programs available, you can not only plan a number of tasks, but can also equip yourself with a printout of track requirements, distances and turning point details. This will allow you to enter co-ordinates into your GPS well in advance of the day of the flight. Some of these task-setting programs will even print a task declaration form for you. Equipping yourself with such a task folder makes task setting easier when you do gather the weather forecast for your day's gliding. It also loses you less time when you arrive at the airfield, or have to make allowance for the weather being better or worse than you had forecast.

Versatility of tasks

Some pilots feel they fly better when there is a task drawn on the map, almost as if this gives a more determined objective for the flight. Despite this, it is surprising how often tasks are cut short or abandoned when soaring conditions are thought to be changing for the worse, or when average speed does not match the ambitions of the task setter. (I am, of course, not referring to competitions, when one has to continue despite, on occasions, the task setter's apparent optimism.)

On the other hand, many pilots cover huge distances without having declared or set a specific task. These pilots, having set off with only a rough idea of an intended route, vary their plan based upon the soaring conditions observed, on weather reports from airfields received on VOLMET, or from listening or talking on the radio to pilots flying in different areas. This flexibility allows them to extend their flights and make full use of the soaring day.

Many gliding clubs would rather impose the restriction of some sort of flight declaration upon cross-country pilots, claiming that not to do so would result in an impossible search and rescue situation. To do so is to restrict one of the basic freedoms enjoyed by soaring pilots. True, if the area a pilot intends flying over can be considered as remote or hostile in the event of an outlanding, then it would be wise for the individual to leave some idea of the intended route or to maintain regular radio contact; but the option to do so should be left to the individual. Even some of the FAI badge rules allow alternative turning points, which can be rounded in any order or not at all. *Soaring is freedom – enjoy it and do not let others take it away!*

Chapter 10

Preparation for Flight

Show me a pilot who has never arrived at the gliding club on a soarable day only to find some essential piece of equipment has been left at home, and I will show you an exceptional person. With any luck, the forgotten item will be a minor inconvenience which can be replaced with a little expense, begging or borrowing, such as a map, battery or final glide calculator. Sometimes it is more serious. I have known pilots leave their glider's winglets or tailplane at home, having forgotten to return the item to the trailer after polishing it. The most serious situation occurs when a pilot forgets to bring his or her brain to the airfield.

A calm, organised approach to preparing for your day's soaring can make it run smoothly. The secret is in preparing for your potential flights in a way that leaves nothing to chance – except for the weather. Such preparation should not only include the glider and its ancillary equipment; it should also include making sure that you, the pilot, are in good shape for the flight and will be comfortable for what may be quite a long time in a small cockpit.

Although preparation for flight includes many navigational considerations, these are not covered in this chapter. This is because aspects such as map and course preparation are dealt with at length in Chapter 16.

Personal preparation

Three important aspects of any soaring flight are that you perform well, that the flight is conducted safely and that you enjoy it. The chance of succeeding in these aims is greatly enhanced if you are both mentally and physically in good shape for the flight. You should do all you can to reduce the chance of being below par on the day you intend going soaring.

If you are in ill health, tired, or suffering from an over-indulgence of alcohol on the previous evening, then your levels of both performance and safety risk being reduced – perhaps to a dangerous level.

Some pilots believe that being physically fit goes further than simply having a medical certificate that states they are fit to fly. Many exercise regularly, jogging or working out in a gymnasium. Whether the resulting fitness benefits their soaring as much as it benefits their general health, only they will know. However, if they feel that they soar better for doing so, then that is what is important.

Ill health

While no pilot is likely to fly when feeling unwell, you may not be sufficiently concerned by lesser ailments. Minor illnesses and allergies, which would not prevent you from going to the office or driving a car, can have serious results if you dare to go flying. Colds, inflamed throat glands, hay fever or sinus problems can prevent pressure equalisation between the air spaces on either side of the eardrum. If such a pressure difference occurs during a descent, it can cause discomfort, pain, temporary deafness, or even permanent damage to the eardrum and hearing. If you are unable to equalise the pressure in your ears by swallowing, do not fly. Similar blockages of the sinuses can also cause facial pain severe enough to disrupt your concentration and prevent you from flying safely.

The symptoms of many minor ailments can be temporarily removed by drugs. However, although many of these remedies are available without a doctor's authority, care should be taken to avoid drugs which may cause drowsiness, dizziness or nausea. If in doubt, seek medical advice or stay on the ground.

Alcohol

The dangers of flying while under the influence of alcohol should be obvious. What is less obvious is the time it takes to reduce the alcohol level in the body to a level at which it is safe to go flying. As most gliding takes place over weekends, and Saturday night is often a social affair, take care not to party too late or consume too much alcohol. To do so will risk either a hangover and the subsequent loss of what may be a good flying day, or, if you do fly, a less than average personal performance. Although hangovers can be painfully obvious, the effect of lesser, residual amounts of alcohol in your body may be less obvious but still detrimental to your performance and safety.

Tiredness

As your flight may involve working quite hard mentally in a hot, cramped cockpit for several hours, it is essential to be well rested in order to avoid fatigue with resulting performance loss. This is another reason why you may have to leave the party early, in order to get a good night's sleep. (Next time you attend a party during a national championship, notice how few of the leading pilots hang around until late!)

Psychology

Your psychological state can also affect your ability to make decisions and soar well. Soaring efficiently, especially when the conditions are difficult, requires huge amounts of concentration. If your mind is preoccupied with events at the office, or still reeling from some pre-flight dispute, your soaring and possibly safety may suffer. Take time to relax and get into the right frame of mind before you launch. Leave your problems outside the airfield gate.

Clothing

Fashion has no place in a glider cockpit. The important thing is that you are protected from the environment and are unrestricted in movement. When it comes to clothing for gliding, practicality and comfort are paramount.

For flight in hot weather, clothing that is light and protects you from the sun is necessary. A sun hat, and a shirt with a collar which protects your neck and possibly has long sleeves, will be of benefit. Light, long trousers (as opposed to shorts) will stop the sun burning your legs. These precautions are even more important if you have light-coloured skin and are flying in a country where the sun's rays are unusually harsh. Footwear should be light, but sturdy enough for what may be a long walk to a telephone after an outlanding. Sandals tend to be bad news if the field in which you have landed is full of stinging nettles or oil-seed rape stubble.

For altitude flying or flight in cold climates, clothing will have to protect you from low temperatures, as well as the sun's rays. A woollen hat or Balaclava is needed to protect your head and especially your ears from the cold. Other areas of the body are best protected by several layers of clothing such as shirts and jumpers. One-piece overalls or flying suits are good as a top layer to protect areas where other clothes are supposed to overlap but often fail to do so when one is seated. Despite the need for several layers of clothing, comfort and mobility should not be compromised.

Footwear for cold-weather flying will have to be more substantial than that suited to warm conditions. It will need to be roomy enough to allow thick socks or layers of socks to be worn. Avoid tight footwear, as this will tend to restrict blood circulation to the feet and toes, adding to the problems of cold rather than preventing them.

Gloves or mittens should be available in the cockpit, if fingers are not to lose feeling and suffer from the cold.

Protection from ultraviolet rays

In whichever climate you are flying, the chances are that you will be directly exposed to the sun's rays. The higher you fly or the hotter the day, the more your skin will suffer from the sun's burning effect. In the short term, such rays can cause painful sunburn. In the long term, they can cause life-threatening skin cancers. Protect against such risks by using high factor sun block creams on any areas of skin which will be exposed to the sun.

Glider preparation

Included under this heading are all of the normal actions that are necessary to prepare a sailplane for flight, such as rigging and inspecting it to ensure that it is serviceable.

There are also other aspects of preparing the glider which are performance related, rather than totally safety related. Some of these performance improvements may require approval by a qualified inspector or engineer. For this reason they are often best done at the same time as the glider is receiving its annual airworthiness inspection. As some of such work may not constitute airworthiness, it may not be carried out automatically by an inspector as part of the Certificate of Airworthiness inspection.

Work to be done annually

Control surface hinge seals

Most control surfaces have some kind of seal between the moveable surface and the fixed part of the aircraft to which they are attached. Such seals are designed to prevent airflow through the hinge gaps. Preventing such unwanted airflow has two beneficial effects on the glider's performance. Firstly, stopping air flowing through such gaps reduces the disruption to the airflow and thus reduces drag. Secondly, if air does escape around control surfaces in this way, the effectiveness of the control surface will be reduced. Subsequently, greater control deflections may be required to achieve the same control response, causing a further increase in drag.

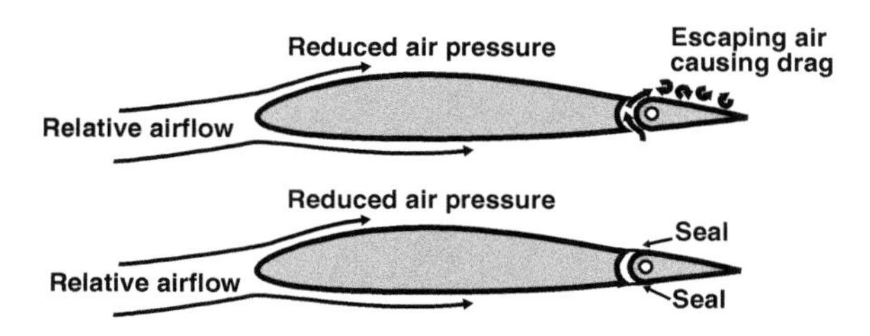

Fig 10.1 Control surface seals. Well-sealed control hinge gaps reduce drag both by preventing air from flowing through the hinge gap and by making control surfaces more effective.

It is important to replace any control surface seals that are torn or missing. The most likely reason for a seal to be missing is if it has not been replaced after maintenance or a repair, or if it has been removed because it was loose.

When fitting a seal to a control surface, make sure it is fitted in such a way that will not hinder control movement in any way. After fitting, check that the control surface still moves to its designed limits, by measuring its deflection.

Pushrod seals

Sealing holes where pushrods pass between the wings and fuselage will help prevent air, introduced into the cockpit for ventilation, from flowing from the fuselage along the inside of the wing and out through the airbrake slots. Air escaping in this way will disturb the airflow over that part of the wing, potentially resulting in a decrease of lift and an increase in drag.

Before embarking on this task, check whether the sailplane manufacturer has already installed such a seal within the wing itself. Doing so might save you some work. If possible, check to see whether such seals are damaged. If they are, and it is practical to do so, have them replaced.

If pushrod seals have not been fitted (as is the case on many early-generation glass fibre gliders) then some improvisation is normally necessary to produce satisfactory seals which do not interfere with control movement or feel. Home-made pushrod seals may be made from polythene, from rubber 'boots' of the type used on car transmission linkages, or from chamois leather – usually accompanied by liberal amounts of adhesive tape.

As with control gap seals, it is essential that no interference or restriction of the controls results from the introduction of pushrod seals.

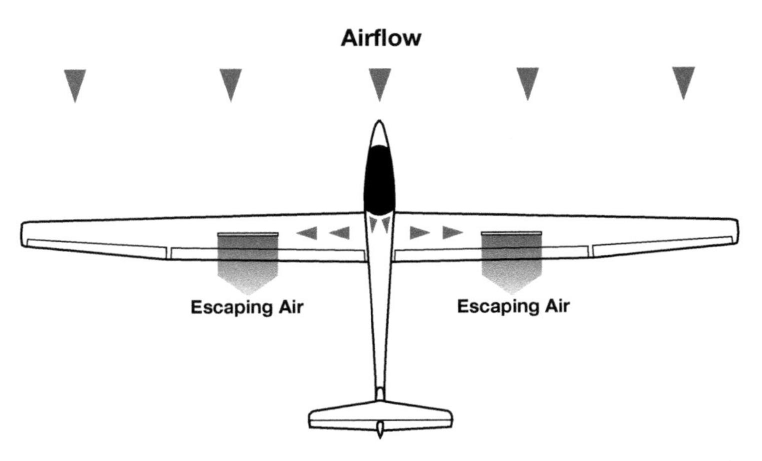

Fig 10.2 Sealing of pushrod access holes. Fitting seals to the pushrods will reduce drag by preventing air from escaping from the fuselage into the wings and out through the airbrake slots.

Airbrake sealing

While we are dealing with airflow over the wing's surface, it is worth checking that the airbrakes, when locked closed, sit flush with the surface of the wing. Badly adjusted airbrake caps will cause discontinuity of the wing's surface, resulting in drag and loss of lift. At high airspeeds they may even rattle, making a disturbing 'machine-gun' noise.

Observing how well the airbrake caps remain flush with the wing is best done when flying at various airspeeds throughout the glider's permitted airspeed range. Often any lifting of the airbrake sealing caps will only occur at higher airspeeds.

Adjustment is normally done using spring-loaded screw adjusters situated beneath the airbrake cap. Caution is necessary when adjusting the airbrake sealing caps. Over-tightening can cause cracking of the surface finish of the caps when the airbrakes are locked closed and the wing is flexed, as occurs in flight.

Turbulators

If turbulator strips are recommended on the type of glider which you fly, make sure they are fitted and are not loose. These simple strips of tape reduce drag and therefore improve performance.

Wheel box sealing

Air escaping from the fuselage through the glider's wheel box will result in noise as well as drag. Whether the glider has a fixed wheel or one that is retractable, it is worth making sure that the lining of the wheel box is well sealed from the inside of the fuselage.

Adhesive tape or bathroom sealing compound can be used, but remember that there may be a need to remove the wheel box lining at some time in the future for routine maintenance. Therefore, it is wise to seek an inspector's advice before using a more permanent method of sealing, such as glass fibre cloth.

Undercarriage doors should be checked to ensure that they close properly. If necessary, the springs or bungees that hold them closed should be renewed. The door hinge lines should be sealed with durable adhesive tape.

Wheel fairings

If your glider has a non-retractable main wheel, make sure that any fairing around the wheel is in place and that the joint between the fairing and the fuselage is sealed with tape. Sadly, the main wheel fairings on some gliders are poorly designed and give insufficient ground clearance. This makes them susceptible to damage, and they are often removed to avoid such damage. Despite this problem, such fairings do have a drag reducing effect and are therefore worth having fitted.

Canopy sealing

A well-sealed canopy is conducive to both reduced noise and reduced drag. Strips of foam draught excluder can be used to cut down the amount of air passing through the gap where the canopy meets the fuselage. This technique is usually necessary on older gliders (Skylarks, Darts, ASK 6s, etc.). Most modern glass fibre gliders have reasonably well-sealed canopies.

Take care not to use foam that is too thick, or you could end up unable to close the canopy without the risk of distorting the frame or cracking the Perspex. Never use force to close a canopy. One problem with canopies is that, as the ambient temperature changes, a Perspex canopy will expand or contract to a different extent from the glass fibre fuselage to which it is fitted. The result is that a canopy that is a good fit in the workshop may be too tight when the glider is parked out in the sun. The sealing of a canopy should be carried out with caution or by an expert.

Wing surface smoothing

Many gliders end up with scratches or chips out of their wings as a result of hangar or trailer damage. Your glider may occasionally be damaged by small stones thrown up by the parachute on the launch cable or blown back at the glider by the tow-plane. Whether your glider is made of wood, metal or glass fibre, it is always worth filling such imperfections. Most stone chips appear on or near the leading edge of the wing – an aerodynamically critical area as far as performance is concerned. To leave them unattended is to fly with a permanent performance handicap.

A few private owners take the improvement of their glass fibre sailplanes to the extreme, and embark on re-profiling the wings. While, at the level of top competitions, the improvement to the glider's performance may give the pilot the edge needed to win, such work is both time-consuming and exacting. You will probably be better spending this time improving your soaring technique. The results of this may be more noticeable.

Water ballast

Fill up both wing and tail ballast tanks (if fitted), and check that there are no leaks and that the dump valves are working efficiently.

Instruments and electrics

Pneumatic instruments should be checked for leaks in the system. Leaks in the total energy system are particularly problematical. Check for perished tubing, or tubes which are restricted as a result of being compressed between instruments or fittings.

The glider's electrical system should be inspected for poor electrical connections or frayed wiring. All batteries should be fully charged and tested.

As an in-flight check on battery state is often useful, a simple, panel-mounted voltmeter is worth fitting. This can give warning of a battery which is losing its charge or failing, and allow you to switch to a standby battery before power reaches too low a level. (Some flight director systems have the ability to display the voltage of the battery supplying them, thus removing the need for a separate voltmeter.)

One feature which is often absent on glider electrical systems is the installation of fuses or circuit breakers. The 12-volt lead-acid batteries commonly used in gliders can supply a large enough current to damage sensitive instruments or cause a fire. As well as installing fuses or circuit breakers which protect individual instruments and systems, it is also wise to install a fuse as close to each battery as possible. (Despite popular belief, once alight, glass fibre burns very well indeed.)

Before the soaring season begins, it is worth taking time to check that the glider is in peak condition and that all of its systems are working well. Starting the soaring season with equipment that needs attention, or with a glider that is below par, can haunt your soaring as you watch those better prepared continually outsoar you.

On completion of the Certificate of Airworthiness inspection and associated work, it pays to check out all of the glider systems and instrumentation.

Pre-flight preparation

Sealing gaps

The design of the wing root/fuselage joint can be critical to climb performance. Any disruption of the airflow in this area can result in the glider failing to climb efficiently, or not climbing at all if the lift is weak. Therefore, the designer tries hard to minimise any disturbance to the airflow at the wing root by carefully designing fillets and using gentle curves.

Despite these efforts, there is inevitably a gap between the wing and the fuselage. No matter how small this gap, air will be able to flow through it – unless you do something to prevent this happening.

As with any unwanted airflow, drag and some loss of lift will result. In this case, sealing of these gaps is easy. Ordinary PVC adhesive tape does the job adequately. Always seal the wing root/fuselage joint. Never fly with this area unsealed.

The same applies to gaps between inboard and outboard sections of wings, and gaps between wing and winglets. All of these gaps need to be sealed if performance is not to suffer.

Where the tailplane fits on to the fin on a T-tail glider, there is normally a gap which needs to be covered with sealing tape. Other types of tail may also benefit from sealing, but whatever type of tail your glider has, be careful not to hinder elevator movement when taping gaps.

It is normally worth renewing sealing tape which was applied for a previous day's flying, as any moisture may cause it to lose its adhesion. Apart from losing some of its performance-saving ability, loose tape can create a very irritating noise.

Making sure the glider is clean

The efficiency of a sailplane relies on the air flowing past its surfaces with the minimum of resistance or disturbance. The manufacturer tries hard to produce as smooth a surface on the sailplane as possible. Any imperfections on the surface of the glider, such as dust, dead insects, or grease marks deposited while rigging, will cause extra drag and, depending on their situation, may reduce lift.

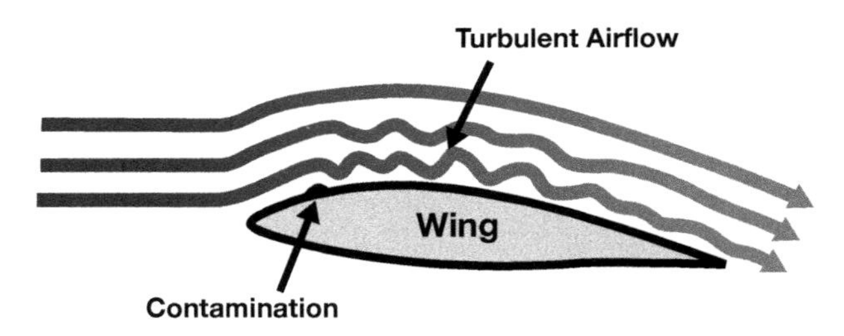

Fig 10.3 Contaminated wings. Even small irregularities on the surface of the glider will disrupt laminar airflow.

Before flying, the glider should be washed, to rid it of such performance reducers. Washing and polishing the glider will ensure that all of the dust accrued overnight, whether the glider was kept in a hangar or a trailer, or parked out in the open, has been removed.

On an airfield that is very dry and dusty, it may be worth leaving wing and canopy covers on the glider until you are almost ready to launch. This will ensure that the dust stirred up by tow-planes has less chance of settling on the glider before launch. Never fly an 'unclean' glider.

Canopy cleaning

A dirty canopy is a dangerous canopy. Dirty or smeared canopies are a liability. With low sun angles or poor visibility, an unclean canopy can result in late sightings of other gliders or aircraft. If a canopy is not clean, then the glider is not fit for flight.

From a performance point of view, marks on a canopy can be a distraction when the need for skyreading and spotting other gliders requires a lot of concentration.

Cockpit preparation

Safety items

The initial cockpit preparation will generally be done as part of the Daily Inspection, and should include all of the major items that are necessary for your safety and comfort.

Ballast weights

As these may not be required by every pilot who flies the glider, they can be regarded as personal equipment. If you require ballast weights, make sure that they are secured to the structure of the glider. Ballast in the form of lead-filled cushions is dangerous and should not be used. Even if it is secured by straps to the glider seat or seat back, such a cushion can cause serious injury in the event of a crash, if it is afforded any movement or it comes loose.

Safety cushions

Recent years have seen an increased awareness of the injuries that can result from the compression characteristics of ordinary foam cockpit cushions during a heavy landing. Energy-absorbing foam cushions are now available which reduce the loads placed on the spine in the event of a heavy landing. Although more expensive, these cushions are well worth fitting, and those composed of household furniture foam should be thrown away.

Parachute

Regarded as an item of personal equipment, a parachute is like the life insurance policy you hope you never have to call upon. Just as your life insurance policy needs to be maintained if it is to be any good, so too does your parachute. Before placing it in the cockpit, check that it is in good condition and is not due for repacking. Check the release pins, the harness and releases, and the general state of the pack.

Operational items

There are few more annoying ways to start a flight than to get airborne, and then realise that you have forgotten to pack some essential item in the cockpit.

If you own your own glider, it may be possible to leave much of the required equipment in the cockpit. Even then, this will only apply to some items and will not include perishable items such as food and drink. It is also necessary to remove maps for flight planning and batteries for recharging.

If you share a glider, or fly various club gliders, then leaving equipment in the cockpit is probably impractical (as well as unwise). One answer is to use a checklist to ensure that everything you need is in the cockpit before you launch. A typical checklist will include reminders of what has to be prepared before the flight, as well as items which must be placed in the cockpit.

You will probably compose your own list of items which you will take with you on every flight. As a result, it is only necessary in this text to briefly mention the importance of some of the essential items. You will almost certainly want to add others.

Task detail sheet

Although you probably know what task you intend attempting, it is amazing how often you hear pilots asking on the radio which ground feature relates to a turning point, or what is the GPS code for the next turning point. A note in the cockpit containing the task details can save this embarrassment. A task sheet can also contain track details, which can be used to check your navigational preparation in case you have made any gross errors. We are all capable of operating a wrong switch occasionally. Should this action delete the settings on your GPS, having the co-ordinates of the intended turning points available on a task sheet may make the difference between using basic navigation or having an easier life thanks to GPS. Other notes which can be written on a task sheet include airspace restrictions and turning point bisectors.

Maps

Apart from the need to navigate, it is a legal requirement in most countries to carry a map which displays controlled and restricted airspace.

Radio frequency list

These days, the restrictions imposed by controlled airspace are considerable. However, despite the reluctance of most glider pilots to talk to air traffic control units, radio is allowing increased access to some airspace from which gliders would otherwise be excluded, or where they would at least be very unwelcome.

In order to exercise the rights of access granted by use of radio, you will need to have available in the cockpit a list of radio frequencies of the various air traffic control units whose airspace you are likely to penetrate. Such a list should be concise and in a format which allows quick reference in flight.

Barograph/datalogger

Many flights which should have qualified for badges, or even records, have not been successful because a barograph was either not carried or not switched on.

These days, it is acceptable to use GPS evidence as proof of completing a task. However, most GPS units will not store data and therefore must be connected to some form of recorder, generally known as a DATA-LOGGER.

A datalogger is a device which stores the GPS position of the glider in its memory, along with the time when you were at each position. Thus, when the data is uploaded into a computer, the position of the glider at any time in the flight can be determined. Some dataloggers also double as barographs, which allows scrutiny of the glider's height at any position to be cross-checked.

Datalogger evidence has been lost because the datalogger's memory had not been cleared before the flight, resulting in it running out of memory during the flight. Failing to record a flight for such reasons is frustrating. Make sure all recording devices are properly prepared, stowed in the cockpit and switched on before the flight.

Final glide calculator

A small item such as a final glide calculator is easily overlooked when stowing equipment in the cockpit. Only when you are nearing final glide range from the airfield are you likely to realise that it is missing. The result is normally an over-cautious, inefficient final glide with resulting loss to your average speed. If you have electronic aids to plan your final glide, a final glide calculator may only be a back-up, but still one which many pilots would prefer not to leave out of the cockpit.

Food and drink

Some sustenance is normally required on long flights. Keeping your energy reserves at a reasonable level is important if fatigue is to be avoided. The choice of food depends on individual tastes but, in general, sticky, sugary foods such as chocolate bars should be avoided. Not only do they melt in the high temperatures of the cockpit, but they also tend to create thirst. Any food to be consumed in flight must be of a nature which allows it to be eaten without half of it ending up spread all over the cockpit.

Dehydration is a major risk during long flights, even in temperate latitudes. The effects of dehydration can be dangerous and result in poor decision making, headaches and illness. It is important to consume enough fluid in flight to avoid such problems. Water (possibly with some dextrose additive) is probably the best drink to take with you in the cockpit. Fruit juices are also favoured by some pilots. Avoid fizzy, sugary drinks as these can actually add to your thirst and dehydration problems. Do not take glass bottles in the cockpit. It is far better to make up a drinking system with a plastic bottle and a tube. In very hot climates, drinking bottles which are insulated with a polystyrene covering can be used to keep drinking water reasonably cool – at least for part of the flight.

Unfortunately drinking enough fluid creates the need to urinate. This may call for some improvisation on the part of the pilot, to avoid discomfort from a full bladder. Whatever means you employ for such relief will have to be stowed in the cockpit and therefore should be on your checklist.

Sunglasses

Sunglasses (or clip-on sun shades if you wear spectacles) are a must. Many pilots have rued the day they have launched only to find that they had left their sunglasses out of the cockpit. Apart from the over-bright environment in which we fly, good sunglasses are an important aid to sky-reading. On a safety note, forgetting to take sunglasses on a flight risks exposing your eyes to the sun's harmful ultraviolet rays.

Hat

Make sure that you have a sun hat in the cockpit. As mentioned earlier, exposure to the sun, even for short periods, can cause sunburn and skin problems unless the skin is properly protected. The head and neck are most likely to be affected and it is necessary to protect these areas with a suitable hat. A well-chosen hat will also help keep you from getting too hot. When altitude flying, a suitable hat may serve the opposite purpose – that is, stop you losing body temperature through your head. Hats with large brims or peaks can reduce upward vision and should be avoided.

Money

Do not forget to carry some money with you for telephone calls to arrange a retrieve if necessary. In some instances, you may need to pay for overnight accommodation while you await your retrieve crew!

Mobile telephone

Stowing a mobile telephone in the cockpit can reduce the time it takes to get your retrieve crew on the road should you land out. It can also allay fears that something has happened to your crew and trailer when your wait in a field is a long one. Providing you have landed in an area with reasonable reception, a mobile telephone can also be an invaluable survival aid in the event of an accident or a landing in a remote area.

Oxygen equipment

If it is anticipated that the flight will reach altitudes above 12,000 feet, then the installation of oxygen will be necessary. The need to check the serviceability of the oxygen system and its contents should become part of the checklist.

Survival equipment

In the event that your flight will take you over terrain devoid of human habitation, such as mountains or deserts, then survival aids will have to feature on your checklist. Even in well-populated areas, location and rescue can take a considerable time.

For areas where low temperatures are the problem, a space blanket or survival bag, and energy-rich food, such as mint cake or chocolate bars, should be carried.

For hot, desert-like areas with little shelter from the sun, a space blanket will also give some protection in the form of shade. As desert areas can also suffer from low night temperatures, the same space blanket will also offer some insulation against cold. For flight over such hot, dry areas, extra water should be carried in the glider and kept in reserve in case of an outlanding.

A strobe light, whistle and some means of lighting a fire are also useful inclusions for flight in places where the arrival of your retrieve crew or help may take some time. As your landing area may not have been conducive to an entirely safe landing, a first-aid kit containing the basics should be part of your equipment. Compact survival packs are available from 'outdoor' sports shops. Some of these may be suitable for storage in the glider.

The inclusion of an Emergency Locator Transmitter (ELT) should also be considered if there is a risk that a search (rather than a retrieve) may be necessary if you land out.

Glider box

By far the best way to be organised, so that you do not have to gather every item you need each time you go gliding, is to keep everything in one place. The easiest way to achieve this is to have a 'glider box' where everything goes after each flight. With such a box, you can simply turn up at the glider and decant all of the necessary equipment into the glider cockpit. The same box can contain non-flight items such as sealing tape and canopy cleaning materials. Attaché-style cases, flight bags, hard plastic tool boxes and even strong cardboard boxes all make suitable glider boxes.

It is best to supplement your glider box with a separate 'navigation bag', or 'map bag', which contains all of the maps and instruments you need for task planning.

Ancillary equipment

Sadly, many pilots tend to think only of the glider and flight equipment when it comes to equipment preparation. A complete glider outfit includes quite a lot of ancillary equipment. This comprises everything from the glider trailer to the towing aids designed to make life easier before launch.

Trailer

An annual maintenance check of the glider's trailer is a wise move. Many glider trailers get little attention and the only work that is carried out on them is running repairs. The trouble with leaving a trailer to get into this state is that running repairs may be necessary in the middle of the season, or even worse, in the middle of a long retrieve.

Check the exterior of the trailer for corrosion of the fittings, loose skin panels and serviceability of the lights. Are the tyres in good condition and are they inflated to the correct pressure? Does the trailer possess a spare wheel that is serviceable, a jack that will safely lift the loaded trailer, and a wheel brace that fits the wheel nuts? If the answer to any of these questions is 'no', then you are in danger of spending an uncomfortable night by the roadside a long way from base.

As far as the interior of the trailer is concerned, make sure that the fittings that secure the glider while it is being transported are in good condition. Poor fittings can cause annoying damage, which in an extreme case may render the glider unserviceable. Wing stands, rigging and tow-out aids to be carried in the trailer should all have secure stowage points, lest such loose equipment causes damage to the glider when on the road.

A spare bulb pack for the trailer lights, along with a wiring diagram for the trailer's electrical plug, are always worth having stowed in any trailer. (It is surprising how many drivers have unhitched the trailer from the car and then driven away with the trailer's electrics still connected to the car. It is at times like these that you will wish you had the wiring diagram.)

Rigging and tow-out aids

Wing stands or trestles have to bear a considerable and expensive weight when rigging. If they are broken or weak, you stand a good chance of the wing or fuselage ending up damaged. Tail dollies, wing dollies and tow-out bars also have to put up with large loads. If the fittings on these are worn or need adjustment, damage to the glider may occur. If any of these items are going to fail, you can rest assured that this will happen when you are trying to get to the launch point on the best soaring day of the year. Check all of this equipment before each season begins and have any suspect items repaired or replaced.

Competition and expedition checklists

The checklists suggested previously contain the essentials for cross-country flying at club level. Once you start flying in competitions (especially competitions away from your home base) or going on expeditions to other gliding sites or foreign countries, you will need to compose checklists that cover a lot more eventualities.

Competitions require paperwork in order to register your entry (entry forms, licences for radios, insurance details, etc.). If you have problems with the glider, such as a punctured tyre or malfunctions of essential instruments (for instance, ASI or altimeter) there may not be spares readily available. Your car, and certainly your trailer, may do more road miles in the space of a competition period than they have done all year, increasing the chance of retrieve problems and the need for a thorough service. Before embarking on any expedition or competition, sit down and plan for every reasonable eventuality. A good checklist can make the difference between having a successful, enjoyable time or a miserable snag-ridden one.

The knowledge that your preparation has been thorough increases confidence, which in turn improves performance.

Chapter 11

Cross-Country in Thermals

Most cross-country flights are carried out using thermals. To achieve this, the glider pilot uses thermals as stepping-stones to keep the glider airborne while covering ground. The classic method of flying cross-country in a glider is to climb in one thermal and, having gained enough height, glide to the next thermal and there regain the height lost in the inter-thermal glide.

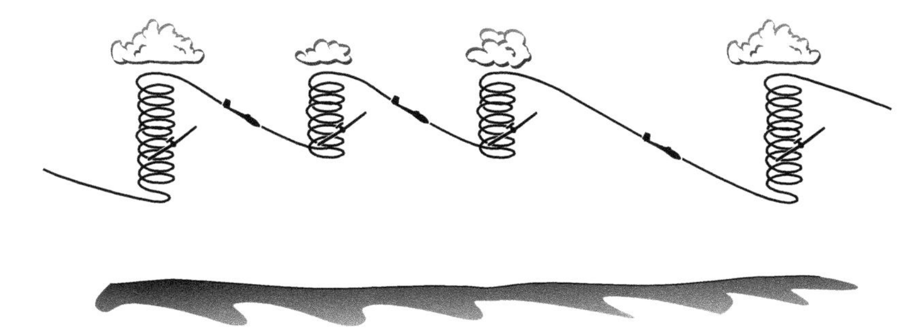

Fig 11.1 Classic climb and glide technique. The classic way to fly cross-country in a glider is to climb in a thermal, then glide to the next thermal.

To be able to do this, weather conditions must be favourable for the formation of thermals which are strong enough to allow the glider to climb, and plentiful enough to enable the glider to glide from one thermal to another.

When thermals are close together, such as when thermal streeting occurs, a glider pilot can cover large distances without circling, simply by moving from thermal to thermal to conserve height, and by adjusting the glider's airspeed to suit the vertical movements of the air. As the distance between thermals is directly linked to convection depth, this 'dolphin flying' technique is often possible early in the day when the cloud base is fairly low but thermals are close together.

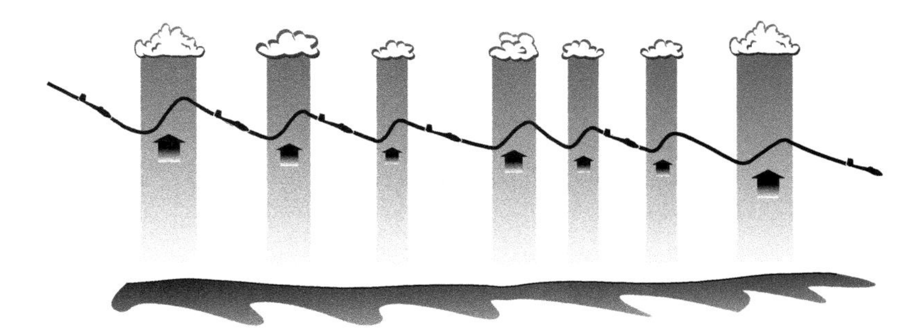

Fig 11.2 Dolphin flight. When conditions permit, it is possible to cover long distances without stopping to thermal, by varying the airspeed continuously to match the vertical movements of the air.

As the temperature rises throughout the day, thermals tend to rise higher and become stronger, but the distance between them increases. As this occurs, classic climb and glide cross-country flight becomes more useful. By the end of the day, although the cumulus base is usually at its highest, thermals are often far apart, requiring high climbs and long glides to achieve cross-country flight. Late in the soaring day, when the amount of surface heating has reduced, it is unwise to allow the glider to get low, as often the thermal circulation which has created the spacing between thermals will result in few new thermals coming up from ground level. Instead, any rising air you find now may be a result of this circulation rather than active ground sources.

Often, for the major part of the soaring day, a combination of classic and dolphin flying techniques is possible. The decisions as to which technique to use will, to some extent, be forced upon you by the soaring conditions. Usually, if you attempt to dolphin fly when thermals are too widely spaced, you will end up low and be forced to regain height by circling. On the other hand, if thermals (either weak or strong) are closely spaced, you may find that you are not losing enough height between thermals to justify stopping regularly to circle to gain height.

The next thermal

If you were crossing a river by stepping from one stone to another, obviously you would be looking ahead for the next stepping-stone. You would not leave the stone you were on until you knew where the next stepping-stone within reach was to be found. Similarly, when thermalling cross-country, you should never leave the thermal you are in until you have a good idea where you are likely to find your next thermal. Like the stepping-stones in the river, this next thermal must be within reach. Just as you would not put your weight on a stone that appeared to be insecure, the expected thermal must also appear good enough to use. Unfortunately, it is not always possible to employ this ideal tactic, as there will be occasions when the prospect of finding the next thermal does not appear good.

The chances are that you will be identifying the probable position of a thermal either by the presence of cumulus clouds or by ground features.

The changing appearance of cumulus clouds as they grow and decay will be your main gauge of whether a particular cloud is indicating the position of lift or sink. Sadly, even cumulus clouds do not always display textbook shapes and characteristics. Often they will fragment or spread out before reaching the expected cauliflower shape. Sometimes, what looks like a decaying cloud will form a new cell or turret, giving the promise of the presence of a new thermal below a cloud which you have already decided to avoid. Such developments will make it more difficult to decide whether lift or sink will be found beneath a cloud.

The only solution to these problems is to constantly scan the clouds and the sky ahead to see which clouds are growing and which are decaying, where wisps of cloud are forming, and in which areas clouds have recently disappeared.

Fortunately, thermalling provides the ideal opportunity for such observation. Each completed circle in a thermal will present a view of the sky on track every fifteen seconds or so. Such repeated observations allow you to build up a frame-by-frame mental picture of the state of the clouds, helping you to judge their stage of development, and which clouds are likely to indicate the presence of a thermal.

As well as looking for clouds to head towards or to avoid, it is important that you take in the whole picture of what is happening to the sky. A large area of sky devoid of cloud, on what is essentially a cumulus day, is unlikely to be an area where thermals are simply not forming cloud. It is more likely that the area is not producing thermals. Such areas will require you either to gain sufficient height in order to be able to glide across them or to make a detour to stay in the soarable conditions suggested by the cumulus clouds elsewhere.

Unlike cumulus clouds, ground features will not normally give an indication of the state of associated thermic activity (the exception to this situation being where smoke is rising from a feature such as a power station). On blue days, features such as towns and hills can often be assumed to be the source of potential thermals, and can be used instead of cumulus clouds to indicate potential stepping-stones across the landscape. On some days, such hot spots may be the only producers of reasonable thermals. If this is the case, your topographical chart becomes a map of probable thermal sources. The result is that you may end up having to climb high and fly long glides to reach the next feature shown on the map, which will hopefully provide your next thermal. Such distribution of thermal sources may necessitate regular detours from the ideal map track.

Lift alignment

When cumulus clouds are present and therefore skyreading is possible, always bear in mind and look for the possibility of LIFT ALIGNMENT. Even on days when cumulus clouds appear to be randomly spaced, look for any cloud distribution that offers even the vaguest line of lift roughly on track. Using such favourable lift distribution to advantage can increase your average cross-country speed considerably.

When cloud streets are present, these form the ultimate high-speed lanes for cross-country flying (the only other contenders for this title being lines of wave lift, or hill lift from long lines of wind-facing ridges).

When cloud streets are aligned with, or close to, your intended track, running along such a street will allow you to sample a large number of thermals of varying size and strength. Often you will be able to run the whole length of a cloud street without having to circle to gain height. Such cloud-street running is usually easier the closer you are to cloud base, and thus it often helps to climb in a strong thermal initially in order to get close to the base of the cloud, and the area where thermals often seem to congregate.

On arriving at a cloud street, turn along it, in the direction of your intended track, until a thermal is encountered. If you have arrived under the cloud street low down, then you may have to climb in the first reasonable thermal you come across until you are at a safe enough height to go looking along the street for a better thermal. If, however, you have reached the cloud street at a safe height, you may be able to fly on track along the cloud street immediately, passing up weaker thermals, until a good thermal is found.

When gliding along a cloud street, you should be experiencing numerous thermals of varying size and strength. If this is not the case, or if, despite being directly beneath the cloud street, you are gliding for a sustained period in sink, then move a few wingspans to the left or right of your present track but remain under the cloud street. The fact that the cloud forming the cloud street may be quite wide does not mean that the line of thermals feeding the cloud is equally wide. You are probably flying along just outside the lift street. Once you establish where the lift street lies relative to the cloud above, this relationship will probably remain the same for neighbouring cloud streets.

End of cloud street tactics

No cloud street goes on for ever, although if you are lucky, one may go at least as far as your goal. If it does not, then you must be prepared to change your tactics as you approach the end of a cloud street.

Occasionally, a cloud street will end in a rain shower. Watch for this and be ready to avoid getting too close to this area. Often when you come to the end of a cloud street, there will be a large gap to be crossed before the next thermal can be reached. Unless you are already close to cloud base, it is wise to take a high climb in a thermal near the end of the cloud street so that the cloudless gap ahead can be crossed. Do not leave the decision to climb in a thermal too late.

If you arrive at the end of a cloud street and do not find a thermal, you will probably have to retrace your track, losing valuable time. Alternatively, if you decide to use what you think will be the last decent thermal before the end of cloud street, and it does not turn out to be as good as you had hoped, then time will be lost in a slow climb or you may have to backtrack to find a reasonable thermal.

A better tactic is to select and climb in what you think will be the second-to-last good thermal of the cloud street. In this way if, having gained the height you require, you do indeed encounter another good thermal before the end of the cloud street, you can use this to accelerate

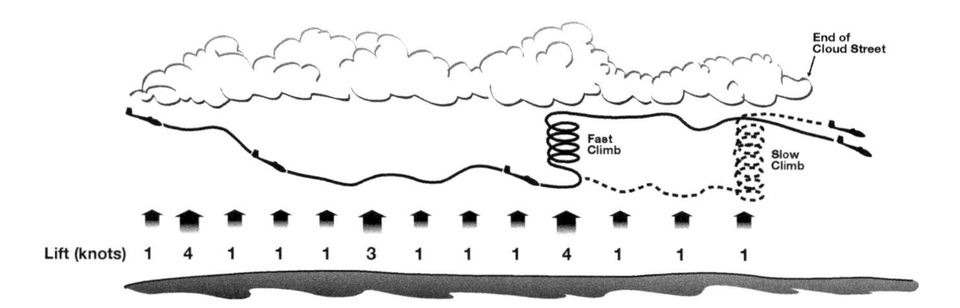

Fig 11.3 Tactics at the end of a cloud street. Counting on the last cloud in a cloud street to give a reasonable thermal may mean having to turn back to find lift. If possible, select a thermal before reaching the end of a cloud street.

before reaching the expected sinking air ahead. If, however, the 'last' thermal of the cloud street does not materialise, you still have the height you need and will not have to waste time searching or climbing in a weak thermal.

Crossing gaps between cloud streets

The areas of blue sky between cloud streets are likely to be areas of sinking air. Avoid straying into these areas except when it is necessary to reach your goal.

If your track crosses a series of cloud streets at or near an angle of 90 degrees, then the advantage of using the cloud streets is greatly reduced. Your task was badly selected. However, the cloud streets will still give you some help in the following ways. Firstly, the cloud streets indicate where good thermals are likely to be found. A short run along a cloud street may well result in a fast climb. Secondly, heading upwind along a cloud street is a useful way to counter the effects of drift experienced when crossing the gaps between cloud streets and when circling in lift.

When leaving the lift below a cloud street, and when approaching another cloud street, try to do so at an approximate angle of 90 degrees to the cloud streets. It is in the area adjacent to where the lift is to be found that the worst sink will probably be found, just as is the case with isolated thermals. On occasions, such sink may be found while the glider is under the edge of a cloud street.

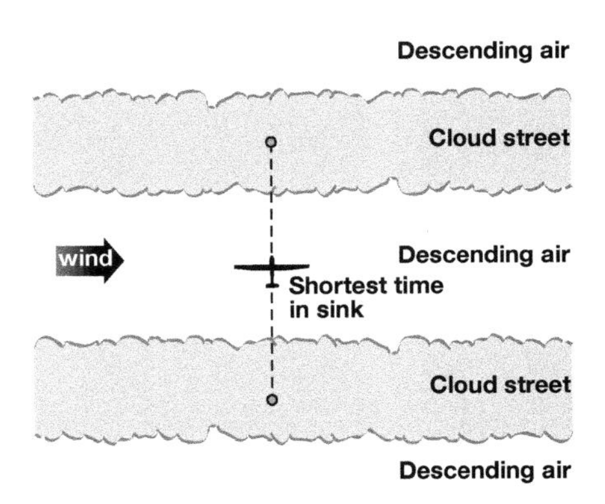

Fig 11.4 Cloud streets at 90 degrees to your track. When crossing between cloud streets which lie at 90 degrees to your track, aim to do so as directly as possible.

Different tactics are necessary when cloud streets lie at a more acute angle to your intended track. If your track has a large into-wind component, then using the cloud streets sensibly can contribute a great deal to your progress and reduce the frustration of battling with a head wind component.

If you were to attempt to fly the direct track from A to C in figure 11.5, although you would be covering the shortest distance possible, you would be spending a great deal of time in the sink zone between the cloud streets and subsequently losing a lot of height.

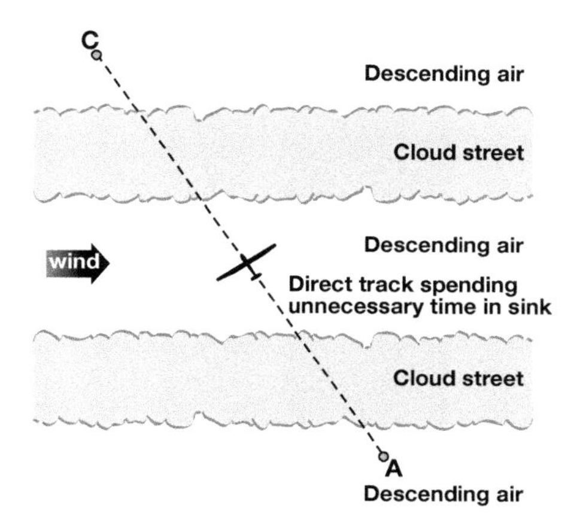

Fig 11.5 Cloud streets at an acute angle to your track. When cloud streets lie at an acute angle to your track, flying directly on track (from A to C) results in the glider remaining in sinking air for longer than is necessary.

Alternatively, flying straight from A to the nearest cloud street, then flying along it to a point abeam C (point B in figure 11.6), and then flying directly to C, will result in your flying a slightly longer distance but spending much more time in the lift zone and less time in the sink zone. Less height will be lost, and therefore less time spent circling to regain height with the resulting drift downwind. Despite the greater distance covered, your average speed should be higher.

Another option would be to use one cloud street for part of the way into wind and then glide across the cloudless gap to the next cloud street. This cloud street can then be used for a further run into wind, taking you closer to your goal (figure 11.7).

In practice, this technique is more commonly used, as several cloud

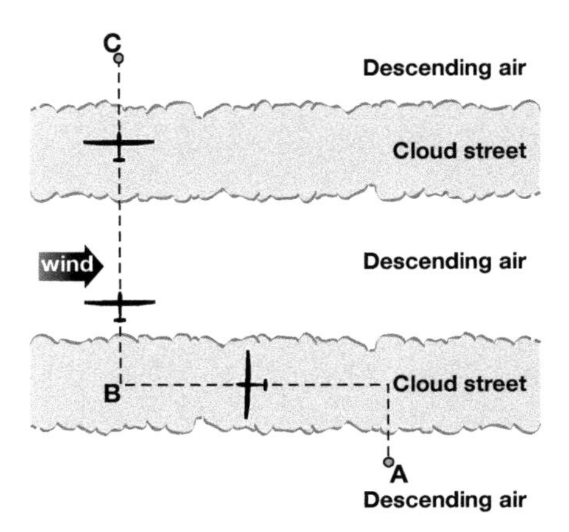

Fig 11.6 Using a dog-leg to cross cloud streets lying at an acute angle to your track. Flying along the cloud street to B, then crossing directly to C, results in the glider spending less time in sinking air.

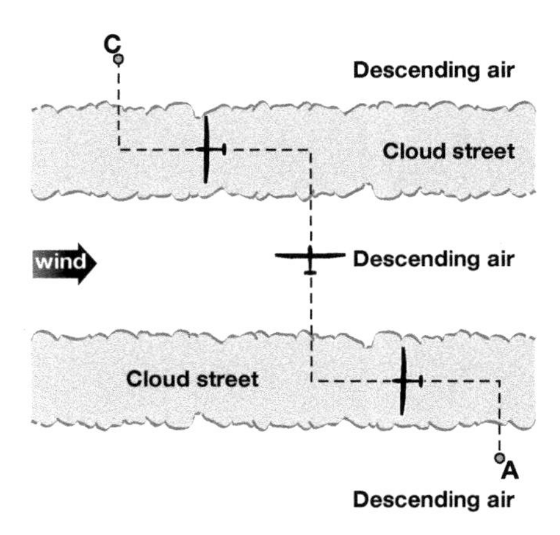

Fig 11.7 Crossing between cloud streets at the most beneficial place. The point where you cross between cloud streets can be chosen to allow you to benefit from the best conditions.

streets are often present. It also has the advantage of keeping you closer to your intended map track, thus making navigation easier; and it has the psychological benefit of making you feel that you are getting closer to your goal.

If, however, you are finding strong and plentiful lift under the cloud street now being used, you may be wise to continue along this cloud street for a while longer, rather than making the jump to a cloud street which might look good but may not actually be as good.

There is one trap which is always set for the sky-watching glider pilot. Any cumulus cloud observed with the sun shining directly upon it (the cloud being down-sun of the glider) may look better structured and more promising as a thermal indicator than if the sun was behind it. Cloud streets, especially, look magnificent when viewed with the sun on them. Such optical trickery, together with most glider pilots' greed for the best lift available, can result in the unwary glider pilot hopping from one cloud street to another, less good, cloud street. Having started a run along this new cloud street, there will be a temptation to jump to the next cloud street down-sun, which will probably look even better. Rather than continually crossing the sink zone between cloud streets to find better lift, it might be better to establish where the lift lies in relation to the cloud street you are under and then exploit it more effectively.

When you decide the time is right to move from one cloud street to another, then the track which you fly will be important to your average speed. You should time your glide to the next cloud street so that you will reach it at or just down-track of what appears to be an active cloud cell. As mentioned before, this is not simply a case of heading out into the blue gap from any position along the present cloud street. Care has to be taken to make your route the most expeditious one through the sink zone, so that you get through it as quickly as possible.

The secret is in making the wind work for you as much as possible. This will mean not only adopting a heading which, because of drift, will result in the shortest track distance across the sink zone. It will also mean flying a heading that will reduce the overall track miles from the point where you leave one cloud street to the point where you plan to arrive at the next.

For instance, in figure 11.8, the glider has simply adopted a heading at 90 degrees to the cloud streets and will drift downwind (to B), requiring a period of flight upwind along the next cloud street to reach a position abeam the departure point of the previous cloud street (C).

The glider in figure 11.9 has adopted a heading which is more into wind, and has a track which will be at 90 degrees to the cloud streets. This glider will arrive under the new cloud street without the need to fly along it for a period to make up any distance lost due to drift.

Taking this tactic one stage further would result in the glider departing the first cloud street on a heading which would cause it to arrive further

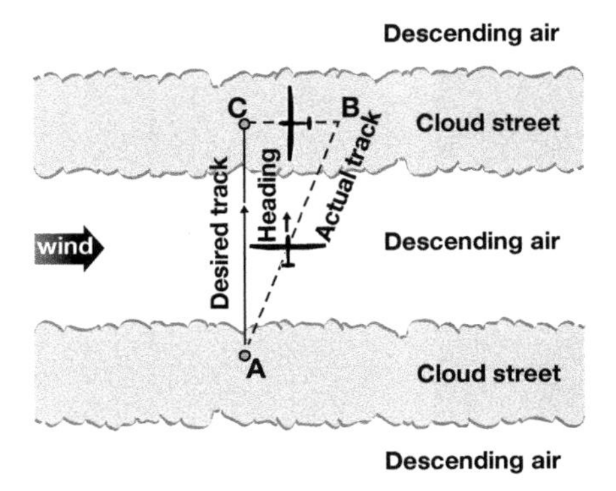

Fig 11.8 Heading versus track when crossing between cloud streets. Pointing the glider directly at the next cloud street will result in the glider drifting downwind.

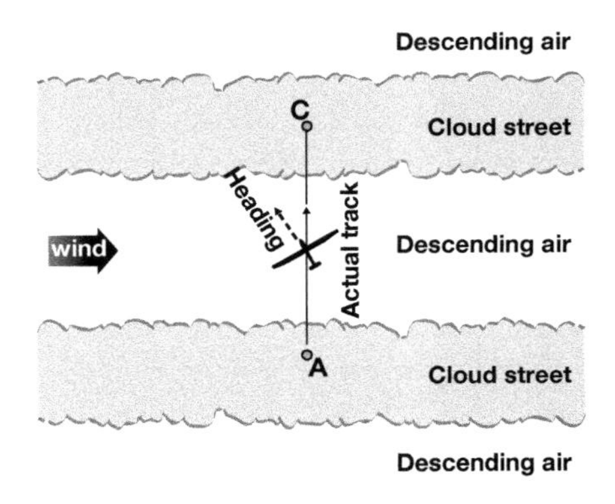

Fig 11.9 Countering drift while crossing between cloud streets. Select a heading which results in the glider flying the track you require.

along the next cloud street in the direction of the goal (at D), thus gaining distance at the expense of a slightly longer flight time in the sink zone (figure 11.10).

When selecting a heading to achieve this, do not be over-adventurous. A heading biased too much into wind will result in the glider spending an excessive amount of time in the sink zone between the cloud streets. The heading you use to cross between cloud streets will have to take account of the wind strength and direction relative to your desired track, and the severity of the sink between the cloud streets.

When flying from one cloud street to another, you will be flying across

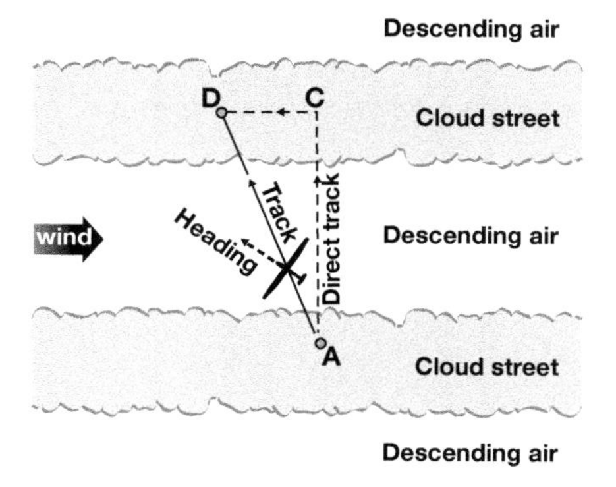

Fig 11.10 Optimum track while crossing cloud streets. If conditions permit, it may be possible to reduce track miles by tracking slightly into wind while crossing between cloud streets.

the direction of the wind. Although you will probably be countering any drift by adjusting your heading, because of the tendency to drift, the visual clues can be misleading. There is often a subconscious desire to apply a small amount of rudder. This will create extra drag and reduce the glider's performance. The best way to detect whether you are flying cleanly is to check the yaw string regularly. After all, you do want to reach that next cloud street with as much height as possible.

If your desired track for the leg has a tail wind component, the above tactics can be modified, as there will be no need to battle your way into

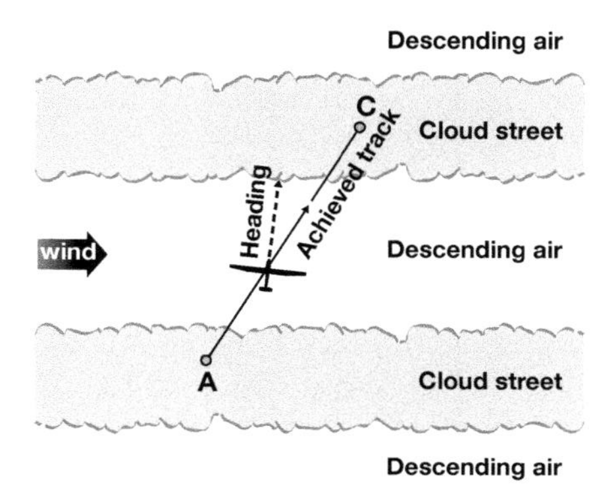

Fig 11.11 Crossing between cloud streets when the desired track has a tail wind component. Choosing the correct track when the leg has a tail wind component can result in increased groundspeed and minimum loss of height.

wind. Instead, you can cross the gaps between cloud streets on a heading that will help you cover task distance more quickly, while still keeping the time the glider is in the sink zone to a minimum.

Whether you choose to use cloud streets at all will depend on how far off your desired track they are aligned. As a general rule, use them if they appear to be helping you to penetrate into wind, or if the lift they provide is allowing you to dolphin fly for long periods. Cross the gaps between them as expeditiously as possible.

Blue thermal streets

Although the tendency is to think of lift streets being indicated by cloud streets, there is no reason why cloudless thermal streets cannot exist. Using these blue streets, as they are known, is not as easy as exploiting the lift under cloud streets.

Identifying their presence on a particular day can in itself be difficult. Staying in the lift without any cloud line to refer to can be near impossible. Apart from knowing the direction of the line of lift, you can never be sure whether you have lost the lift by straying to the left or right of the lift zone, or whether you have flown off the end of that particular lift street.

However, there are ways of deducing the line of a possible blue street. For instance, if you are climbing in a thermal and there is another glider climbing in a thermal more or less directly upwind or downwind of your position, then it is not unreasonable (if a little unscientific) to assume that a probable lift street lies between the two thermals. The same theory can be applied if a good thermal source is visible directly upwind of your present thermal.

Once such a cloudless lift street is discovered, the same tactics used when there are cloud streets around can be employed (that is, turning upwind or downwind when lift is encountered and crossing cloud gaps at an expeditious angle to the perceived lines of lift).

Stratocumulus, large cumulus clouds and showers

Many days that begin with well-spaced cumulus cloud can suffer from spreadout of the cumulus cloud into stratocumulus, over either the whole task area or parts of it. On other days, what may have begun as a day with promising-looking cumulus clouds can turn into a day of large cumulus clouds, cumulonimbus and showers.

The problem with stratocumulus is that it can cut off the sun's heating of the ground for long periods. Despite this, thermals already triggered may continue to feed the clouds for some time. The sky close to cloud base may continue to provide reasonable, if somewhat weaker thermals.

Watch for such spreadout developing and take a high climb while you still can, as new thermals are likely to be scarce until the stratocumulus starts to break up. Look for any breaks in the cloud, or sunshine reaching the ground. It is where either of these events appears that new thermals are likely to occur. Alternatively, if no sunlit ground is within easy gliding range, look for darker areas of the cloud, which may still be being fed by thermals. In such conditions, exercising caution (that is, flying conservatively) often pays off and is normally the best tactic.

The build-up of large cumulus clouds poses different problems. Lift will usually be good under active clouds, but the thermic circulation which is part of such clouds may suppress other thermals in the vicinity. This is particularly true when clouds become large enough to produce showers. Not only does the sink from such clouds ruin other thermals in the area, but also the soaking of the ground in their path can delay the rebirth of thermal activity for some time.

When clouds start to build excessively, be prepared to make long glides across areas of sky that contain no lift and are potentially filled with heavy sink, or to make large diversions to stay in favourable air. It is important that you learn to appreciate the glide performance of your glider and can estimate the size of gap you can cross without risk of landing out. This is an area where many pilots seriously underestimate their glider's performance, leading them to hang around at the edge of a cloudless gap unnecessarily, thus losing them time which may be needed to complete the task.

If a shower is observed to be blocking your track, it may indeed be necessary to 'park' the glider in a thermal and await an improvement in the conditions. Such tactics may be unavoidable if the shower in question is obscuring a turning point.

Frontal overcast

When a weather front (typically a warm front) is approaching your task area, the overcast it creates brings more serious problems with it. The thickening cloud of an approaching warm front is unlikely to break up to give gaps, in the way that stratocumulus does. Such overcast is likely to be the herald of the end of your soaring day.

Once the cloud cover is complete and thick enough to stop the sun's heat reaching the ground, few, if any, new thermals will form. However, if the overcast reaches the area after the ground has warmed sufficiently, some thermal sources, above which incipient thermals have formed already, may still release their thermals. Thus for a period after the cloud cover has become total, some thermals may still be found.

In such conditions, using blue thermal techniques, and looking for the best potential thermal source features, may help you find usable thermals. Do not expect the thermals these provide to be of a comparable strength to the thermals present before cloud cut off the sun. Instead, regard them as get-you-home thermals, and reduce your airspeed between thermals more towards your best glide speed. With luck, you will be near the end of your task by the time the cloud cover kills the last thermals, and you will be able to reach a point from which you can final glide.

Unfortunately, the weather forecast was correct. The warm front was approaching fast, and as I approached the last turning point in the company of many other gliders flying the competition task, the thickening stratus cloud had cut off the sun completely. The sky looked dead for the last leg.

I decided to break away from the pack and divert south, well off track, to the site of a disused steel works, while everyone else continued towards the turning point. There I found a thermal and climbed at a steady 2 knots to cloud base before setting course for the turning point.

As I approached the turning point, I could see gliders in fields all around it. I rounded the turning point at 2,000 feet. My gamble had paid off.

I did not have enough height to make it home and eventually landed in rain in a field 10 miles short of the finish. That thermal had given me second place for the day.

Planning ahead

The success of a thermal cross-country flight depends on much more than your ability to use a thermal into which you have stumbled. A major part of thermalling cross-country is judging what is likely to happen – not just in the thermal you think you are heading for, but for some distance ahead. All of your stepping-stones have to be in the correct place. If you cannot find a thermal to bridge a gap, then you will have to plan some other route. If an expected thermal does not materialise, you will have to be able to reach an alternative one. Always have a contingency plan and do not wait until you are low with no lift and no ideas. Watch the sky in general and the sky on your route in particular. Watch for changes in the weather pattern, the cloud cover, the type of cloud and the state of the cumulus. Only by doing so will you be able to adopt the best tactics for the conditions ahead. Thermals behind you are history. Judging what the future holds is an important aspect of thermalling cross-country.

Other types of lift

Lastly, although you set off with a thermal cross-country flight in mind and have probably been using thermals successfully for some distance, keep your mind open to the possibilities of other types of lift.

When low on a thermal day, hill lift, not thermal, may be available or may even be responsible for the climb you are achieving. Higher up, wave may be waiting to be exploited if you show a little patience. Sea breeze fronts and squall lines may offer a fast run for part of the route. The atmosphere is ever changing and is full of surprises. Keep your eyes open and use your imagination and your soaring will benefit.

The first leg of the competition flight paralleled the southern coast of England in a moderate southerly wind. Having got low, I had drifted north of track and had to push forward into wind to reach the first turning point, which was a prominent church on a small hill.

Thermals had been blue and not very strong on this leg, but further inland the cumulus clouds looked promising for the second and third legs, if only I could get around the first turning point and glide northwards to reach them.

Getting low again as I approached the turning point, I saw three other gliders sitting in a field. Just as I turned, I hit the only decent lift I had encountered for some time. I banked tightly but after half a turn lost the lift and was in strong sink. I straightened up towards where the lift had been but did not have enough height to reach it. The result was a landing in the same field as my colleagues. They too had not switched their brains from thermal mode to hill lift mode quickly enough, and had encountered the same curl over.

Chapter 12

Speed Flying

'Slow down and you will fly faster.' That was the invaluable advice given to the author by a former world champion.

How can such an illogical statement be of value? The secret lies in the conservation of energy – in this case, energy in the form of height.

The important point to note is that 'flying fast' in cross-country flying means achieving a high *average* speed over the entire task and not simply a high airspeed on the glide sections of the flight.

Another important aspect always to bear in mind is that when we talk about lift, we are talking about a *rate* of climb. Therefore, one knot (or 100 feet per minute) is not simply a gain of height; it is a gain of height of 100 feet *every* minute. Unfortunately, sink indications are also *rates* of sink. What this implies is that to gain height will take precious time. Similarly, the longer you spend in sink the greater the amount of height lost – height which will take time to regain.

Optimising your flying by only climbing in strong lift will reduce the time spent circling (a time when, generally speaking, you are not covering ground).

On the face of it, spending less time in sinking air would seem to suggest flying as fast as possible through sink. This will get you through the area of sinking air quicker, but will also throw away more height due to the glider's reduced efficiency at higher airspeeds.

To illustrate this, let us look at the POLAR CURVE of a typical Standard Class glider. The polar curve of a glider is a graph of the glider's expected rate of descent through the air at various airspeeds. A polar curve is only a theoretical guide to a glider's performance, as it will not allow for the movements of the air mass in which the glider is flying. That is, it assumes that the glider is flying through air which is neither ascending nor descending, or, to put it another way, that the glider is flying in still air.

An inspection of the polar curve of a typical glider shown at figure 12.1 will highlight some basic points.

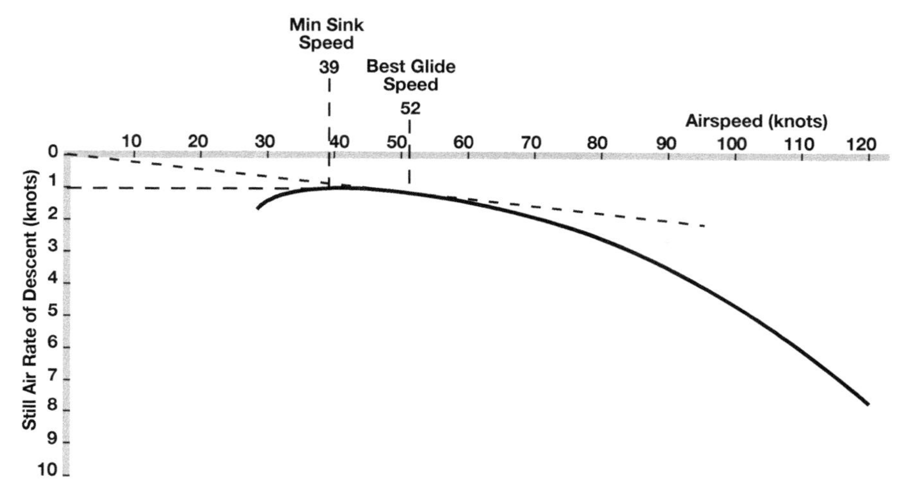

Fig 12.1 The polar curve. A glider's polar curve shows the still air rate of descent that can be expected at any given airspeed and therefore is a measure of a glider's performance.

1. As the airspeed increases, so does the rate of descent.
2. The airspeed at which the rate of descent is least (known as the MINIMUM SINK SPEED) appears vertically above the highest point of the curve (39 knots).
3. The airspeed at which the best glide angle is achieved (best lift/drag ratio in still air) is vertically above the point where the tangent from the graph's origin touches the polar curve. This airspeed is known as the BEST GLIDE SPEED and is greater than the minimum sink speed (52 knots).
4. At low airspeeds the curve is very flat, showing that there is a reasonable airspeed range over which the glider is very efficient and will not lose excessive height. At the higher end of the airspeed range, the rate of descent increases much more rapidly as airspeed increases, resulting in a steepening of the curve as the glider's performance reduces quickly.
5. The glide angle (lift/drag ratio) at any airspeed can be calculated simply by dividing the airspeed by the corresponding rate of descent.

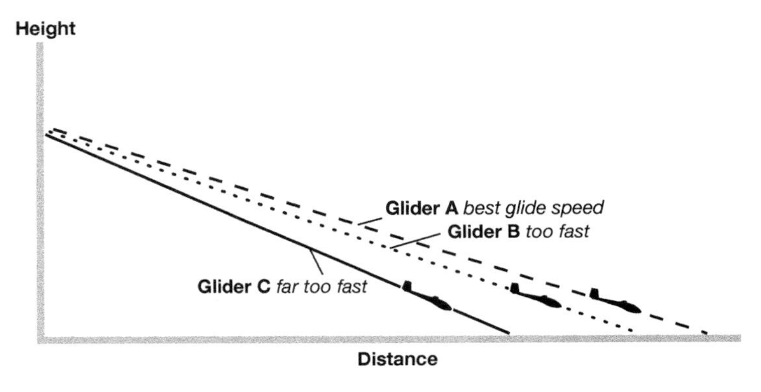

Fig 12.2 Flying at the optimum airspeed. Flying at too high an airspeed wastes height and reduces the distance which can be covered from a given height.

The fact that, in still air, as airspeed increases so does the glider's rate of descent means that flying too fast will throw away height. Time will be used up regaining any height that is lost, whether it is lost unavoidably or wastefully. The best airspeed at which to fly through still air is the airspeed at which the best glide angle is achieved. By flying at this airspeed, the glider will cover the maximum distance with the height available (glider A in figure 12.2).

Flying faster will result in an increased rate of descent, a poorer glide angle and less distance covered (glider B in figure 12.2). Glider C, flown by a pilot who missed out on flying Concorde, flies far too fast, covers least distance and is also first on the ground!

If the air through which our three identical gliders are flying is descending, then it will benefit them to fly faster, in order to get through the sink faster. Again, there is an optimum airspeed at which to fly through any sinking air. This will change depending on the rate at which the air is sinking. The faster the air is sinking, then the faster the glider needs to be flown.

The necessary airspeed to fly through sinking air can be deduced from the glider's polar curve as follows (figure 12.3).

1. Extend the vertical axis of the graph upwards, using the same scale as below the origin (zero knots).
2. Draw the tangent to the curve from the value on the upper part of the vertical axis corresponding to the rate of sink (in this case 2 knots).

3. Read off the required airspeed on the horizontal axis vertically above where the tangent touches the performance curve (70 knots).

(There is nothing magical about this jiggery-pokery. It is just an easy way of moving the whole curve down by the amount of sink being experienced, since it is easier to move the horizontal axis of the graph upwards than to re-draw the whole curve lower down.)

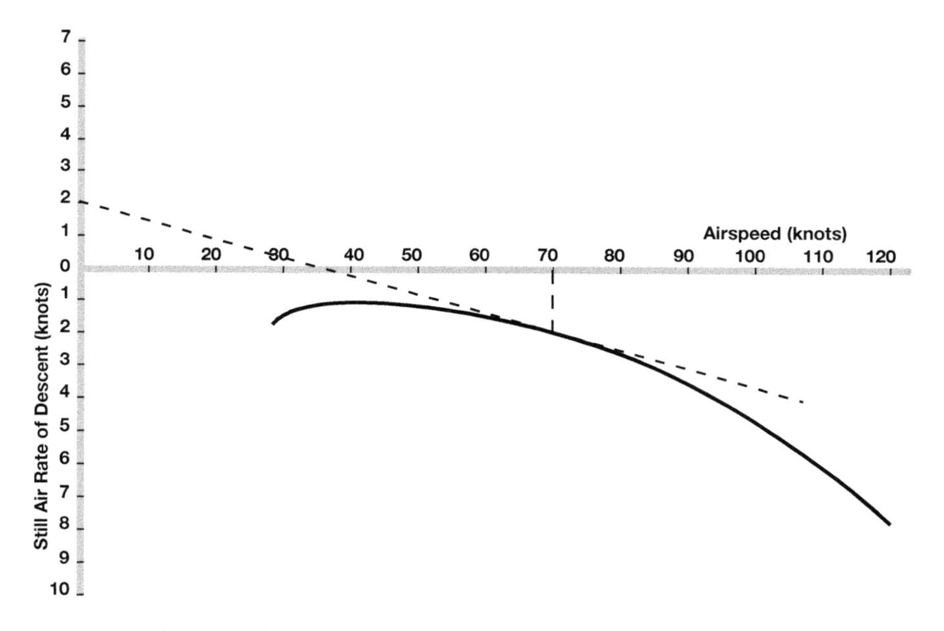

Fig 12.3 Calculating the optimum airspeed to fly through sinking air.

Figure 12.4 shows the importance of flying at the correct airspeed through sinking air. It assumes that our three gliders are crossing an area 5 nautical miles wide where the air is descending at 2 knots (200 feet/minute).

Glider B flying at the airspeed for best glide angle (52 knots), is in the sink longer and, as a result, loses more height. Glider A flies at the higher, optimum airspeed (70 knots) and gets through the sink quicker. The reduced time spent in sinking air means less height is lost. Glider C goes to the other extreme and dives through the sink at too high an airspeed (90 knots) and wastes height.

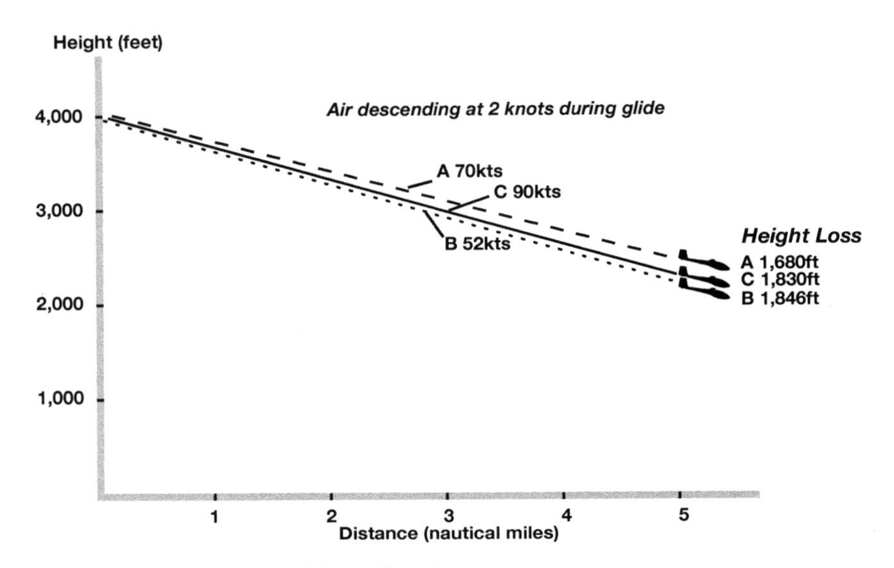

Fig 12.4 Optimum airspeed through sinking air.

Table 3 shows the performance achieved by each of the gliders. The rate of descent displayed in the table is the combined rate of descent caused by the descending air mass plus the glider's still air rate of descent at the airspeed flown.

TABLE 3

GLIDER	AIRSPEED (knots)	TIME (min:sec)	RATE OF DESCENT (knots)	HEIGHT LOSS (feet)
A	70	4:17	3.92	1,680
B	52	5:46	3.20	1,846
C	90	3:20	5.49	1,830

The aim of these somewhat simplified examples is to help you see the importance of flying at the correct airspeed for the prevailing conditions.

There is only one problem. This technique will give the best results in terms of the maximum distance that can be flown assuming that there is no more lift and therefore no further climb to be had. It can be used if you are trying to cover the maximum distance at the end of the day or if no more lift is expected. (Gliding for maximum distance will be covered in more detail in Chapter 18 on THE FINAL GLIDE.)

However, if there is another thermal within reach, you can be a bit more adventurous. In order to increase your average speed, you can afford to fly through the sinking air at an even higher airspeed, as the extra height lost can be regained. The degree to which you increase your airspeed when gliding between thermals will depend on how fast you think you can regain the extra height lost. Therefore, the determining factor as to how fast you should fly is the rate of climb you expect in the *next* thermal.

Again, the glider's polar curve can be used to determine the required airspeed to use when flying between thermals. Here is how we do it (figure 12.5).

Let us assume that the next thermal will give us an average rate of climb of 3 knots (300 feet/minute) and that our glider is flying through still air.

1. Using the upper part of the vertical axis, re-draw the tangent from the 3 knots point to the polar curve.
2. Vertically above where the tangent touches the curve, read off the airspeed on the horizontal axis. This is the optimum airspeed at which to fly in these conditions (75 knots).

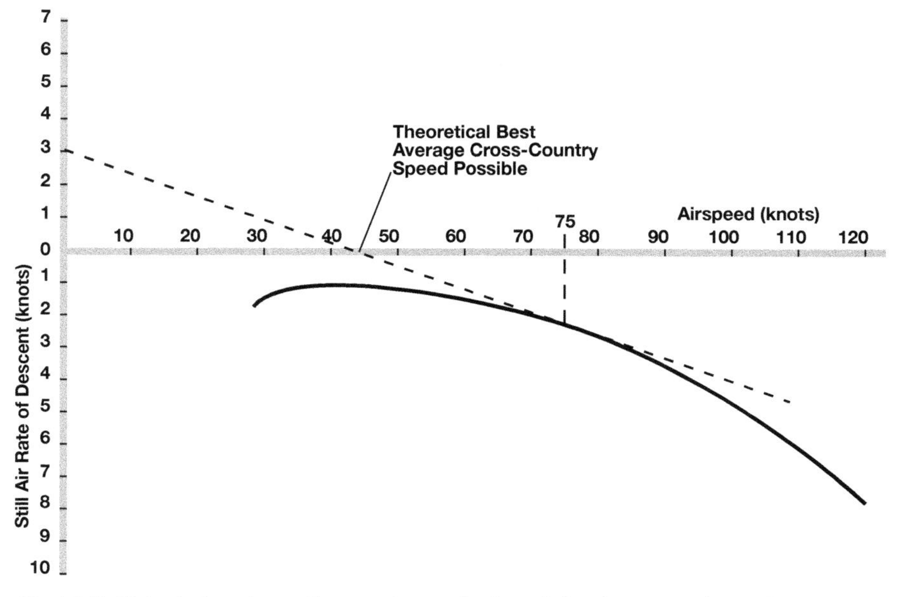

Fig 12.5 Calculating the optimum airspeed when flying between thermals.

(Incidentally, the airspeed value where the tangent crosses the horizontal axis is theoretically the average cross-country speed achievable with this average rate of climb. This will depend on the conditions not changing, and cross-country being carried out in the classic climb and glide style, rather than by dolphin flight. However, this figure tends to be optimistic, as it cannot take into account the condition of the glider or any performance-losing mistakes made by the pilot.)

Now let us look at the effect this has on our three competing gliders, crossing the same 5 nautical mile gap between thermals.

At first glance, this time pilot A seems to have lost out in height to pilot B, but remember that they have arrived at a thermal which will give an average rate of climb of 3 knots. Pilot A has arrived 1 minute and 46 seconds ahead of pilot B, but 200 feet lower. Thanks to the thermal strength, pilot A will regain the height lost in 2 minutes and 58 seconds, which will mean that the total time for the glide and climb will be 6 minutes and 58 seconds. By the time that pilot B regains the height lost, he will have taken 8 minutes and 4 seconds for the same glide and climb.

Pilot C still does not have the idea and is again flying at too high an airspeed. He ends up low and takes longer to regain the height lost, giving an overall glide and climb time of 7 minutes and 38 seconds, that is, 40 seconds longer than pilot A. Table 4 summarises these performances.

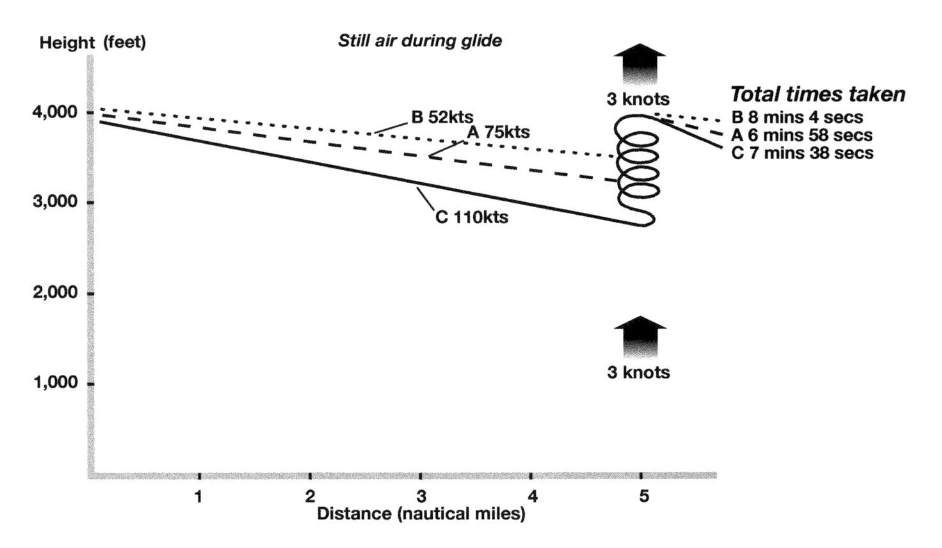

Fig 12.6 Optimum airspeed between thermals.

TABLE 4

GLIDER	SPEED (knots)	TIME (min:sec)	RATE OF DESCENT (knots)	HEIGHT LOSS (feet)	CLIMB TIME (min:sec)	TOTAL TIME (min:sec)
A	75	4:00	2.23	892	2:58	6:58
B	52	5:46	1.20	692	2:18	8:04
C	100	3:00	4.63	1389	4:38	7:38

Note, however, that on this occasion, thanks to the rate of climb available, pilot C (flying faster) has done better than pilot B, who flew too slowly.

Now it is unlikely, if there are thermals around, that the air in between them will be still air. The chances are that it will be descending. At what airspeed do we have to fly through descending air if we are heading for a thermal? Let us look to the polar curve for the answer (figure 12.7). Assume that the thermal for which our pilots are heading is 5 nautical miles away and will give them 3 knots average rate of climb, and that the air through which they are flying is descending at 2 knots.

1. This time we need to move the polar curve down by the value of the sinking air; that is, 2 knots. As we have already seen, we can achieve the same result by moving the horizontal axis upwards by 2 knots.
2. Now move the horizontal axis upwards again by the expected rate of climb of 3 knots (2 + 3 = 5 knots).
3. Therefore, from the 5 knot mark on the upper part of the vertical axis (the new origin), draw the tangent to the curve.
4. Vertically above where the tangent touches the curve, read off the optimum airspeed on the horizontal axis (87 knots).

Which of our three pilots achieved the best results on this 5 nautical mile section of the flight?

Again, pilot A flew at the optimum airspeed for the conditions (2 knots sink and 3 knots expected rate of climb). Pilot B, flying too slowly, spent longer in the sink, lost more height and took longer to regain it, losing 2 minutes and 29 seconds to pilot A. Pilot C, our speed merchant, once again beat pilot B's performance – but the time spent regaining the extra lost height resulted in his losing 36 seconds to pilot A on the combined glide and climb. See figure 12.8 and table 5.

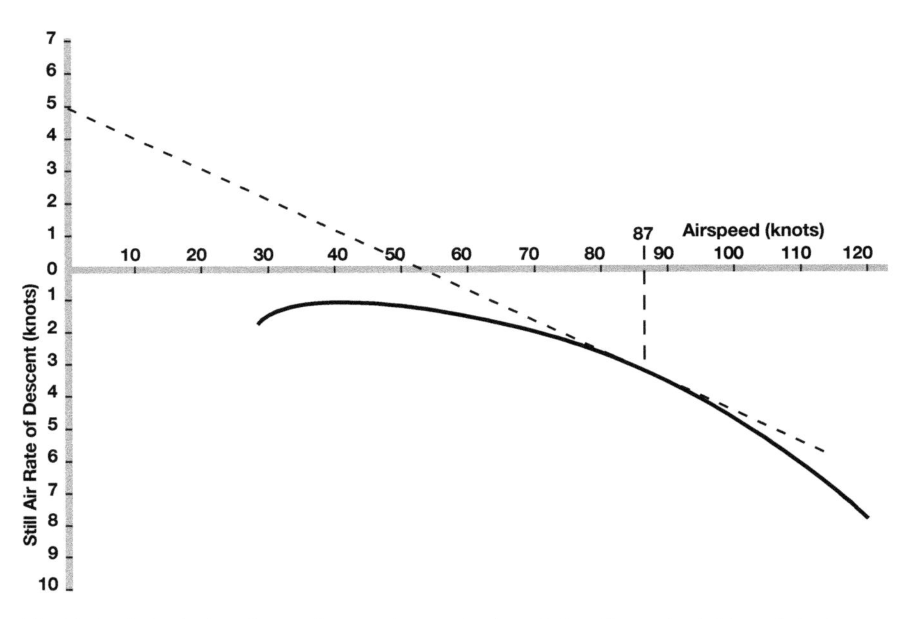

Fig 12.7 Calculating the optimum airspeed when flying through sinking air between thermals.

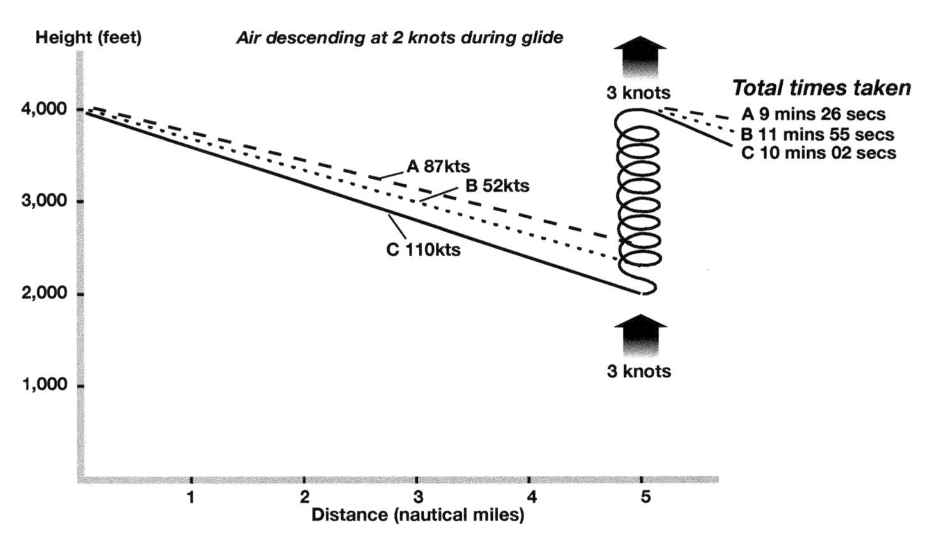

Fig 12.8 Optimum airspeed when flying through sinking air between thermals.

TABLE 5						
GLIDER	**SPEED (knots)**	**TIME (min:sec)**	**RATE OF DESCENT (knots)**	**HEIGHT LOSS (feet)**	**CLIMB TIME (min:sec)**	**TOTAL TIME (min:sec)**
A	87	3:27	5.20	1,794	5:59	9:26
B	52	5:46	3.20	1,845	6:09	11:55
C	110	2:44	8.03	2,190	7:18	10:02

Now although the thought of losing 36 seconds would make a competition pilot cringe, as an average cross-country pilot you may think that such a time loss is insignificant. But hold on! This time loss is over a 5 nautical mile section of the flight. Given that you will probably need around 35 such climbs to complete a 300 kilometre flight, the likely total time lost over such a flight would be 21 minutes for pilot C and a staggering 1 hour and 27 minutes for pilot B!

Many cross-country pilots disregard such facts, saying, 'I am interested in distance, not speed'. If the task is a long one, or if the length of the soaring day is shortened, perhaps by the approach of a weather front, then this lost time may be the difference between covering the distance and making it home, or ending up in a field. *(On a 500 kilometre task the total time lost to pilot A would be 35 minutes for pilot C, and 2 hours 25 minutes for pilot B!)*

As mentioned, it is the average rate of climb that will be achieved in the next thermal which dictates the optimum glide airspeed. This statement needs some qualification, as the first thermal you come across need not be the thermal you use for a climb. With modern sailplanes, you often have the performance to pass up weaker thermals, stopping only to climb in a thermal offering a good rate of climb. Only by such selection (and rejection) of thermals will you achieve high climb rates and thus high average cross-country speeds.

This does not mean that you should fly straight through weaker thermals at high airspeed. It is often possible to extract some energy and height even from these rejected thermals by reducing airspeed, so as to remain in their lift longer. This technique will be discussed in more detail in Chapter 13 on DOLPHIN FLYING.

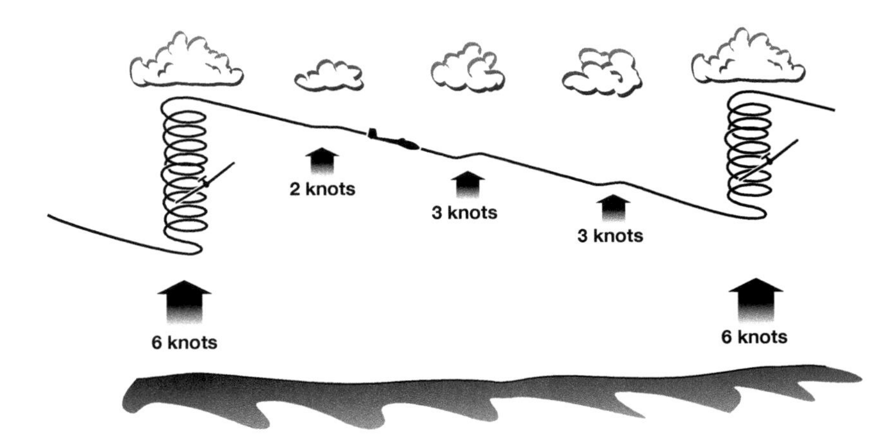

Fig 12.9 Selection of strongest thermals. The highest possible average cross-country speeds will be achieved only if the strongest thermals are used for climbs.

Average rate of climb

Many pilots overestimate the rates of climb being achieved in thermals. It is essential that you know how to assess the average rate of climb that you are achieving and be honest with yourself. If you do not, you will end up flying too fast between thermals and throwing away height. Overestimating rates of climb is the kind of mistake that pilot C may have been making.

If you were to use a stopwatch and time how long you spend in a thermal, and then calculate the average rate of climb achieved, you would probably be disappointed with the result. Most pilots only achieve about one-half, or possibly even one-third, of the rate of climb indicated on the variometer.

Modern flight director/variometer systems normally offer average rate-of-climb indications as a standard part of their display. However, many of these show the average climb rate over a pre-set period (normally set by the pilot before flight) and will not give the average over the whole climb. The 'running average' that such instruments give does help, as it tends to smooth out the sometimes erratic indications of the variometer, giving a more realistic, if still optimistic, impression of your average climb rate.

Other variometer systems allow you to start the averager at the beginning of the climb and stop it at the end of the climb, thus giving you the average rate of climb for the whole of the timed period. This may be an improvement over the running average, but is dependent on your decision on when to start and stop the averager.

The practice of many pilots which causes the greatest over-estimation of the average rate of climb being achieved (and therefore expected in the next thermal) is that they either wait until they are established (or centred) in the thermal or until they have started to climb, before they 'start the stopwatch', either physically or mentally.

To correct this tendency, we have to define 'climbing' not as the period when the glider is actually ascending, but instead as *any time that the glider is 'not flying on track'*. This 'starts the stopwatch' as soon as airspeed is reduced from the inter-thermal cruise airspeed to begin thermal centring AND keeps the 'stopwatch running' until the increased airspeed is re-established after leaving the thermal on the glide to the next thermal.

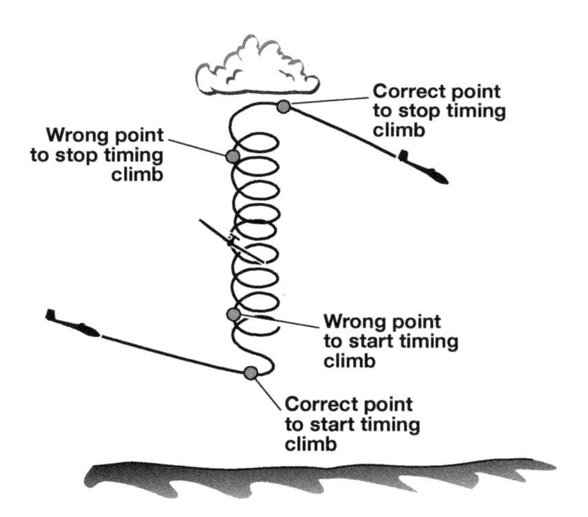

Fig 12.10 Defining average rate of climb. Climb time begins when the glider slows down before entering the thermal and ends when it achieves cruise airspeed after leaving it.

From figure 12.10, you can see that our new climb time now includes the slowing down period as you enter the thermal, the initial centring period, the climb phase and the time to accelerate on leaving the thermal, instead of just the climb phase. This means that the average rate of climb will suffer if any of these phases is badly executed. This should help you to better estimate the likely average rate of climb in the next thermal, thus allowing you to fly at a more realistic inter-thermal airspeed.

This improvement in mental attitude can be achieved to some extent automatically if the variometer system has an 'airspeed switch' which starts the averager timer as the airspeed falls below a certain pre-set value, and stops it as the airspeed increases as the glider leaves the thermal. Alternatively, some systems are triggered by selection of the flap setting used for thermalling.

The speed-to-fly ring

As you can imagine, drawing or trying to follow pre-drawn lines on a polar curve while flying a glider is impractical. The solution to this problem came in the form of an adjustable ring fitted on a bezel on the circumference of the variometer. This ring is known as the MacCREADY SPEED-TO-FLY RING, named after its inventor and former World Champion, Dr Paul MacCready. Each ring is calibrated for a specific type of glider, and printed on it are an index arrow and airspeed marks which are usually at intervals of 5 knots.

When the speed-to-fly ring is set, the variometer needle, as well as indicating the rate of descent as usual, will indicate the airspeed at which the glider should be flown for optimum performance.

For instance, when the speed-to-fly ring is set with its index arrow opposite the zero lift/sink mark on the variometer dial, the airspeed indicated opposite the variometer needle will be the airspeed that will result in the maximum distance possible, whatever the vertical characteristics of the air through which the glider is flying (figure 12.11).

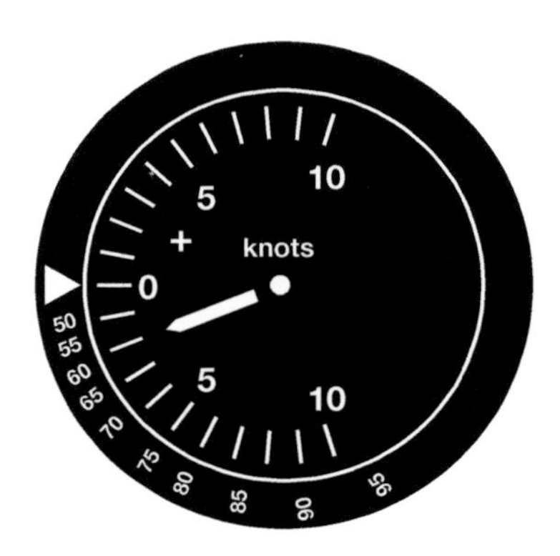

Fig 12.11 MacCready speed-to-fly ring. With the index arrow set at zero, the speed-to-fly ring will show the airspeed required to cover the maximum distance.

If you set the index arrow of the speed-to-fly ring opposite the average rate of climb expected in the next thermal (in figure 12.12, this is 3 knots), then the airspeed indicated on the speed-to-fly ring by the variometer needle will again indicate the optimum airspeed at which to fly through the air, whatever the rate of descent of the air mass. However, this time it will allow for the fact that the glider will be able to regain any height lost at a rate of 3 knots. The average rate of climb in the next thermal, and therefore the setting used on the speed-to-fly ring, is generally known as the MacCREADY SETTING or MacCREADY VALUE.

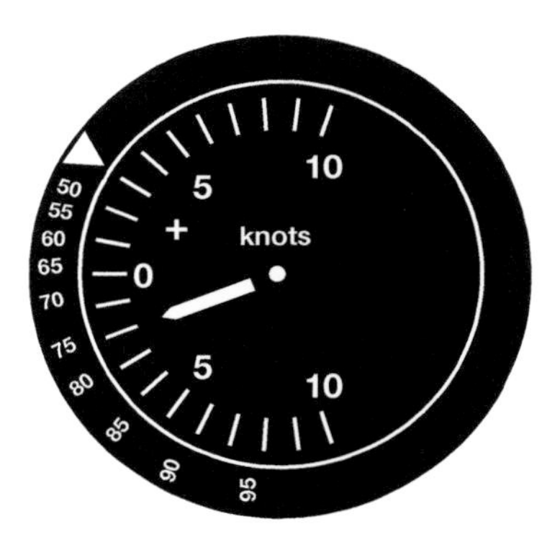

Fig 12.12 Speed-to-fly ring setting. The speed-to-fly ring can be set to the expected average rate of climb in the next thermal. The airspeed required is then read off next to the variometer needle.

To arrive at the optimum airspeed will require increasing the airspeed gradually. As the airspeed increases, so will the glider's rate of descent. This will result in the variometer needle indicating an increase in the optimum airspeed. Responding to these indications will, after a series of airspeed increases, result in the glider flying at the optimum airspeed for the flight conditions present and expected.

Incidentally, a speed-to-fly ring can only be fitted to variometers which have a linear scale. Fortunately, most variometers have linear scales and only a few have logarithmic scales.

The effect of incorrect MacCready settings

As discussed earlier, it is difficult to estimate an accurate value for the average rate of climb being achieved, let alone that which is expected in the next thermal. Fortunately, such errors of estimation do not, in theory, lead to a large loss of average speed, providing the speed-to-fly ring setting is not wildly inaccurate.

However, care should still be exercised when selecting a MacCready setting. Too high a MacCready setting will result in flying too fast, which will reduce the glider's glide angle. This could result in your failing to reach the next thermal or a thermal of the expected strength. Too low a MacCready setting is a safer option, as the inter-thermal height loss will be less and there will be an increased chance of reaching good lift, although there will be some reduction in average speed.

Despite this apparently safer route to cross-country soaring, reducing the MacCready setting to zero, when there are good thermals to be had, is disastrous to average cross-country speed. Although you will probably have a better chance of remaining airborne longer with a zero MacCready setting, you may not achieve a high enough average cross-country speed to complete the task. A zero MacCready setting should only be used in an emergency, such as when you are getting low, or if conditions ahead look as if a long glide will be required to reach the next thermal or your destination.

Air mass variometer

There are several disadvantages in using a standard variometer. As it indicates the descent rate of the sailplane, and not just the air mass, it tends to give the pilot the impression of extreme sink, when at higher airspeeds much of the indicated rate of descent is the result of the glider's inefficiency at such airspeeds. A far greater problem is that it is difficult to determine the rate at which the air is rising when flying at any airspeed much above best glide speed. In addition, as airspeed is adjusted, as dictated by your speed-to-fly device, the indicated rate of descent will change, requiring a further airspeed adjustment. This creates the need for tedious 'needle chasing'.

The answer to these problems is to introduce a calibrated 'leak' of air from the pitot system, which introduces air into the variometer system (between the variometer and the capacity flask). Because this leak will vary proportionately as the glider's airspeed varies, it will offset the flow of air into the capacity flask that would otherwise flow through the variometer. If the system is well set up, the indication on the variometer is the rate of descent (or ascent) of the air mass and not the glider, irrespective of the glider's airspeed. Many pilots find this information much more helpful. A variometer calibrated in this way is known as an AIR MASS VARIOMETER, or NETTO.

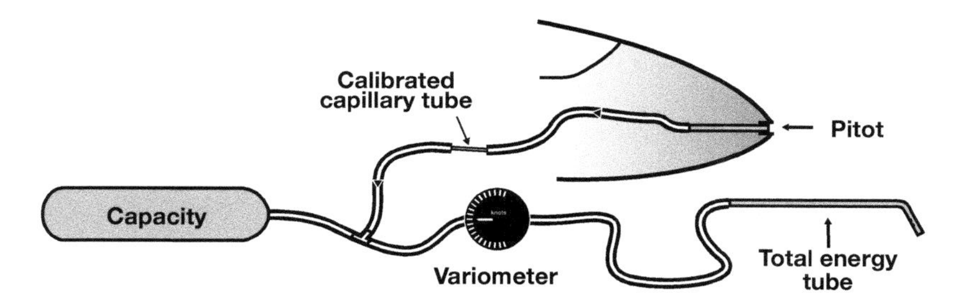

Fig 12.13 Air mass variometer. By allowing a calibrated flow of air from the glider's pitot system into the pipe between the variometer and the capacity flask, the variometer will show the rate of descent of the air mass, not the glider, irrespective of the glider's airspeed.

A speed-to-fly ring can be fitted to such an air mass system. However, it is necessary to re-calibrate the standard speed-to-fly ring, otherwise the indications will be meaningless.

If an air mass variometer is installed, it is normally desirable to incorporate a pneumatic change-over switch, which will return the variometer to read the 'normal' rate of descent of the glider rather than the air mass. This is necessary when thermalling so that the actual rate of climb of the glider is displayed.

Such a switch should be carefully labelled and its setting checked regularly. If this is not done, one can find oneself 'thermalling downwards' with the variometer showing that the air mass is rising, but obscuring the truth that the sink rate of the glider is greater than the rate of ascent of the air mass. This is a particular problem when thermals are weak and workload is high.

Electronic variometers and flight director systems

To a great extent, technology has come to our rescue. The introduction of printed circuit variometers and microchips means that all of these functions can be presented to the pilot in a variety of different forms, along with audio warning tones for lift, sink, flying too fast or flying too slow.

Although it was initially developed as a pneumatic system, perhaps the most useful aid to speed-to-fly technique is the 'zero reader' display, or SPEED DIRECTOR, which is now common on most flight director systems.

With this system, all the pilot has to do, having set the MacCready value, is fly at the airspeed that keeps the variometer/speed director needle on zero. If the needle goes up, the pilot should reduce airspeed, as the glider is being flown too fast with respect to the air through which it is flying. If the needle indicates down, the nose of the glider should be lowered to fly faster through air that is descending faster. When the optimum airspeed is achieved, the needle will again indicate zero.

Many flight director systems incorporate a separate indicator either within the main variometer display or as a separate unit mounted on the instrument panel. This may simply give a prompt to 'pull' or 'push' on the control column. Such a system was once given the impressive title 'cruise control'.

Normally the audio output of such systems has a silent band, which occurs when the glider is being flown at the optimum airspeed for the conditions. Thus, if the optimum airspeed is being maintained, the pilot literally has a quiet life!

On electronic systems, the change-over from variometer mode to speed director is normally achieved by an airspeed sensor which operates as the glider adopts or leaves thermalling airspeeds. This reduces the risks of being in the wrong mode while thermalling. Another option usually available is to use a microswitch connected to the flap system to trigger the mode changes, in a similar way to that in which electronic averagers can be switched on and off automatically. In addition to these automatic devices, there is normally a switch with which the pilot can override the automatic change-over from variometer to speed director.

Even if an electronic variometer system is fitted to your glider, it is wise to have a back-up mechanical variometer which is fitted with a speed-to-fly ring in case of failure of the electrical system or the instrument. For this reason, the method of construction of a speed-to-fly ring is explained in Appendix 1.

Speed-to-fly simplified

The workload on a cross-country flight is often high. Many pilots find that slavishly altering airspeed as dictated by a speed-to-fly device adds unnecessarily to the workload. As the penalties for using inaccurate MacCready settings are not excessive, many pilots adopt more general MacCready settings, depending on how soaring conditions are shaping up. For example, a pilot in the UK may use the following settings in the following soaring conditions.

Soaring Conditions	MacCready Setting
Survival day	0
Below average	1
Good	2
Excellent	3

Different countries or regions will require different scales according to what quality of average soaring conditions they enjoy.

An even simpler way to use speed-to-fly technique would be to adopt target airspeeds when cruising between thermals. These airspeeds will depend not only on the soaring conditions in your region but also on the performance of your glider. In this case, you may use something resembling the following guidelines.

Soaring Conditions	Inter-thermal Airspeed
Survival day	Best glide speed
Below average	70
Good	80
Excellent	90

By reducing the workload, the adoption of such techniques allows more time to be spent skyreading and making decisions which will help you achieve high average cross-country speeds.

The importance of achieving high rates of climb

As the rates of climb being achieved dictate the average inter-thermal airspeed, achieving the best possible climb rate in each thermal is essential. This can be achieved firstly by using only the strongest thermals available, and secondly by climbing in each thermal as efficiently as possible. This means that your initial centring must be efficient, and subsequent centring must maintain the glider's position in the best area of lift for the whole time that you are in the thermal.

Since the average rate of climb is calculated from before you enter the thermal until after you leave it, it is also important to conduct these parts of the flight efficiently. Do not reduce the airspeed unnecessarily before you reach the thermal. The pilot who reduces airspeed to thermalling speed when approaching a likely cloud is the pilot who loses a lot of height in the sinking air around the thermal. Keep the airspeed at the required penetration speed until lift is contacted. Many pilots seem to have learned this bad habit of slowing down and going into a 'search mode' because their instruments are badly compensated for total energy. These problems, which often haunt club gliders, make it difficult for the pilot to know what the air mass is doing. If a little attention were paid to the calibration of such instrument systems, then pilots might avoid gaining such bad habits.

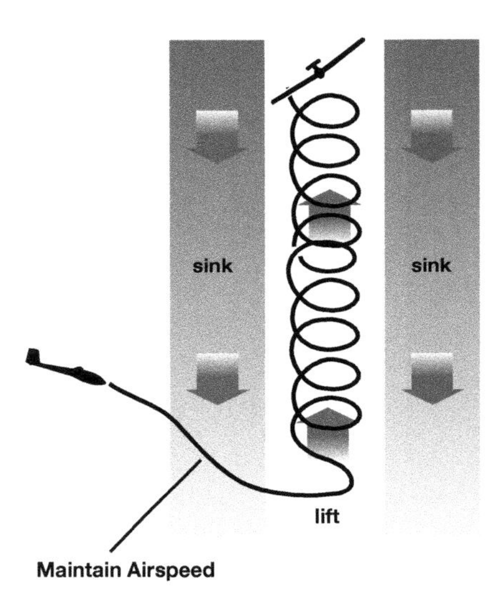

Fig 12.14 Entering a thermal. Maintain airspeed until lift is contacted. Avoid the temptation to slow down as you approach a thermal. Doing so will mean that more height will be lost in the sink which surrounds the lift.

Another bad habit, this time at the top of the thermal, is the tendency to hang around trying to gain more height after the climb rate has decreased. Unless there is some good reason to be there, such as needing more height to reach the next thermal or because the sky ahead looks dead, you should not be floating around reducing your average climb rate.

This phenomenon can be seen on blue days, often in competitions, where gliders gaggle at the top of each thermal waiting for someone to lead the way. Some might call it caution; others would call it under-confidence.

The place to break such habits and generally practise thermalling is near your home airfield – where you can be ruthless about which thermal you will take and which you will reject, when you will leave each thermal, and to what height you are willing to descend before selecting the next one. All of this can be done safely in the knowledge that you can land back on the airfield for another launch should you get it wrong. The confidence that such exercises give will reduce the doubts and tension when you are trying to fly fast on a cross-country flight.

One classic example of climbing too high in weak lift is after a near-outlanding, when a weak thermal has saved the glider from ending up in a field. After the initial relief of climbing away, it is very tempting to stay with the thermal until you are much higher than is necessary to reach the next good thermal. The time to leave this weak thermal is when you believe you can safely reach better lift. After all, you have probably ruined your average speed enough for one day by falling into such a hole!

Theoretically, the time to leave any thermal is when the *actual* rate of climb being achieved falls below the rate of climb you expect in the next thermal. I use the word 'theoretically' because, although we are dealing with a known rate of climb in the thermal being used, the strength of the lift in the next thermal has to be guessed on the basis of experience gained on the day, or on visible indications, such as clouds or the behaviour of other gliders. Certainly, once the actual rate of climb has fallen below the overall average being achieved for the thermal up to that time, staying with the thermal will obviously be reducing your average rate of climb and subsequently your overall average speed.

Of course, if you have doubts about the quality or even the existence of the 'next' thermal, then common sense should override theory and it may be better to take extra height in the current thermal. At any rate, never leave a thermal unless you have a good idea where the next thermal is to be found.

As achieving a high rate of climb is so important, nothing should distract you from this goal. To this end, navigation should be carried out when cruising – not when thermalling. Before reaching a thermal, you should already have worked out the likely direction in which you will leave it once the climb has been completed. If controlled airspace is a consideration, then you should have worked out the height to which you can climb without infringing such airspace while in the glide and not when climbing in the thermal.

The manner in which you leave the thermal also has an effect on your 'average rate of climb'. The same area of sinking air surrounding the thermal, which potentially caused excess height loss when you arrived at the thermal, is again waiting to rob you of some of your hard-gained height. Leaving the thermal's lift at too low an airspeed will waste height as you transit the sink. Increasing airspeed as you meet the sink will also result in a depressing unwinding of the altimeter. The only answer is to accelerate while the glider is still in lift.

By doing so, you will be flying at a reasonable airspeed when you hit the sink. Unfortunately you will have to use an educated guess, based on the day's previous thermals, as to how fast the air will be descending and what airspeed you will require. After crossing the sink on the outside of the thermal, you can adjust your airspeed more accurately as dictated by your speed-to-fly ring or speed director.

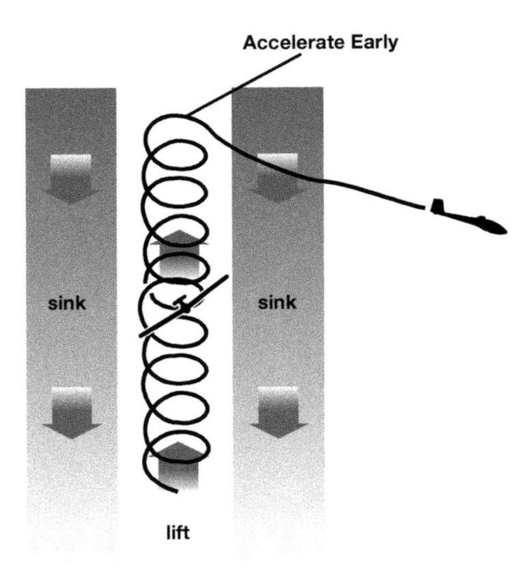

Fig 12.15 Exiting a thermal. Accelerating to inter-thermal airspeed before you leave the lift will mean that you will spend less time in the sink surrounding the lift.

Techniques used for increasing airspeed before exiting a thermal vary. Some pilots favour increasing airspeed while flying the last circle in the thermal; others advocate tightening the turn on the side of the thermal opposite the exit track and then accelerating as they fly through the thermal's core. Both these techniques will ensure against excessive height loss, but may have to be modified if there are other gliders in the thermal.

Height band technique

Thermal selection and rejection is not always as simple as having a minimum average rate of climb as a target and disregarding any thermals which give lift that is weaker than this. The pilot who doggedly presses on to low heights, flying at airspeeds dictated by a high MacCready setting and awaiting the ideal thermal, may occasionally be lucky and find it. More often than not, such a pilot will end up low and have to climb in a very weak thermal to survive (with the resultant loss of time), or even worse, will end up in a field.

The factor that must be borne in mind is the probability of finding a suitable thermal. Mathematicians will produce great formulae and calculations to prove their probability theories. As a glider pilot whose only aim is finding a good thermal, the only factor that you need to bear in mind is that the lower you get, the less search time you have to find a thermal and the less fussy you can afford to be.

Let us look at how this works in practice by examining a section of a cross-country flight and what is known as the technique of using HEIGHT BANDS (figure 12.16).

Let us assume that, so far, the flight has resulted in regular thermals giving average climb rates of 4 knots. We have just left such a thermal near cloud base (at 4,500 feet) and are gliding out on track (A). Our MacCready setting is 4 knots.

After gliding a short distance, the glider encounters another 4 knot thermal. We can either climb in this by circling or, if we are still relatively near cloud base, then simply pulling up in the lift may regain us the height which was lost in the short glide (B).

In fact, while flying close to the upper limit of the height band, pulling up is often a better option than starting to turn in the thermal, as inevitably some time will be lost centring. If possible, it is better to avoid such short, circling climbs for this reason, as the total time lost over a whole flight can be considerable.

Being at or near the top of our OPERATING HEIGHT BAND (in this case 4,500 feet), we continue on the next glide.

Next, we encounter a 3 knot thermal (C), but since the sky ahead looks good, we do not stop to circle in its lift, but merely pull back the airspeed to stay in the lift longer. Further on, a 2 knot thermal tempts us (D), but since the glider is still at a safe height, we can again continue until we find the expected thermal with a strength of 4 knots (E). As long as the average rate of climb equals or exceeds 4 knots, we continue climbing in this thermal until we reach the top of our operating height band.

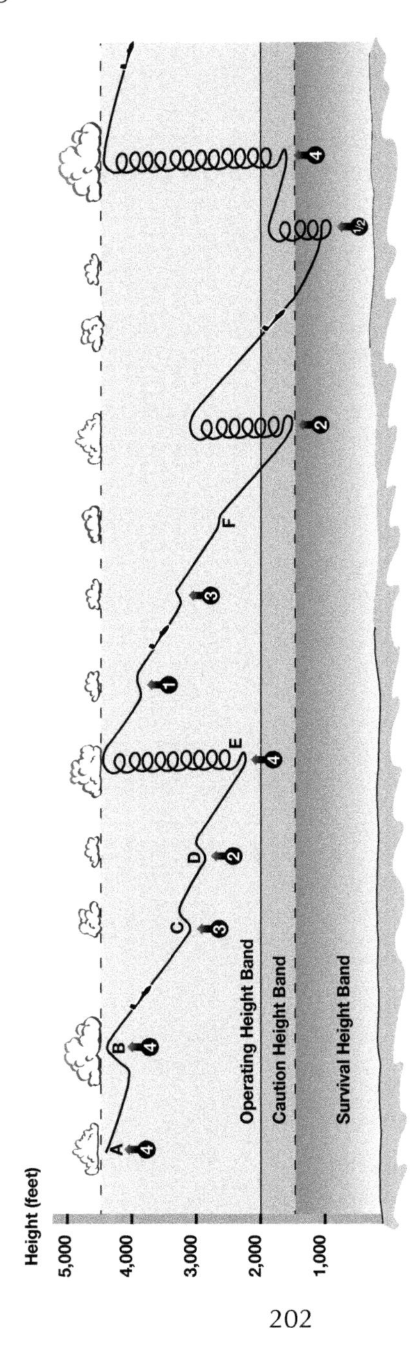

Fig 12.16 Height band technique. Selecting and using reasonable height bands increases average cross-country speed and reduces the chance of landing out.

After leaving this thermal, we treat the glide to the next thermal in the same way. However, should the glide continue with the glider only encountering weaker thermals and getting down towards 2,000 feet, the search time available to find a 4 knot thermal is becoming limited (F). It is now necessary to select a lower MacCready setting, based on the sky ahead (say 2 knots).

Now, if a thermal offering a 2 knot average rate of climb is encountered, then we will use it, BUT ONLY AS HIGH AS TO GIVE US ENOUGH HEIGHT TO SAFELY REACH A POSSIBLE 4 KNOT THERMAL. The glide to this potential 4 knot thermal should be made using a MacCready setting of 4 knots.

Should a glide take us below 1,500 feet, then we will have to go into survival mode and reduce our MacCready setting to ‘zero’. Now we will have to accept any thermal to get the glider to the 1,500 feet level, where we can again set our hopes (and MacCready setting) for a 2 knot thermal.

In this example, the height band where the chosen MacCready setting is 4 knots (between 2,000 feet and 4,500 feet) is called the operating height band. The band between 1,500 feet and 2,000 feet I would call the CAUTION HEIGHT BAND (2 knots MacCready value), and that below 1,500 feet, the SURVIVAL HEIGHT BAND.

If you can keep the glider in the operating height band, you have the best chance of achieving a reasonable average speed. However, as any competition pilot will tell you, getting low and falling into the caution or the survival height bands usually has a disastrous effect on the average cross-country speed achieved.

The heights selected for these height bands will vary from day to day and even during a day, depending on such factors as the soaring conditions, the cloud base, the sky ahead, the terrain being overflown and the base of controlled airspace. On a good day, with a high cloud base, or in countries with a deep convection layer, the operating height band may be very much higher and deeper. On some days, it may pay to operate nearer cloud base. On blue days it may also be safer to stay high.

One problem of using cloud base as the upper limit of the operating height band is that close to cloud base the visibility is poor and the cloud immediately above the glider often obscures the sky ahead. This makes deciding on the best track difficult, and prevents skyreading immediately before leaving a thermal. For these reasons it is often better to use a height band upper limit which is a few hundred feet below cloud base.

Very early in the day, cloud base is often low but thermals are closely spaced. This will give a narrow operating height band. Later in the day, thermals rise and are more widely spaced, allowing a deeper operating height band.

Wetlands or irrigation areas ahead may require temporary raising of the lower height bands, whereas airspace may require limiting the operating ceiling and occasionally the lower limits of the height bands.

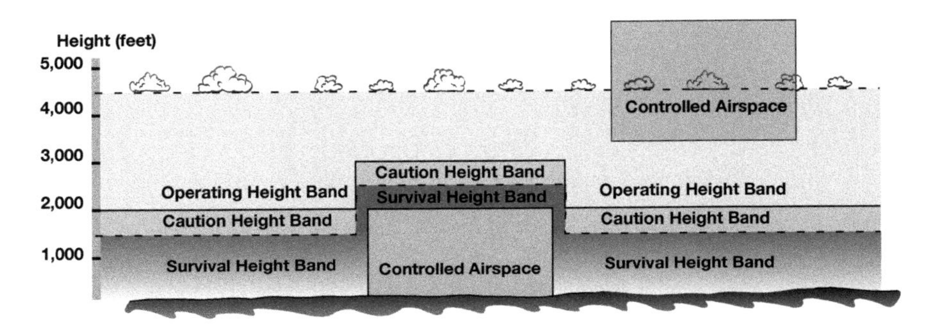

Fig 12.17 The effect of controlled airspace or a bad soaring area on the level of height bands. It is necessary to modify the level of the height bands when crossing airspace or poor soaring country.

Whatever limits you set for the height bands, and whatever MacCready setting you choose, you must be willing to modify these to suit conditions. Such judgement will be based on the above factors, but usually the determining criterion will be the sky ahead and changing soaring conditions. However, this does not give you an excuse for indecisiveness. Decision making is the essence of speed flying. Such decisions should be carefully made and based on sound reasoning. The only other requirement is self-discipline. In fact, I believe that one important difference between a good pilot and a great pilot is having the discipline to fly in the correct manner based on calculated decisions, without allowing influences, such as nervousness of outlanding or getting low, to distract or to affect the decisions made.

In the final analysis, although flying at the correct theoretical airspeeds for the air mass through which the glider is travelling will enhance your average cross-country speed, it should be remembered that these improvements are only minor relative to the need to keep the glider in favourable air. Avoiding areas of sinking air, and spending more time during the cruising phases of the flight in air which is rising, is just as important as achieving fast climb rates. As you will see in the next chapter, this fact is displayed vividly when you achieve successful 'dolphin flight'.

Chapter 13

Dolphin Flying

While discussing the high standard of some of the younger competition glider pilots, one senior pilot stated, 'The trouble with them is that they have always flown gliders in which they have not had to circle too often to gain height'.

This sarcastic statement does contain one hidden truth. It lies in the fact that if you could fly a glider on track without having to turn, your average cross-country speed would benefit.

Most of the previous chapter dealt with the need to adjust airspeed to suit the vertical movement of the air through which the glider is flying. At some points during the discussion of height bands and thermal selection, we gained height in thermals by pulling up while flying through them, rather than circling. Such calculated variations of the glider's airspeed can give prolonged glides (and even height gains) using this technique.

This DOLPHIN FLYING (named after the porpoising nature of the flight path) is an extension of speed-to-fly technique. As much of the technique depends on rejecting thermals, it only developed as the performance of gliders improved. Pilots in earlier, low-performance gliders could not be choosy about which thermals they would use. As glider performance improved, glider pilots could sample more thermals as a result of the increased glide angles of modern gliders.

Dolphin flying still utilises all of the speed-to-fly theories of the climb-and-glide style of flight regarded as the 'classic' way to fly cross-country, only more so. With dolphin flight, the airspeed is reduced and increased in unison with most of the noticeable changes of air mass. I say most, because if you were to change the attitude and airspeed for every little change of air mass, then not only would you be flying a rather jerky trajectory, but also you would be reducing your performance to an extent that might make dolphin flying of less value. With every single control input there is a drag penalty.

The success of dolphin flying depends on two important factors. The first is smooth flying of the glider. The second is trying to keep the glider flying in the best available air mass.

The first of these, achieving smooth flight, comes with practice. The important point is to anticipate and thus be ready for the need to change airspeed. When other gliders are thermalling on track, it is easier to judge when a pull-up in lift is likely to be required. In this instance, especially if a thermal is crowded, airmanship is of prime importance if you are to avoid a collision or incurring the wrath of the other pilots. At other times, 'seat of the pants' information may tell you when lift is near – but be careful not to react to imagination or sensations caused by control inputs (so-called 'stick lift').

If thermals are strong and very sharp edged, or if you are using high inter-thermal airspeeds, then your pull-ups should be positive. The quicker the transition from sink to rising air, the harder you should pull up. The smoother the air mass changes, the gentler should be your control inputs. While advocating firm pull-ups in thermals, even these should be practised with care. Although most gliders are immensely strong, you will not be doing the airframe any good by imposing large loads on the structure. From a performance outlook, pulling up sharply at higher airspeeds is not so much of a problem, but pulling g at low airspeeds will result in an increased amount of induced drag and a reduction in aerodynamic efficiency.

As most gliders are not designed to fly under negative or reduced g loads, sustaining such loads is wasteful of energy. For this reason, do not be over-enthusiastic on the control column when you push forward to increase airspeed. Easing forward on the control column should be done before exiting the lift, so that the airspeed is not being increased in sinking air.

Even more important than the timing and the manner of controlling airspeed is keeping the glider flying in the best air mass available. Therefore, dolphin flying can only be done if lift is favourably distributed with no large areas of sink. Dolphin flying is therefore easier if there are cumulus clouds or cloud streets indicating the position of lift. However, it can also be used effectively in blue thermal conditions.

Keeping the glider in the best air mass usually involves track changes to stay under lines of cumulus clouds, or simply flying under small cumulus while *en route* to better-looking clouds. It involves avoiding crossing large areas of blue sky when there is a viable route using cumulus clouds as indicators of thermals as if they were stepping-stones across a lake. Although the tendency is to think of dolphin flight only in the vertical sense, viewed from above the flight path would often appear as a horizontal zigzag.

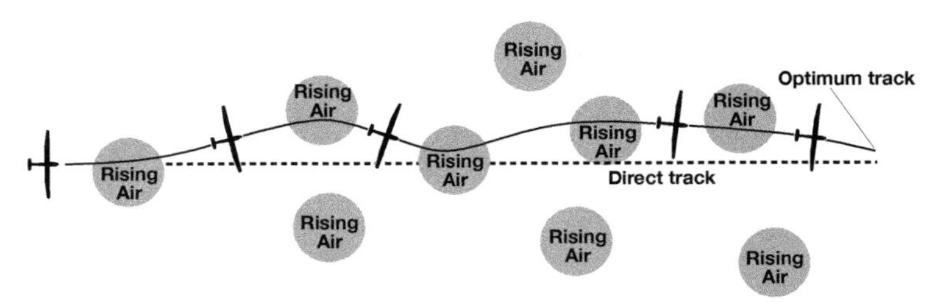

Fig 13.1 Following the best track. Flying a route which keeps the glider in favourable air is an essential part of dolphin flight.

With this in mind, do not be obsessed with following the track on the map or the exact track required according to your GPS. Impressive as it may appear to land out exactly on track, it is more impressive to stay airborne – and even more so to achieve a high average speed home.

The need for total energy and an air mass variometer or speed director

It is possible to dolphin fly, after a fashion, with a normal variometer, but it is not ideal. One essential requirement of any variometer, if it is to be useful in dolphin flight, is that it incorporates some form of total energy device which is well set up and gives smooth readings across the whole of the glider's normal airspeed range.

Dolphin flying is made much easier if the glider is also equipped with an air mass variometer (which is compensated for total energy, of course) or a speed director system. As you will normally have other duties to perform while trying to match the airspeed to the vertical movements of the air, it is a great help to have an audio system fitted to your speed director.

Whatever type of variometer is being used, it is important that it gives smooth indications and is not over-sensitive. While a slow variometer is a nuisance when dolphin flying, an over-sensitive variometer or speed director can be more of a distraction than a help, leading to over-controlling and confusion. Most electronic variometer systems present the pilot with the option of dampening down the indicator's response time. Be careful to select a response rate that is still fast enough to be useful.

The advantage of dolphin flying

If dolphin flying can be carried out successfully, average cross-country speeds far greater than those theoretically possible by classic climb and glide methods can be achieved.

Analysis of the barograph trace shown below for a 456 kilometre quadrilateral showed that the average rate of climb in the 16 thermals which were used during the flight was 4.3 knots. Using classic climb and glide technique, the best achievable average cross-country speed should have been 94.5 kph. As much of the task distance was completed using dolphin flying techniques, the achieved speed for the task was 104.6 kph.

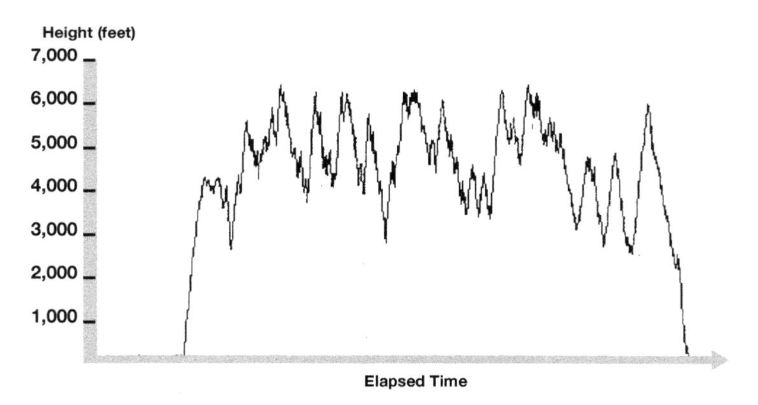

Fig 13.2 Barograph trace showing dolphin flight.

However, attempting to adopt dolphin flight techniques in unsuitable conditions (for example, when strong thermals are spaced well apart with little or no rising air in between) will invariably lead to a loss of average speed, probably as a result of the glider ending up low and having to take a weak thermal climb. This is usually the result of rejecting thermals in which a circling climb would have been preferable.

The technique of dolphin flying

The following section of a hypothetical flight is an example of how to use dolphin flying technique.

Let us assume that thermals with average rates of climb of 4 knots are common, and that there are frequent thermals of lesser strengths around. We have therefore decided that it is reasonable to select a MacCready setting of 4 knots.

Having left a thermal after a climb in a 4 knot thermal, we glide towards our next likely cloud. Should we have to vary our course slightly under wisps of cumulus or small cumulus clouds, we do so, providing the course deviations are within 30 degrees at most. We fly the airspeeds indicated by our speed-to-fly device. If we encounter sink, then we fly at the appropriate airspeed. Should we fly into lift, we reduce the airspeed accordingly.

If the thermal we encounter appears to be a 2 knot thermal, we bring back the airspeed to the value corresponding to that thermal strength, but we *do not stop to circle*. As the glider gains height from the pull-up in lift, we anticipate the point where the glider will exit the lift, being ready to increase the airspeed *before* the glider enters the sinking air around the thermal. If there are clouds around, this can best be judged by observing the area of cloud above and increasing airspeed before leaving the lift area under the cloud.

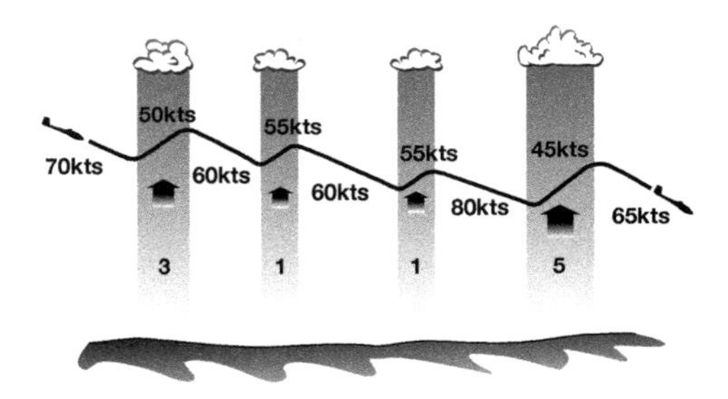

Fig 13.3 Dolphin flight utilises speed-to-fly theory.

We continue this system either until we hit a thermal giving 4 knots or better (in which case we climb by circling in the thermal) or until we reach a height where we are prepared to be less selective about the thermal strength acceptable (that is, the bottom limit of our operating height band).

Conditions near cloud base are often better

On some days, with a high cloud base, the conditions may make dolphin flight difficult. On these days the thermal circulation pattern may allow successful dolphin flying when close to cloud base, even though lower down the thermal spacing does not favour dolphin technique. The answer in such conditions is to take a climb to cloud base and attempt to stay high, to benefit from the better lift distribution near cloud base. It may be necessary to top up your height regularly by taking short climbs in the best lift available by circling in it.

Maintaining height (not gaining height)

On many occasions, when airspace in which gliders are prohibited has a lower limiting level that is below the height achievable in thermals, it is necessary to limit any climbs in order to remain clear of such airspace and within the law. (A similar situation may occur when near cloud base but not wishing to enter cloud.) At such times, if the glider is flown on track when the base of the controlled airspace is reached, subsequent thermals may make it impossible to stay below the airspace if normal speed-to-fly or dolphin flight is continued. In order to deal with this situation, it is necessary to increase airspeed to a point where no more height is gained.

If the desire is to remain at this height (perhaps to stay in better conditions close to cloud base), then you might find yourself increasing airspeed in lift (so as not to gain height) and decreasing airspeed in sink (thus using your excess airspeed to conserve height). Such a technique goes against the speed-to-fly theory discussed so far. A better way to deal with the situation is to increase your MacCready setting to a level that keeps your height below the imposed limit, but still allows dolphin flight. Finding the correct MacCready setting to achieve this will be by trial and error, but will probably involve initially undersetting the desired value and then increasing it until the glider is flying at airspeeds that will prevent it from gaining height. If you find that conditions change for the worse, or that you are throwing away more height than you intend, decrease the MacCready setting back towards its original value.

Maintaining height (or gaining height)

There are other instances where you may be required to maintain, or even gain height, while flying on track. Such instances do not normally involve controlled airspace, as it is a brave pilot who crosses over restricted airspace with only a small margin of height above its top. It is more likely that this desire to conserve height will arise when attempting to arrive at the next thermal at the same height as gliders already circling in the thermal, or occasionally at a turning point that must not be turned below a specific height.

These circumstances require an approach to dolphin flying that really concerns decision making when using speed-to-fly techniques. On the face of it, what is required is to decrease the MacCready setting so as not to lose height. This will result in lower cruising airspeeds and some reduction in average cross-country speed. This technique risks the use of 'forced' dolphin flying – and as such, one has to ask oneself if it would be better taking a climb by circling rather than 'stretching the glide'. A lot depends on the circumstances!

Having stated this case for classical (climbing and gliding) style cross-country, it should be appreciated that circumstances, especially the nature of the lift being used, may dictate different tactics. For instance, where fairly constant height can be maintained in hill lift, such as that produced by long lines of hills or mountains, or along long bands of wave lift, the chance to convert lift into airspeed arises. Providing such lift is strong, constant and predictable, then high airspeeds and record-breaking flights become possible. In these circumstances, remaining close to the ridges or lenticular clouds may benefit from some 'negative dolphin flying', where airspeed is increased in lift in order to maintain a relatively constant height and a high average cross-country speed.

Lastly, as well as the need to have a favourable air mass, the ability to benefit from dolphin flying requires not only a lot of practice but also the right glider. The higher the performance of the glider, the better the chance of useful dolphin flight. Fortunately, the designer's tendency is for bigger, heavier sailplanes with higher cruising airspeeds. All of this lends itself to an increased chance of successful dolphin flight.

Chapter 14

Water Ballast

Many pilots seem to treat water ballast as some kind of magical juice which, when carried in a glider, will propel it at great speed around a task. Well, sorry folks, analysis shows that apart from containing tolerable amounts of various pollutants, water ballast is only water. It does not create some sort of chemical thrust, but can, for aerodynamic reasons, increase a sailplane's average cross-country speed – but only if used properly.

The advantages gained from the ability to carry water ballast can be lost, and it can become a liability if it is not used sensibly. Gaining the maximum performance advantage from a glider is not simply a case of filling up the water ballast tanks to the point of overflow on every day on which you fly cross-country.

An understanding of why water ballast increases performance, when to carry it, and when to jettison it, is essential if you are to get the best performance from your glider.

How water ballast increases performance

If a glider is made heavier, the wing will have to produce more lift to counter this increased weight. The way that this is achieved is by flying at a higher airspeed.

To gain a wider understanding of what happens to the glider's performance when water ballast is carried, look at the polar curves shown at figure 14.1. The solid line is for a glider without water ballast. The broken line shows the polar curve for the same glider, but this time with water ballast being carried.

1. Note how the whole curve is moved towards the right and downwards when water ballast is carried, that is, towards the high airspeed end of the graph.
2. Note also that with the increased airspeeds comes an increased descent rate. Water ballast does not improve the overall glide angle achieved.

3. The advantage of carrying water ballast on a glider is that the airspeed for any given glide angle is increased. For instance, from the graph it can be seen that the airspeed which gives the best glide angle is higher when water ballast is carried.

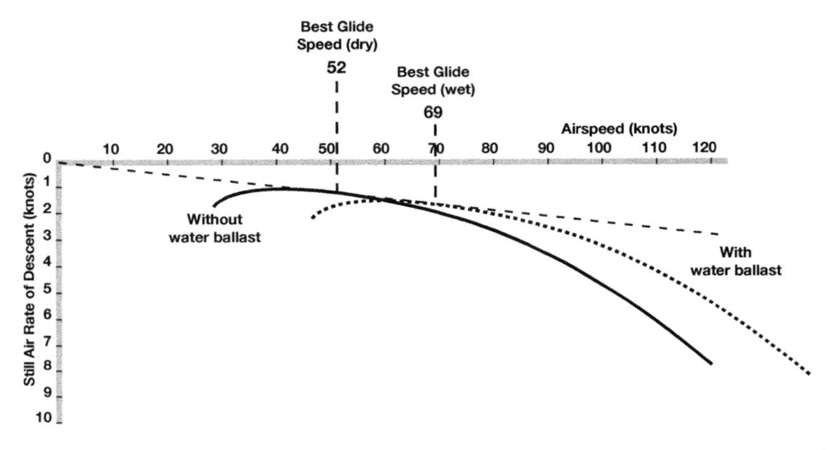

Fig 14.1 Water ballast. Comparison of the polar curves of the same glider with and without water ballast shows that the airspeed at which any one glide angle is achieved is increased when water ballast is carried.

The disadvantage is that when water ballast is carried the increased descent rate in still air means that the rate at which the glider will climb in rising air will be diminished, compared with the same glider without water ballast on board.

When to carry water ballast

As a result of its effect on climb performance, it is only worth carrying water ballast if soaring conditions are reasonably good.

If the lift is moderate or strong (that is, average climb rates in excess of 3 knots are common) then the carrying of water ballast could be beneficial. However, lift strength is not the only factor that determines whether carrying water ballast in a glider is worthwhile.

As water ballast is carried in tanks in the wings, rate of roll will normally suffer. This reduced roll response means that manoeuvring the glider regularly, as when centring in thermals, will be more difficult. Because of this, when thermals are broken or distorted and there is a constant need for re-centring, the carrying of water ballast can be a liability.

The increase in the glider's still air rate of descent as the angle of bank is increased was illustrated in the chapter on thermal soaring. When water ballast is carried, the glider's circling polar is adversely affected, such that if thermals are narrow and high angles of bank are required to stay in good lift, then the glider would probably be better not carrying water ballast. On the other hand, if areas of lift are large as well as strong, having water ballast on board will not incur such penalties.

If the need to climb by circling is reduced by plentiful lift, as is the case when individual thermals are closely spaced or thermal streets are present, then the circling flight disadvantages do not apply and water ballast can be a distinct advantage. In these conditions, the greater mass of a sailplane loaded with water ballast can make dolphin flight more effective. Therefore, if dolphin flying is likely to be possible, then carrying water ballast is worthwhile as long as overall climb performance is not severely hindered when climbing by circling becomes necessary.

As the only reason for carrying water ballast is to increase your average cross-country speed, it can be advantageous to carry it on take-off even on days when conditions do not promise to be good enough to make it beneficial when flying on the task. On such days, the extra airspeed on the initial glide on track, either direct from release from aerotow or after going through the start zone, may marginally decrease the time it takes to fly the first leg of the task. However, whereas such seconds gained may be of importance in competition flying, many pilots would prefer to avoid the extra work involved in filling up the glider with water ballast for so little gain. After all, if soaring conditions are likely to be so weak that you will have to dump your water ballast early in the flight then your task is unlikely to be a lengthy one requiring a high average speed over a long day's flying.

The higher airspeeds used when the glider is carrying water ballast make penetration into a head wind easier. Therefore, if a leg of your task looks like being flown into a head wind, it is helpful to have water ballast on board. However, this definitely does not justify struggling around a task and having to accept low rates of climb, just for the advantage of having water ballast on board for the final glide.

When to get rid of water ballast

If the average rate of climb being achieved is consistently below 3 knots, then water ballast should be jettisoned. Because of the slower climbs achievable, not getting rid of your water ballast could actually result in average speeds lower than those achieved by a similar glider that is not carrying water ballast.

If you get low and are in danger of having to land, then it is wise to dump your water ballast. Not to do so will mean having the higher rate of descent that comes with the glider being heavy. This will lead to less time airborne and less time to search for lift. Although the glider's glide angle will not be altered, in such a situation what is needed is time airborne and not speed across the ground. Getting rid of your water ballast may just give you the extra time you need to find a thermal.

When forced to adopt such survival tactics, the thermal which saves your day will often be weak. In fact, it may be so weak that the glider will only climb if water ballast has been jettisoned. There will always be the exceptional case when, having dumped your water ballast, you immediately encounter the best thermal of the day. Annoying as this is – that's life!

Dumping water ballast

Some gliders may be capable of carrying 60 gallons (600lb) or more of water ballast. Dumping this amount of water through one or two three-centimetre drain holes can take a considerable time. As it is wise to land with the water ballast tanks empty, it is necessary to open the drain valves several minutes before landing.

As it may be desirable to lighten the glider by dumping some, but not all, of the water ballast in order to 'fine tune' the glider to the soaring conditions being encountered, it is necessary to know the rate at which your glider loses its water ballast when the drain valves are open. To work this out, fill up the water ballast tanks by using containers that hold a known quantity of water. Once full, make a note of the total amount the glider's ballast tanks can hold. With the wings held level, open the dump valves on both wings. Using a stopwatch, time how long it takes for all of the water ballast to drain from the tanks. By simple calculation, you can now work out the rate at which your glider can dump its water ballast. (If your glider has multiple tanks requiring water ballast to be dumped in a specific sequence, then some modification to this experiment may be necessary.)

Now that you know your glider's water ballast dump rate, as long as you know how much water ballast you had to start with, you can dump a measured amount simply by opening the water ballast dump valves for a specific period. This is useful when soaring conditions deteriorate, or you realise that you have launched with more water ballast on board than can be justified by the nature and distribution of lift.

In order to be able to dump a specific amount of water ballast without losing all of it, it is necessary to have drain valves on the water ballast tanks that are reliable and will form a good seal once they are returned to the closed position. Smearing the valve seats with a small amount of grease often achieves such reliability.

If you should decide to jettison some or all of the glider's water ballast, it is necessary to reset your speed-to-fly ring or the wing-loading setting on your flight director system. Not to do so will mean flying too fast between thermals and effectively throwing away height. In order to know what adjustments to make to your flight director, it is necessary to calculate the wing loading of your glider for various amounts of water ballast. It is handy to display this information in the cockpit for quick reference.

When jettisoning water ballast, make sure that no other gliders are flying below you. It is bad airmanship to drop water ballast in crowded thermals. Contaminating the wings of another glider with water can seriously hinder its climb rate to the extent that if lift is weak, it may fail to climb at all.

Knowing how long it will take to rid the glider of all of its water ballast is essential, especially if a landing is to be made in an area which is marginal in size or may have a rough surface.

Landing with water ballast on board

Some gliders are designed to be strong enough to allow a landing to be made with water ballast on board without risking damage to the glider. This is an asset if you have to land back for another launch, as it saves you wasting time reloading with water ballast or having to rush off without it on a day when it should be carried. However, landing with water ballast on board requires some special safety considerations.

The stalling speed of a glider with water ballast on board will be higher than the same glider when it is not carrying water ballast. To allow for this fact, extra airspeed will be required for the approach and round out, if a stall or a heavy landing is to be avoided.

As the glider will have more mass and more airspeed on touchdown if it is landed with water ballast on board, it will take longer to stop once on the ground. Always allow for a longer than normal ground run after landing, and adopt a landing run which gives lots of room for error. Never land in a marginal area without first jettisoning your water ballast.

Remember that lateral control suffers when the wings are full of water. This may result in loss of aileron control on the ground run earlier than would normally be the case. In such circumstances, land well clear of obstructions and other aircraft. Do not try to land close to the launch point or you may find yourself ground looping into the line of waiting gliders.

As the touchdown airspeed will be higher and the ground run will be longer if you land without jettisoning your water ballast, the stresses placed on the glider will be greater. If the landing area is rough, you will be wise to land with empty water ballast tanks.

When not to launch with water ballast on board

There is no point in launching with water ballast on board if the soaring conditions are likely to be weak or marginal and the task does not require a high average speed.

When flying with water ballast, one has to consider the effect of low ambient temperatures. Even although it is unlikely that the relatively large amount of water in the wing tanks will freeze during the flight (unless the temperatures are very low and the flight is a long one), drain valves may become jammed with ice, leading to the inability to dump the water ballast before landing. The small amount of water that can be contained in tail ballast tanks is much more likely to suffer from freezing problems. The smaller-diameter pipes and valves used in tail ballast systems also increase the risk of blockage by ice.

While on the subject of tail ballast tanks, these normally drain automatically when the drain valves for the wing tanks are opened. This means that a partial dump of water ballast may result in a large proportion or possibly all of the tail ballast being lost. As many gliders benefit from carrying tail ballast even when the wing ballast tanks are not full (subject to the centre of gravity of the glider remaining within its design limitations), taking off with too much wing ballast can be bad news. Having to dump water ballast (and as a result tail ballast) may leave you flying a glider which does not handle or climb as well as you would wish.

To launch with water ballast or not – that is the question

Lastly, as an exercise to demonstrate the complexities involved when deciding to carry water ballast, consider this scenario.

The intended flight is to be a long-distance thermal flight. In order to cover the distance, you will have to get airborne and set off on track as soon as it appears to be remotely soarable. As you are starting early, the first hour or so will have weak, possibly blue, thermals. As the day progresses you can expect the lift to become strong, weakening again as you enter evening.

Do you take off with water ballast on board, even though you may initially have to struggle to stay airborne, or do you reduce the risk of an early outlanding by not loading water? Even if you are achieving slow climbs and a low average speed in the early part of the day, is it worth accepting this so that you can benefit from having a heavy glider when thermal strengths increase?

I have probably not given you enough clues to make an infallible decision. Other factors that could be added are glider type and how well it performs with water ballast on board, whether it carries tail ballast, the size and nature of the thermals, wind strength and the nature of the task – that is, is it a triangle or a downwind attempt? Are there likely to be lift streets?

The above exercise is designed to make you think. Add as many variables as you wish and see what conclusions you reach. Water ballast can be great to have on board, but can also be a liability.

Chapter 15

Cross-Country in Wave

When cross-country soaring is mentioned, most pilots tend to think of thermalling cross-country. However, there is no reason why wave flying should not also come to mind. Although in most countries thermals provide the reliable energy for cross-country flight, some countries look on wave flying as the main way to cover ground fast. New Zealand is a classic case in point, having been the venue for successful World Distance Record attempts utilising wave lift. In many other countries, wave soaring offers the potential for cross-country flying.

As well as being possibly the sole type of lift for cross-country flight, wave also has the potential to increase performance on thermal cross-country flying days – by increasing either the average speed or the distance flown.

On an essentially thermal cross-country flight, the degree of difficulty in gaining entry into the wave system will determine how advantageous it is to use wave lift to increase average cross-country speed. The decision to spend time attempting to enter the wave will depend on how strong and extensive the wave lift appears to be, and on how difficult its effect is making thermalling. Often, the time lost getting into the wave lift will reduce the average speed so much that it makes it worthless in terms of speed advantage.

On the other hand, when wave lift can be contacted, it may create the opportunity of covering a greater distance. Wave soaring may be possible before thermals begin to form and after the last thermals have died. Flights may be planned with early starts, or continued until sunset, when usable wave lift is present.

Cross-country flights planned using only wave lift require all of the techniques covered in the chapter on wave soaring, although the nature of the wave may be different. For altitude flying, a wave system with only one or two waves may be sufficient to achieve the heights desired, providing the amplitude of the waves is large enough.

For cross-country flying, the wave train (the number of waves downwind of the source) ideally needs to be numerous. This will allow the possibility of flight both upwind and downwind, as well as along the line of the waves.

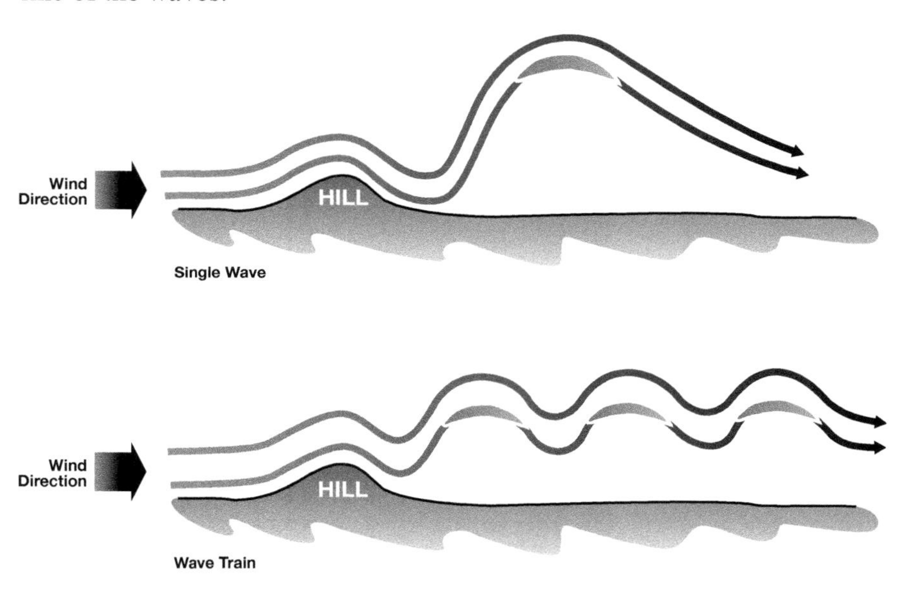

Fig 15.1 Comparison of wave trains. For cross-country in wave, ideally there will be a large number of waves downwind of the primary wave.

For this latter distribution to occur, the wave need not go to a great altitude. Therefore, days when the wave is conducive to cross-country flying need not necessarily be conducive to altitude flying, and vice versa.

The profile of such a flight might resemble the climb and glide pattern of a thermal cross-country. However, the nature of the lift will mean little or no climbing by circling and more opportunity for straight-line cruising.

Getting started

The initial climb in wave will probably require the typical cautious climb, involving tracking back and forth along the front of a lenticular cloud. Once established at a safe height, you can decide the best cruising height band to use. This decision will be based on the level at which the best lift was encountered, and that which gives the best chance of skyreading. Operating at a level close to which there are lenticular clouds can make staying in lift easier. Operating well above the level of lenticular clouds often makes skyreading easier.

In the case of wave flying, skyreading means working out the wave pattern, and the likelihood of the presence of good lift, from the appearance and position of lenticular clouds. A wave system can change quickly, resulting in movement, enhancement or collapse of the wave. Lenticular clouds can appear and disappear a lot quicker than cumulus clouds. Therefore, it is essential that you are vigilant if you are to avoid being caught out.

In some ways, skyreading while wave flying is like skyreading when thermalling cross-country. You must always be looking for the most well-defined clouds, which are likely to indicate the best wave lift and the pattern of the wave. The main difference is that you may often be skyreading from above the clouds instead of below them.

Cruising

Progressing cross-country involves running along the bands of wave lift, jumping from one area of lift to another by crossing the areas of sink or, more commonly, a combination of both of these techniques.

Whenever possible, the best progress will be made by running along bands of wave lift in front of lenticular clouds – providing, that is, that these bands of lift run more or less in the direction of your desired track. Unless you are either very lucky or know the wave system you are working very well, the chances are that no one wave bar will run the whole distance of the leg of the task which you are flying. Therefore, the need will arise to move from one area of lift to the next one, upwind or downwind. Although this will inevitably involve some height loss, whether this loss is large or small will depend on how you cross the zone of descending air between the areas of lift.

Moving from wave to wave

There are two aspects to crossing between lift zones efficiently. The first is crossing at the optimum place; and the second is flying through the sink at the optimum airspeed for the rate of descent being experienced, that is, the best speed-to-fly.

Leaving one area of lift and heading upwind or downwind across the sink to reach the next area of lift needs careful planning. Time will have to be spent regaining any height lost – whether such height loss was necessary, or excessive due to careless flying. It is necessary to find a crossing place where the sink is least severe. This can often be done by deduction. Because we are dealing with a wave effect, a place where the lift is strong is probably upwind or downwind of an area where the sink is also severe. Conversely, an area where the lift is weak will probably be next to an area where the sink is also weak. Therefore, it would make sense to cross from wave to wave at such points of weak lift and sink. If you are climbing in strong lift in one position, do not cross directly upwind or downwind to the next area of lift from that position. Instead, turn to run along the line of the wave lift (probably gaining height as you do so) until the lift reduces, and then head off for the next wave. Once you have reached where you expect the next area of lift to be, turn along the wave until you establish the glider in the stronger lift. This will probably be directly upwind or downwind of the previous area of good lift, although you should not necessarily expect the same ascent rate.

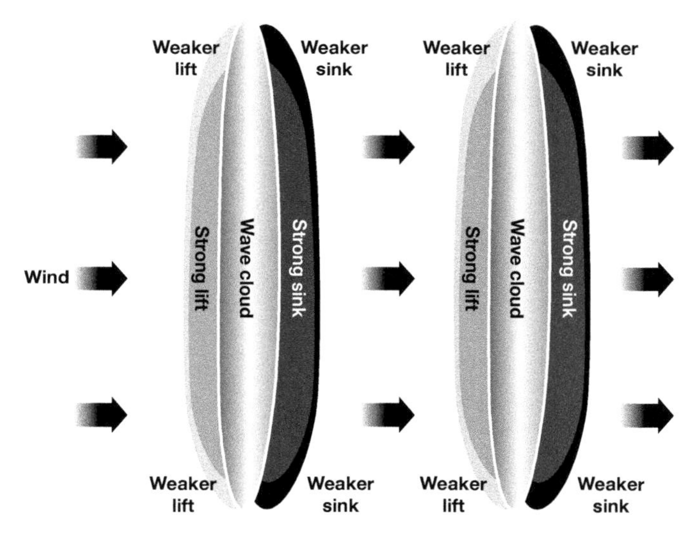

Fig 15.2 Crossing wave gaps. Cross wave gaps at the point where lift is at its weakest. This is where the sink between lift zones should also be at its weakest.

Often the streamlines of the air will show on the surface of any lenticular clouds. Such clues, combined with the vertical nature of the leading edge of a lenticular cloud and the size of the cloud gaps, help establish where the air is ascending (or descending) most, thus helping you to find the best lift (or avoid the area of worst sink).

Flying into wind, I found that the glider's lack of performance, along with the strong head wind and the sinking air between areas of wave lift, made progress into wind frustrating – but this was the way I had to go.

Each climb to 13,000 feet was followed by a glide which lost 8,000 feet to cover 5 nautical miles of ground. Only after this early attempt at using wave lift to fly cross-country did anyone tell me about the more efficient ways to cross wave-gaps.

On the other hand, the high performance of a modern sailplane may allow you to jump wave gaps directly, thus braving the strongest sink to arrive at the strongest lift. Even with the performance available from such sailplanes, there will often be a significant height loss, and therefore ample height must be available before embarking on this type of windward leap, otherwise you may not be high enough to contact and use the next area of wave lift.

Whichever method of crossing wave gaps you decide to use, once you have started your crossing, do not allow the sink encountered to persuade you to turn back. Such action will cause you to lose a lot of height and will prevent you from reaching your goal.

Flying at the best speed-to-fly involves all of the same considerations as when thermalling cross-country, plus a few more.

When thermalling cross-country we assume that the thermals, being in the same air mass as the glider, are, like the glider, affected by any wind. Therefore, except when trying to reach a point on the ground, such as when final gliding, we do not make allowance for head winds or tail winds. We simply select a MacCready value which corresponds to the expected average rate of climb in the next thermal and do as our speed-to-fly device tells us.

As we generally assume that most of the wave used for cross-country flying is of the *standing wave* type, we can assume that it is stationary above a point on the ground and not moving due to the wind. Therefore, when flying to the next area of wave lift up or downwind, some allowance will have to be made for the effect of the wind, which will prolong the flight to the next area of lift upwind and will reduce the time it takes to reach the next area of lift downwind.

In order to reduce the time spent in the sinking part of the wave, it is necessary to increase airspeed above the normal speed-to-fly airspeed when flying upwind. Conversely, when flying downwind, it will pay to fly at a lower airspeed than indicated by your speed-to-fly device, thus benefiting from the lower still air descent rate of the glider at lower airspeeds.

As glider performance varies greatly, and estimating the wind strength at altitude is never easy, it is difficult to come up with any hard and fast rules as to how much to adjust the airspeed to allow for head or tail winds. (Many modern flight director systems, coupled to GPS, will give a read-out of the head or tail wind affecting the glider, thus removing the need for guesswork.) As the wind at altitude may be considerable, perhaps greater than 50 or 60 knots, an increase of airspeed by as much as 40 knots above the indicated MacCready speed may be necessary to penetrate into a head wind in order to transit the sink zone and reach the next area of lift. In such wind strengths, any jump to a wave downwind will occur much quicker; and although it will pay to reduce airspeed, the amount by which this can usefully be done will be limited by the fact that, below a certain point, any further reduction in airspeed will not significantly reduce the glider's sink rate.

The polar curves in figure 15.3 illustrate this point. They are both drawn allowing for a sink rate of 5 knots, with figure 15.3(a) showing the construction for a 40 knot head wind and figure 15.3(b) for a 40 knot tail wind. The vertical axis is displaced by the value of the wind component – to the right for a head wind and to the left for a tail wind. A tangent is drawn from the displaced origin of the graphs to the glider's performance curve. The optimum airspeed for flight in each of these conditions is read off the airspeed axis vertically above where the tangent touches the polar curve. Notice that the flatness of the polar curve at lower airspeeds means that with a strong tail wind there is limited value in reducing the airspeed by any great amount.

Any decision to increase airspeed to improve penetration through sink must be tempered with caution. At the altitudes at which you may be flying, you will need to consider the true airspeed of the glider and take care that this does not become excessive.

If your aim is a high average cross-country speed, then speed-to-fly technique will have to take second place to keeping the glider in the height band that gives the best lift. This may mean slowing down at times to top up height, even when the lift could be better. Dropping out of the optimum height band may waste so much time that completion of the task may become impossible.

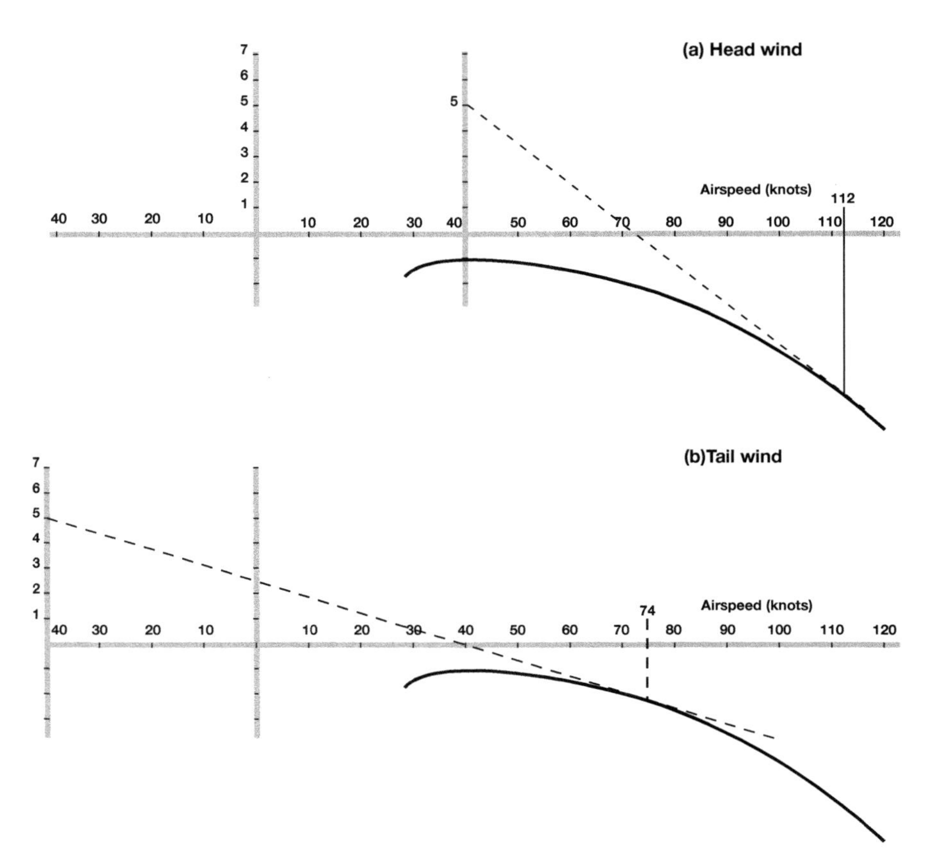

Fig 15.3 Speed-to-fly when crossing wave gaps. Whereas an increased airspeed will be beneficial when penetrating a head wind, the advantage of reducing airspeed in a tail wind is much less significant.

Another problem which may affect the airspeeds you decide to use is that not all variometers give reliable indications at higher altitudes. This, coupled with the fact that altitude affects the airspeed indicator to a different extent, means that any speed-to-fly information will be inaccurate.

Most mechanical variometers will not be affected by altitude and, if fitted with a speed-to-fly ring, will indicate the optimum airspeed at which to fly. The tendency will be to select this airspeed on the airspeed indicator. However, as the airspeed indicator will under-read at altitude, you will actually end up flying too fast. Some electric variometers under-read at altitude. Despite the tendency of the airspeed indicator to under-read, the effect of these variometer errors is to demand a slower speed-to-fly than is required.

In a situation where the sink zone you are crossing may be wide and give a consistent rate of sink, an accurate speed-to-fly indication may be more desirable than when thermalling cross-country. The errors mentioned will almost certainly deny you such accuracy. As the errors caused by your mechanical variometer's speed-to-fly indications are perhaps more predictable, it may be better to rely on this for speed-to-fly guidance, and make an allowance for the errors by flying slower than its MacCready ring suggests. Again, reference should be made to the table comparing indicated airspeed to true airspeed, but as a conservative rule of thumb, reduce speed from whatever your speed-to-fly ring suggests by one knot for every 1,000 feet of altitude.

The fact that lee waves lie parallel to the high ground which triggered them means that a line of wave lift is often staggered or even bent. Different wave systems, caused by different lines of mountains, may meet at an angle. Such characteristics create the possibility of 'jumping' to a

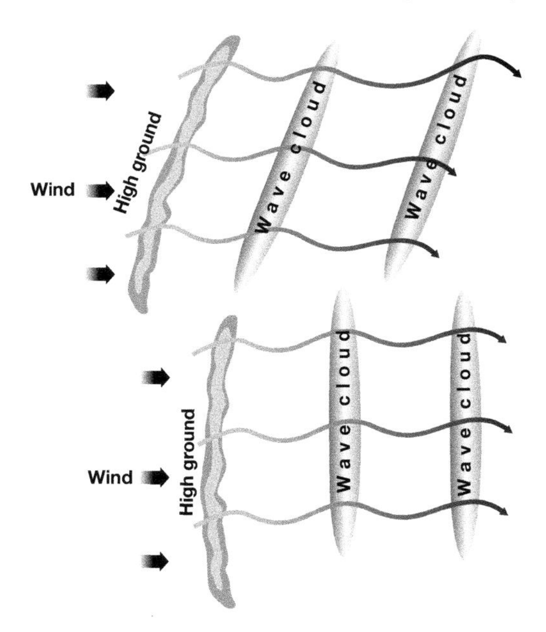

Fig 15.4 Bent, herringbone and staggered waves. The shape of the high ground causing the wave means that not all waves are straight and parallel.

different line of lift without flying across a sink zone. By studious sky-reading and careful track planning, it may be possible to progress from wave to wave without much, if any, height loss. In fact, if you are careful and conditions permit, you may be able to maintain both a high cruising speed and a near-constant altitude while clocking up track miles. However, the optimum altitude for cruising in the best lift will often mean being close to lenticular clouds. This will make skyreading more difficult and may necessitate occasional climbs to a level where, although lift is weaker, the wave pattern can be seen and assessed more easily.

Task selection

When wave is forecast, the direction of the expected wind is normally known. This will allow task setting to be accomplished with some degree of certainty – providing, that is, that the wave lives up to expectations and cloud cover does not make conditions too hazardous.

With a reliable forecast and the benefit of local knowledge, the ideal task will be either a goal-and-return or a narrow triangle which is aligned more or less parallel with the waves (and therefore more or less at 90 degrees to the wind direction). Such a task will give the best chance of

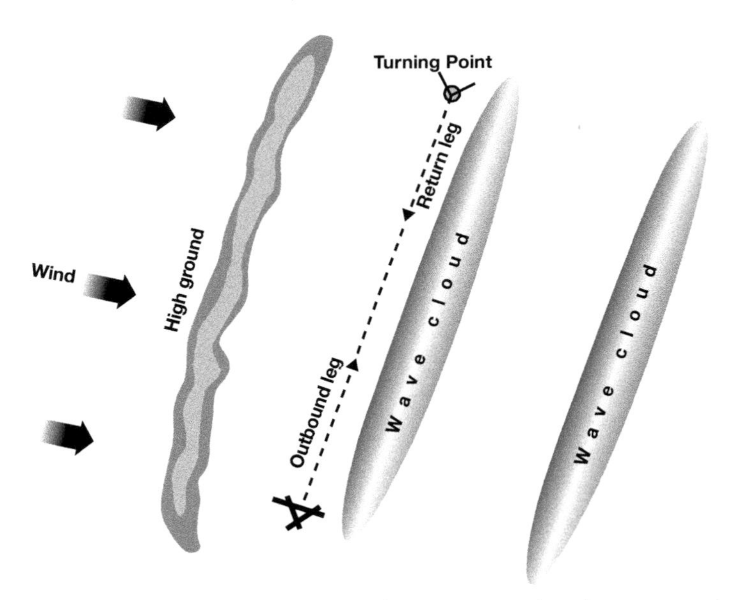

Fig 15.5 Goal-and-return or narrow triangles as wave tasks. These types of task are ideal when flying for speed.

running along wave bars without continually having to penetrate into wind.

Should the wave system be extensive, with many waves downwind of the source mountains, then larger and more adventurous tasks can be set, with perhaps one or more turning points being set upwind or downwind. This option may be forced upon a task setter if prohibited airspace or a coastline precludes flight in the ideal direction.

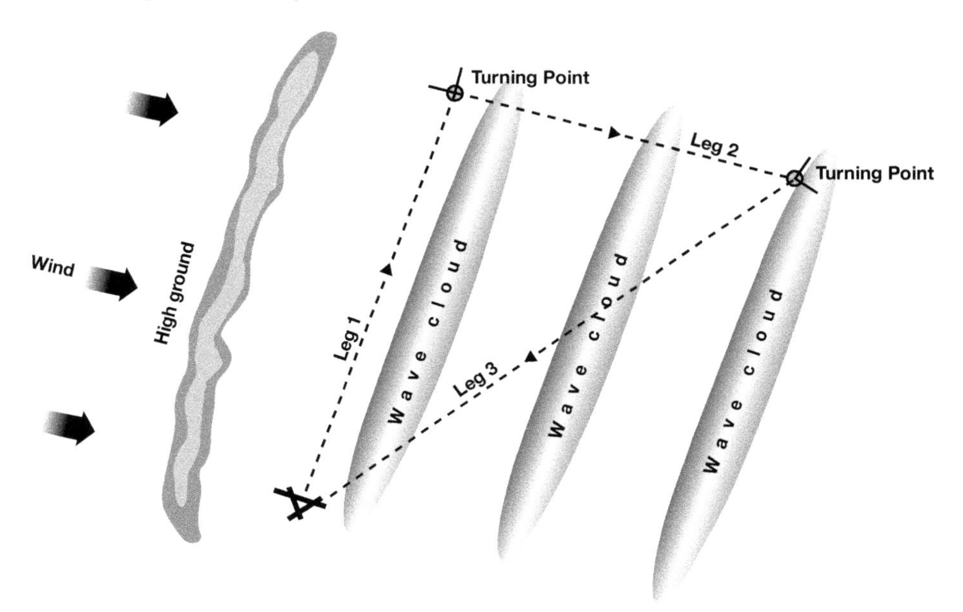

Fig 15.6 Triangular tasks in wave. These tasks require the wave system to cover a larger area.

Navigation

As with any cross-country flying, there is always the extra workload introduced by the need to navigate. Unlike wave flying in the local area, navigating on a wave cross-country flight will involve more than simply keeping a few known landmarks in view.

As well as the problem of identifying ground features from what may be a great height, there is a good chance that many of the ground features which you expected to see may be obscured by cloud below the level of the glider. This in turn can create doubts about your position and your estimate of the wind strength. Such concerns can adversely affect your concentration and soaring performance.

Applying all the skills of basic navigation (as covered in the following chapter), avoiding jumping to conclusions and appreciating that your groundspeed may vary from being very high to fairly pathetic should keep you from getting lost. Sensible use of GPS will enhance your navigation immensely and even allow you to find places hidden by cloud, but you must take care not to be lured into an area where a descent through cloud into unlandable terrain may become unavoidable.

Final gliding

The idea of setting up a final glide from a height of 20,000 feet or so sounds like one of those luxuries that only occurs in daydreams. From a high cloud climb on a thermic day, in theory the distance covered during such a glide would be impressive. Unfortunately, unlike a final glide in the still air conditions that often exist after the last thermals have died, wave lift and sink will continue beyond the end of the day. This makes planning and judging a final glide in wave conditions less than a science.

Any final glide in wave conditions will inevitably encounter large areas of lift and sink, varying (usually decreasing) wind strengths and possibly turbulence. Cloud at low levels may make it unwise to press on with a final glide.

Turbulence is often encountered near the level of the top of any high ground in the vicinity, but may occur at any level. This, together with the problems of making sure that true airspeed does not become excessive, means that it may be wise to fly at conservative airspeeds.

The operating heights involved may easily make it possible to begin a final glide from a point on a leg other than the last leg. In this way, the final glide may involve flying around one or more turning points without the need to top up height in further lift.

As always on a final glide, detours should be made to use any lines of lift which are roughly on track. If achieving a high average speed is important, care must be taken to avoid arriving overhead the home airfield with an excess of height. Unlike final gliding in solely thermic conditions, there may even be the need to use some airbrake to increase the rate of descent over the last few miles, if strong lift or turbulence means that excess height cannot be safely lost by converting it to airspeed.

Chapter 16

Navigation

In many ways, navigation for the glider pilot may seem to be a relatively straightforward business. However, when compared to the navigation of a light, powered aeroplane, navigating while flying a glider can be more difficult than it first appears, especially for the pilot new to cross-country.

The first problem is that, unlike the aeroplane pilot, the glider pilot has no engine to keep the glider airborne. This will not only mean circling regularly to regain height, but will also mean making detours away from the desired track to find and exploit rising air. Secondly, if there is any significant wind, a glider will drift a considerable distance from track while it circles, for what may be long periods, in weak lift. This in turn will mean that the glider cannot achieve a constant groundspeed, so easy calculations of the distance covered and estimates of times for any one leg are not normally possible.

Add to these points the fact that the power pilot often has an array of radio navigation aids and occasionally radar assistance from air traffic control to help maintain course, and it can be seen that glider navigation needs to rely on basic navigation with map and compass. The skill to use these basic techniques comes from practice. This chapter is aimed at giving you some idea of the navigational errors we can all make, and the ways to avoid them.

The availability of relatively cheap, portable, satellite navigation systems (GPS) has at last offered the glider pilot some degree of assistance with navigation. However, even these wonderful pieces of equipment have their limitations and, like all electrically powered gadgets, are capable of failure in a variety of ways. Therefore, it is essential that every glider pilot aspiring to cross-country soaring masters the elements of basic navigation. For this reason, the use of GPS and its limitations will be dealt with separately, later in this chapter.

Flying a glider cross-country requires navigation in two planes. The first is the horizontal plane. This is the part that tells you where you are, where you are going, how far it is to the turning point or goal airfield, and what that airfield or town is called. The second is the vertical plane. This part lets you know how high you can climb before you are up against the base of an airway, or how high you have to be to cross a restricted or prohibited area. This vertical navigation relies on horizontal navigation to tell you the position on the ground above which the airspace boundary lies. Safe, legal cross-country flight depends on relating the glider's position in both these planes.

Equipment required

The essential equipment for glider navigation is a map and a compass. With these, you can fly cross-country in anything from an ASK 8 to a Nimbus 4. The map should be marked with the latest airspace and topographical information. The compass should have been checked by 'swinging', to ensure that it will read accurately on all headings.

In the United Kingdom, the essential map to have is the half-million-scale map (1:500,000) as this is the only one showing all of the current controlled airspace information. (For a while, such maps appeared marked only with airspace below 5,000 feet. These maps are inadequate for gliding and, should they reappear in future, they should be avoided.)

In addition to this, quarter-million-scale sectional charts (with a scale of 1:250,000) can be useful to identify specific features, especially turning points or airfields. These are also more useful for giving your crew directions to your landing point should you land out. In some countries (such as Australia), one-million-scale charts (1:1,000,000) are considered the norm, as obvious ground features may be far apart.

Although, in the United Kingdom, you are legally required to have the 1:500,000 scale map for airspace reasons, some pilots prefer to navigate using 1:250,000 scale maps. The problem appears to be that for some pilots the smaller-scale map (1:500,000) does not give enough ground detail, whereas, for other pilots, the larger-scale maps (1:250,000) give too much. One advantage of learning to use the smaller-scale maps from an early stage is that, once you are attempting longer-distance flights, you will not need as many maps in the cockpit. The one that you do have, the 1:500,000 scale map, will be the one that you are used to.

Whatever map you decide on, the more you study it, the easier it will be to use in the limited space and time available in flight. Get to know its legend and the way it depicts airspace, and generally become familiar with what distances its scale represents.

Pre-flight preparation

Much of the navigation work necessary can be done on the ground before the flight. The task can be drawn on the map, the wind direction assessed and the headings for each leg worked out. Major ground features and airspace restrictions near the track can be noted.

The more preparation you can do on the ground before the flight, the more you can concentrate on keeping the glider airborne and flying as efficiently as possible.

One way of achieving this is to put as much information as possible in an easily visible place. As you will have to use the map regularly in flight, this is the easiest place to display the information required. The typical glider cockpit does not lend itself to having many pieces of paper containing the different information required. By the time the first turning point is rounded, the cockpit can resemble an airborne litter bin.

Once the task has been decided, draw the legs of the route on the 1:500,000 scale map. (Using a map with a glossy, protective covering means that you can use 'permanent' marker pens for this work. These can be cleaned off after each flight, leaving the map ready for the next task. Chinagraph pencils can also be used for this, but as Chinagraph tends to rub off on the hands, the lines will have to be covered with clear tape to avoid accidental erasure while flying.) Make sure that the lines you draw are of a colour that stands out from the printed lines on the chart and that, while being wide enough to see at a glance, they do not obscure any of the map's features or information. This is especially important at turning points.

Mark on the first, or perhaps all of these legs, an arrowhead in the direction in which the task is to be flown. (Do not laugh – even some competition pilots have flown around a task in the reverse direction!)

Next, mark on the map the expected wind velocity at the mean flying level. This can be done simply by an arrow representing the wind direction and a number for its strength. Again, this should be marked in a prominent place, but not so close to the track lines as to obscure route details.

Work out the track of each leg. To do this, place a protractor with its centre on the task line so that the north arrow on the protractor lies parallel to the north/south grid lines on the map. From the protractor, read off and note the track of each leg.

Next, you need to make allowance for the effect of the wind. As the amount of drift due to the wind will depend on the glider's speed, which in turn depends on how good soaring conditions turn out to be, this wind allowance can only be a very rough guess. The important point is to bias the final heading into any wind and not end up a long way downwind of track. If you are very mathematical, you may wish to make assumptions of what your average speed will be and work out 'accurate' headings for each leg. Personally, I would rather get on the launch grid, as when you are in flight, the chances are that the best lift will not be on track, and therefore you should not be on track either. The headings marked on the map will be used mainly to exit thermals in roughly the desired direction or for 'dead reckoning' glides in poor visibility. (DEAD RECKONING is the name given to navigating to a point by referring only to the elapsed time and heading flown from a previous known point.)

Table 6 will give you some idea of the value of the crosswind component (in knots) that you will experience, given a wind speed and the angle of the wind relative to your track. It can be useful to keep such a table with your flight planning equipment.

TABLE 6

WIND SPEED (kts)	**ANGLE BETWEEN TRACK AND WIND (degrees)**								
	10	**20**	**30**	**40**	**50**	**60**	**70**	**80**	**90**
5	1	2	2	3	4	4	4	5	5
10	2	3	5	6	7	8	9	9	10
15	3	5	7	9	11	13	14	14	15
20	3	7	10	13	15	17	18	19	20
25	4	8	12	16	19	22	23	24	25
30	5	10	15	19	23	26	28	29	30

EXAMPLE: Track = 342°
Wind = 290°/15 knots
Angle between Track and Wind Direction = 52°
Crosswind Component = 11 knots

As a 'rule of thumb', once you have worked out the crosswind component (in knots), double this figure and use this number of degrees as a drift correction. This should be enough to keep you on or even slightly upwind of track, despite the drifting which will be inevitable while thermalling. Make sure your correction for drift changes your heading towards the wind and not away from it.

EXAMPLE: Track = 342°
Crosswind Component = 11 knots from the left
Correction from Drift = 11 (knots) × 2 = 22°
Heading Required = 342–22 = 320°

If you do not allow for drift during the planning stage, and simply try to fly from one ground feature to another along the map track, the result will be that you will continually be trying to get back into wind.

The final requirement is to work out the compass heading you will have to fly in order to make good the track required. Here there is a slight problem. The MAGNETIC NORTH POLE is to where the needle of a magnetic compass points and is dependent on the earth's magnetic field. This point is not in the same place as the TRUE NORTH POLE through which the earth's rotational axis runs. The heading you will steer on your compass is relative to MAGNETIC NORTH, whereas the grid lines on your map (and therefore the heading you have worked out from the above) will refer to TRUE NORTH. Magnetic north varies from true north by a differing amount depending on the area in which you are flying. This amount even changes from year to year. The amount of this MAGNETIC VARIATION, as it is known, for the area in which your task is set is shown on most maps as a pecked line annotated with the number of degrees of variation and whether magnetic north is east or west of true north.

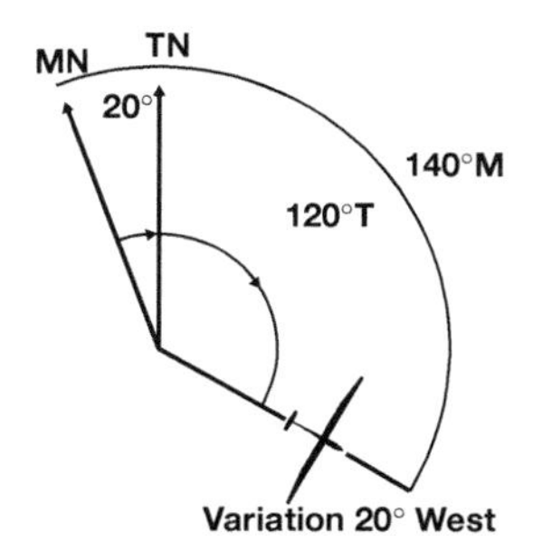

Fig 16.1 Magnetic north. A compass needle points towards magnetic north, which is displaced from true north. Allowance must be made for this difference at the planning stage.

To obtain the magnetic heading for a leg of the route, add (or subtract) the magnetic variation for the area to (or from) the true heading, after correction for wind drift has been applied. Whether you add the variation or subtract it depends on whether magnetic north is west of true north (in which case you add the number of degrees of variation to the heading you have calculated) or east of true north (in which case you subtract it). Some pilots remember whether to add or subtract the variation using the mnemonic '*East is least* (subtract) – *West is best* (add)'. What you are left with is the magnetic heading that you will steer on your compass.

An alternative to this technique is to measure the direction of the track by aligning your protractor's north arrow with the north arrow of a neighbouring VOR compass rose marked on the map. If the track direction is measured in this way, no calculation to allow for magnetic variation is necessary, as VOR compass roses are already orientated towards magnetic north. (VORs are radio aids used by power pilots. Check the map's legend to see how they are represented on the map.) It is still necessary to make an allowance for any expected wind.

Once you have ascertained the final magnetic heading that will give you the desired track, mark this on the map next to the appropriate leg of the task. Repeat the above procedure for the remaining legs of the task, bearing in mind that the wind will have a different effect on each leg.

The following is a summary of this procedure:

1. Draw the track on the map.
2. Mark the wind direction and strength on the map.
3. Measure the direction of the track (°T).
4. Make allowance for the effect of the wind.
5. Add or subtract the magnetic variation as appropriate.
6. Mark the magnetic heading next to the track (°M).

You should now have a map marked as shown in figure 16.2.

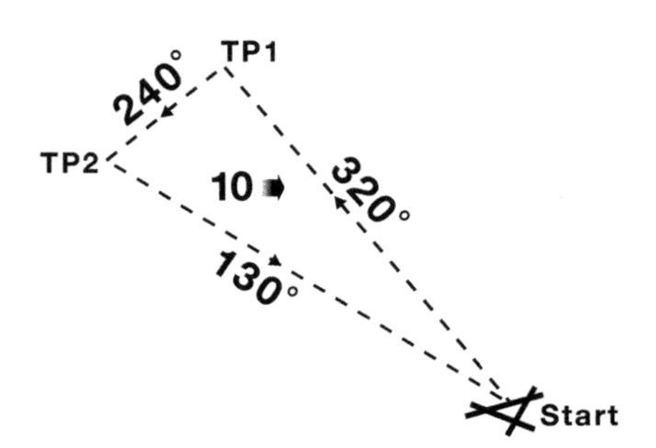

Fig 16.2 Map prepared for flight. Most of the information required to fly a task can be displayed on the map.

If the task is a very large one, it may be helpful to mark on the map the times at which you expect or need to be at the turning points or half-distance points. This is more to check that you are achieving a high enough average speed, rather than for navigation purposes. Such checks may help you decide on the tactics that need to be employed for the rest of the flight (e.g. whether you need to speed up and take more chances or if the time you have in hand allows you to play safe).

Often the general clutter of information on a map can make it difficult to find pertinent airspace information at a glance. If this is the case, it may be useful to highlight relevant heights, such as the base of controlled airspace, by marking it more clearly on the map.

The only other information that you may require on the small-scale map is some turning point or final glide details. In general, the more detailed, larger-scale maps are better used for this purpose. (Preparation for these parts of the flight is covered in the relevant chapters.)

If you decide to take larger-scale maps with you in the cockpit, you may wish to mark these with track lines and headings.

Folding the map

After you have marked the map as desired, you will have to fold it so that it is of a manageable size for use in a glider cockpit. There are two common ways of doing this.

One is to fold the map in a concertina fashion (often done immediately after the map is purchased). If this method is employed, any part of the map can be viewed simply by an action similar to turning a page of a book or turning the map over.

Another method is to fold the map so that only the task area is visible on the front and, if necessary, spreads to the exposed rear side of the map. This method normally involves taping the map into this new shape. It has the advantage of re-shaping the map in such a way that turning points or the goal airfield do not end up obscured by folds in the map.

Whichever method you use must:

* Result in a map size small enough to be usable in the cockpit
* Show the whole task area
* Show enough map detail outside of the task area to allow for possible detours from track, whether intentional (to find lift or a safe airfield at which to land) or as a result of getting lost

Initial study of the route

Next, study the route. Take note of the major features which you will overfly or pass close to.

One technique which may be helpful when you are new to cross-country flying is to make a table of features that you expect to encounter as you fly each leg, and their position relative to your track. Such a table for the section of map shown at figure 16.3 would appear as follows.

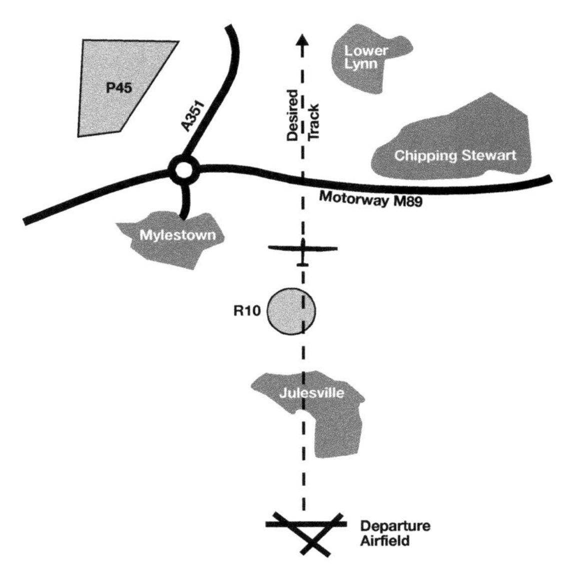

Fig 16.3 Map showing part of a cross-country task.

TABLE 7

LEFT OF TRACK	ON TRACK	RIGHT OF TRACK
	JULESVILLE	
	RESTRICTED AREA R10	
MYLESTOWN		
	MOTORWAY – M89	
		CHIPPING STEWART
'A' ROAD – A351		LOWER LYNN
PROHIBITED AREA P45		

Do not try to list every little village or town. Use only major features that you are unlikely to miss – even in poor visibility.

Once you are more experienced, you will be able to visualise the position of ground features relative to your track by a quick glance at the map and so will no longer require such an *aide-mémoire*. In addition, when you begin to do a lot of cross-country flying, longer flights and possibly even competition flying, you simply will not have the time for such elaborate route planning. However, in the early stages, when flights tend to be shorter, such a table can reduce what will be a fairly high in-flight workload, and as a result will increase your chances of a successful task.

Having studied the main navigation features on the route, it is worth spending a few minutes considering the topographical features over or near which you will fly. This has more to do with soaring than navigation. Look for areas of wetland, river valleys or irrigation, which might cause poor soaring conditions and require caution or re-routing. See if the track takes you close to high ground, where a small diversion may result in better thermals for that part of the task. Such features may cause diversions that can be pre-planned, thus reducing the chances of becoming lost or disorientated because of last-minute detours.

With the exception of turning point and final glide information, your map is now ready for stowing in the cockpit.

Navigation immediately after launch

The best time to start thinking about navigation in flight is as soon as you are sure you can stay airborne. The first priority after launch is to find a thermal and climb in it until you are at a safe height. Only after you can safely remain in the air should you seriously start thinking about the navigational aspects of your chosen task. After all, if you cannot manage to soar then you are not going anywhere.

Even before you set off, you can carry out some observations and planning which will help once you decide the time is right to set off on track.

The first of these is to check that the compass is giving sensible indications. This is easily accomplished by reference to the airfield's runway, the orientation of which is known. Flying parallel to it should result in the compass indicating the expected heading.

As the compass may be influenced by metal objects in the cockpit or the magnetic field set up by electrical instruments, it is necessary to carry out this check in at least four directions, for instance, north, south, east and west. All of these can be judged by reference to the runway or the bearing of local features whose relative position can be estimated from the map.

Such DEVIATION, as this magnetic interference of the compass is known, can affect the compass when flying in one direction even when the compass gives correct indications on other headings. After you turn the first turning point or begin the final glide is a bad time to discover that your compass is wildly out on crucial headings.

Normally, if you are flying in your own glider, major deviation problems will not arise unless you have made changes to your instrument layout or have stowed steel pickets or the like in the cockpit. On the other hand, compasses in club gliders often lack the attention such instruments need after re-installation of instruments or modifications to the glider's electrical system. Obviously any deviation from the expected headings needs to be noted before setting off on task if you are to avoid navigational difficulties.

I was flying a club SF27, which I had never flown before, on a 300 kilometre lead-and-follow exercise. With a Cobra and a SB11 following behind, we had rounded the last turning point. Having started the last leg, I could not make sense of the landmarks I saw ahead. My compass told me that I was heading south, but the lakes which appeared ahead were not on the map. The only lakes shown on the map were to the west in controlled airspace.

When I asked the Cobra pilot, who was following me closely, what his compass was indicating, he said that we were heading more west than south. The compass, which had been reliable on other headings, was almost 50 degrees out when indicating south.

While local soaring before you start the task, check the map for the first features you will pass on the first leg of the task. If visibility is good, you may see several of these features (for instance, in Australia one can often see the first turning point of a 500 kilometre triangle from overhead the start point). On the other hand, if visibility is poor or bad, you may only see the first feature, if that. Check from the map (or your features table) to which side you will pass this feature. This will give you some idea of the direction in which you will go and save you having to worry about headings and drift.

En route navigation

Once you begin the task, use ground features and the map to navigate whenever possible, and use the compass mainly as a regular back-up to confirm that you are roughly on the correct heading.

If ever you cannot see the next feature on track, fly on the compass heading for the leg until a definite position fix is obtained. Never forget that staying exactly on track has a low priority compared to the need to follow the best areas of rising air.

Navigation will be a continual chore while you are flying cross-country. The best time to be checking your position or the map is when you are gliding in the upper half of your operating height band. Below this, the need to find lift requires much of your attention. As the average cross-country speed depends on the climb rates being achieved, when climbing in thermals you need to concentrate on staying in the best lift. Therefore, it is wise to leave the navigation until you leave the thermal. To do this means that when you accept that a thermal is good enough to use, before entering it you should know your position and how high you can climb without infringing controlled airspace.

Do not try to follow every little feature on the route. Most villages and many small towns do not appear on small-scale maps. Use only major features such as larger towns. On approaching a turning point, you will have to use smaller features to ensure that you are in the turning point zone, but such detailed navigation will be covered in Chapter 17 on TURNING POINTS. Some of the features that can safely be used for general route navigation, along with advice on their pitfalls, are as follows.

Sizeable towns

Reasonably large towns make good navigation features when expected on or near track. When you are new to cross-country flying or are low, even a fairly small town can look like a city, giving rise to doubts about your position. Towns are not usually so abundant that you can mistake one for another if you know that you are more or less on track. However, in industrial areas, numerous towns may appear to merge together, making it difficult to distinguish any particular town. One confusing feature of towns is that they often change shape because of the building of new housing developments or industrial areas. This can often occur faster than maps can be updated.

Motorways and major roads

These are generally excellent features if they lie more or less along a section of your track. However, a particular motorway or road may be missed if your track crosses it at right angles and visibility is poor. The angle at which your track crosses a particular road may help confirm your position.

Lakes and reservoirs

Large lakes and reservoirs are usually easy to spot. However, in times of drought, these features may not retain the shape shown on the map. In hot countries, they may disappear completely. In some seasons, flooding may expand the size of lakes and obscure other local features.

Forests

Large forests show up well. Small wooded areas are not worth using for general route navigation, but their relative position is often useful to confirm another feature's identity.

Rivers, canals and irrigation areas

Only large rivers make good navigation features. Smaller rivers are often obscured by the foliage along their banks and can easily be missed. In dry countries, such foliage may be the only indication of the river's existence and may outline the feature enough to make it obvious. In wet areas, many small rivers may be present, making the identification of any particular one difficult. However, such an area will still serve as a position fix, as it will appear on most maps. Irrigation areas have the same characteristics. Canals often show up well, and are not so common as to be mistaken for each other.

Railway tracks

Railway tracks show up remarkably well. This applies both to tracks currently in use and, surprisingly enough, to many disused tracks, even where the rails have been removed decades ago. Much of this is probably because they tend to form straight lines or long curves and therefore catch the eye, while most other features on the landscape turn and twist. Like motorways, they will be easier to see if your track does not cross them at right angles and visibility is reasonably good. Also, as with roads, the direction in which a railway track runs relative to your track will help confirm which railway line you are looking at, and thus your position. Watch out, however, as occasionally a railway line disappears into a long tunnel!

Airfields
Large civilian airfields are usually unmistakable. However, as they are usually embedded in a block of controlled airspace, you will seldom be close enough to use them for navigation. Military airfields, both used and disused, can be useful for fixing your position. The biggest problem, certainly in some parts of Britain, is that their abundance can make it difficult to determine which airfield you are looking at. This is where the 1:250,000 scale map is very useful. These maps not only show airfields, but also show the runway pattern of hard-surfaced airfields. They even give this information for disused aerodromes. One note of caution is necessary here. Many of the disused airfields shown on maps have had the runways dug up and have been returned to farmland. Despite this, most still show the runway pattern through the soil. Do not expect these airfields to show up at a distance. Many active airfields, both military and civil, display their identity in the form of a couple of letters in the airfield's signal square, or even their full name on a hangar roof.

Power stations
Power stations act as beacons in the landscape. Even when the visibility is poor, the plume of smoke or steam emitted by these stations can often be seen when the actual cooling towers or chimneys cannot. The only problem you may have in some industrial areas is deciding at which particular power station you are looking.

Obvious landmarks
Definite features such as large viaducts, lines of hills and bridges over estuaries can make good navigation features. In England, ancient civilisations saw fit to carve large white horses or giants on hillsides, and to erect stone circles, to help glider pilots of the future to navigate. (Can you come up with a better reason why they created them?) These are useful, although in some areas the number of chalk horses can lead to confusion. As many of these man-made features can be concealed in the folds of a hill, do not waste too much time looking for them in order to confirm your position.

Useful practices
Many pilots find it helps them visualise their track and the relative position of expected ground features if they orientate the map so that the map direction of the leg being flown points towards the nose of the glider. By doing this, features which should appear on the left or right of track

appear on the correct side of the track drawn on the map. The problem with this technique is that, at some point in the flight, the pilot will need to be able to read upside down or keep turning the map around to read the names of places.

Others like to note the times at which specific points on the ground were passed. This can be done either on a kneeboard or, more commonly, on the map next to the map feature. The reason for making such notes is to give an elapsed time from the last position fix, should the next expected ground feature not be sighted. The elapsed time from the previous fix gives some idea of the distance travelled along the track or on a rough heading. The result is an increased chance of knowing the glider's position, and whether or not the expected navigation feature has been missed or will shortly be reached. The same system can be used by starting a stopwatch on passing a recognised ground feature.

The more cross-country flying you do, the easier navigation will become. Not only will you get to know specific ground features on the routes you commonly fly, but you will also gain the ability to glance at the map or the ground and quickly relate a feature's position to your track.

The first of these acquired skills – recognising specific features – will result in your building up a mental map of the areas over which you fly regularly. The result will be that, even if you are uncertain of your position, the sight of such a feature will trigger a mental picture for comparison and identification, thus allowing a definite position fix to be confirmed using the map.

The second – the ability to use the relative position of ground and map features to fix your position quickly – is the key to successful navigation, whether on track or hopelessly 'off track'.

All of the time while you are flying cross-country, make a note of the sun's relative position to your position and your track. Although the sun will move gradually as the day progresses (and your track will change after turning points are rounded), an awareness of the sun's relative position can be a useful aid.

Firstly, it confirms what your compass and ground references are telling you. Secondly, and probably more importantly, even when circling in thermals, a rough estimate of the direction of the required track on exit is always available by reference to the sun's relative position. This is important when you consider that most compasses used in gliders are wildly inaccurate at the rates of turn that gliders achieve when thermalling. One great advantage of being aware of the approximate track you will take when the time comes to leave the thermal is that, even while thermalling, you can constantly be scanning the sky on track for your next source of lift or the development of any cloud streets. Knowledge of the sun's relative position gives you all of this without ruining your rate of climb or risking losing the thermal by looking at a map.

Despite these uses for the sun as a navigation aid, its usefulness will obviously be limited by large amounts of cloud cover or thick haze. Surprising as it may seem, the sun is often more useful in this role in countries lying in temperate latitudes than in those closer to the equator. This is because in tropical and equatorial regions, the sun spends more time closer to the overhead than in more northerly or southerly latitudes. Thus, its angle to the observer is not as great, and therefore it is not as useful as a reference for navigation in the middle of the day.

Compasses and their limitations

The compass is an essential instrument for cross-country soaring. Used sensibly, it can be more than a device that merely indicates the direction in which to fly. Used wrongly, it can result in total confusion, disorientation and undeserved mistrust of the instrument.

As well as showing what heading you are flying and (by making allowance for the wind) giving you some idea of your track, the compass also allows more general orientation. By remaining aware of the direction of north, you can orientate your map and compare the relative positions of ground features, as well as defining the direction in which line features, such as railway lines and major roads, run.

However, magnetic compasses do have some limitations. The problems of magnetic variation and deviation are not the only ones that haunt the magnetic compass.

The earth is surrounded by a magnetic field. This magnetic field has a North Pole and a South Pole. The magnetic compass works because the north-seeking end of its magnet points towards the magnetic north pole. The more you move away from the equator towards the north or south poles, the more the earth's magnetic field curves downwards towards the vertical, or 'dips'. As the needle of the compass will align itself with the lines of the earth's magnetic field, the compass needle will also tend to dip towards the nearest magnetic pole.

As it is the horizontal component of the local magnetic field that gives the compass (and, in turn, the pilot) the glider's heading, the tendency of the compass magnet to dip reduces the accuracy or even the usefulness of the compass. To reduce the amount of dip in a compass, the manufacturer's use a pendulum system, whereby the pivot point is higher than

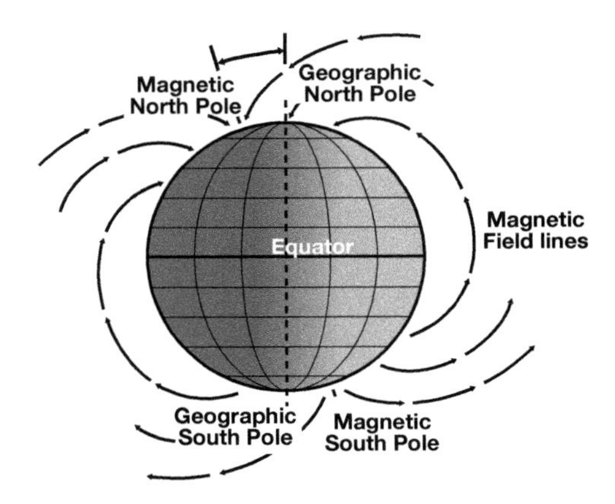

Fig 16.4 Compass needle dip. A compass needle will dip as it aligns itself with the earth's magnetic field, thus reducing the accuracy of the instrument as latitude increases.

the compass's magnets. This means that the weight of the magnet assembly acts as a pendulum and so counters the tendency of the magnets to dip.

The use of such a pendulum system means that the basic magnetic compass used in gliders (and as a back-up compass on larger aircraft) suffers from ACCELERATION and TURNING ERRORS.

When the glider accelerates, the compass will tend to show a turn towards the nearest magnetic pole. Deceleration will have the opposite effect. This error is at its worst when the glider is changing speed on an east or west heading, and reduces as the heading becomes more northerly or southerly. On a heading of north or south, acceleration errors do not exist.

When the glider is turning towards or through the direction of the nearest magnetic pole (for instance, through north when flying in the northern hemisphere) the compass will be slow to read the correct heading. When turning through the direction of the furthest magnetic pole (south when flying in the northern hemisphere) the compass will move too rapidly. What this means is that at the high rates of turn normally used in gliders, the basic magnetic compass is of little use when turning.

I could become all scientific here, and give you rules to apply when trying to straighten up on a compass heading using a basic magnetic compass, but my advice is not to bother. It is far easier to straighten up in what you think is the correct direction, wait until the airspeed is constant,

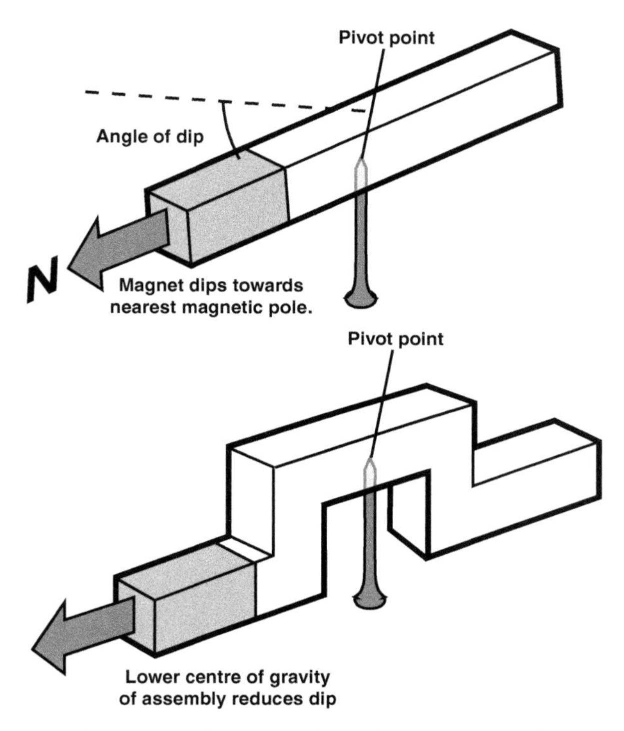

Fig 16.5 Compass design to reduce dip. Magnetic compasses often use a pendulum effect to counter needle dip.

and then check the heading on the compass. If it is, say, 30 degrees to the left of the desired heading, turn right 30 degrees, judging the angle by reference to the ground and sky ahead. Re-check the compass once manoeuvring has finished and its reading has steadied. If necessary, make further smaller corrections.

Cook compass

In the COOK COMPASS, the whole compass is set on a mount, which allows the compass to be adjusted to hang with its rotational axis vertical at all times. This maintains the sensing magnet horizontal. Providing the axis of the compass is kept vertical, turning and acceleration errors are eliminated.

Bohli compass

The BOHLI COMPASS also uses a gimbal system (rather than a pendulous one), which allows the sensing magnet to self-align with the earth's magnetic field, thus avoiding turning and acceleration errors, and the errors caused by dip. The compass rose is marked on a transparent plastic dome at the top of the compass and is viewed through a mirror. A brightly coloured bead is mounted at the end of a thin needle, which is attached by its end to the sensing magnet. The bead is viewed through the transparent dome and its position indicates the glider's heading. The upper part of the compass can be tilted from the vertical to the left or right to allow for the banking of the glider and to keep the compass rose horizontal.

Some reasons why pilots get lost

Many inexperienced and some experienced cross-country pilots worry about getting lost, especially if there is controlled airspace near their track. Such concern can play on the mind to the extent that their soaring performance is badly affected, as is their ability to navigate. Any serious worry can reduce the ability to think rationally.

Should you become 'uncertain of your position' when in this frame of mind, you may start making hasty judgements about where you are or where you have gone wrong. In fact, you may not 'have gone wrong' at all. You may not be lost. You may just think you are! A pilot in this situation may well jump to conclusions about what the town in view is called, or whether the track made good has drifted to the left or the right of the desired track. Once unsound conclusions are reached, then a pilot is beginning the process of getting lost.

For instance, if the town ahead and to the left is not the town that should have been directly on track, but is wrongly assumed to be, then this could lead to a change of track. ('The wind must be from a different direction' may be the pilot's reasoning.) Once the heading is deliberately changed in this way, the pilot has lost all definite contact with the required track and, as a result, is fast on the way to becoming disorientated and lost. Now being off track, it is unlikely that the next expected feature will be seen, and when, after a period of time, this has not materialised, our pilot may either make another disorientating track change or give another town a name which is not its own.

One common reason why such a scenario might occur could be the fact that many glider pilots overestimate the average groundspeed that they are achieving. This results in the expected ground feature still being some distance ahead. The problem would not have developed if the pilot had simply continued on track for a while longer and not jumped to conclusions.

Another influence that can lead to a pilot varying track and losing touch with his or her position is if the track lies close to controlled airspace. A pilot who is under-confident about navigation may subconsciously vary the glider's track away more and more from such airspace, out of fear of infringing it. If such a track change is unplanned, such wanderings may well lead to the pilot not knowing how far off track the glider has strayed.

Several factors may combine to make navigation difficult to the point of causing uncertainty of a pilot's position.

If the visibility is poor, the distance from which you can see and identify ground features may be less than half of the distance between recognisable ground features. This can result in a situation where neither the feature that you used as your last position fix, nor the next feature, is visible. If you are heading directly into sun, especially a lowering sun, this will reduce forward visibility or compound the problems of already poor visibility. Flying over terrain with few landmarks creates the same potential problem.

Large showers or lines of showers have the ability to obscure large areas of ground and deny you sight of the navigational features upon which you are relying.

In these situations, it is essential to have a compass that you can trust and to steer a sensible heading without wandering aimlessly about the sky. Any heading changes that you make to find lift must be noted, so that you have a good idea as to which side of track you have moved and by how much.

Even on days when the in-flight visibility is good, cloud shadows can make ground features less than obvious, and can even give the impression of forested areas on track. Fortunately, shadows cast by cumulus clouds are fairly localised, and other nearby features should guide your eye to the obscured feature that you are expecting – providing you search the terrain systematically.

Ground features that lie at the foot of hills, ridges or mountains may be impossible to see if viewed with the high ground between you and the feature. Reservoirs and lakes, which normally make good navigation features, can often be obscured from view by the hills in which they lie. If an expected feature cannot be seen, study the contours on the map to see if this may be the reason for the failure to spot the feature.

Surprisingly, navigation can often suffer from too many landmarks. In some areas, particularly industrial areas, towns, power stations, railways and roads may be so abundant that distinguishing one from another may be difficult. As the air in industrial areas tends to be full of smoke particles, the resultant poor visibility in and downwind of such an area may compound the navigation problem. The point to remember here is that knowing *exactly* where you are is never essential in general route navigation, except when you are close to the boundary of controlled airspace. Such problems, however, do tend to make the finding of turning points difficult.

Re-establishing your position when lost

If you are one of those pilots who worries about getting lost and you miss an expected ground feature or two, then here is what you should do. Providing you are not near an area of controlled airspace, take a deep breath, find a thermal and climb as high as you can. The deep breath is always a good way to relax. The thermal will give you height and reduce the workload which comes with being low. The circling involved in using the thermal will stop you heading off in some uncertain direction. The extra height gained will give you a better view of your surroundings and allow you to see more remote ground features.

If there is controlled airspace near your estimated position then, should you discover that you have strayed into it, either glide out of it immediately or, if you feel that you are too far inside it to be able to get clear of it, then you should pick a field and land. *Concentrate and do not rush the field landing.* Your Chief Flying Instructor would rather have an apologetic pilot than an injured one with a broken glider.

The old adage, that 'one is never lost, simply temporarily uncertain of one's position' is probably more truthful than euphemistic. After all, if, say, ten minutes have passed since your last definite position fix and you have been gliding at an average airspeed of around 60 knots, then, assuming there is not a howling tail wind, you will only have covered 10 nautical miles from that last known position. In all probability, you will have been flying in one general direction. This limits the possible area in which you might be. (So calm down – all is not lost!) Look around you for any ground features, and see if they match any on the map in that area.

It is important that you should look for ground features first, and only then look at the map to see which they might be. If you look at the map first and try to find ground features to match what you see on the map, there is a very great risk that you will twist the facts to make the ground features match what you want to see.

When confirming your position, it is essential to use more than one feature, and to compare the relative position of whatever features you use with each other and relative to your compass heading.

For instance, you may find that your only definite ground feature is a town. It is still feasible to use it to re-establish your position, as long as you check that it matches its map representation accurately.

Personally, if I am using a town to establish my position, I try to identify five characteristics that tie in with the map. For instance, if we take the town shown in figure 16.6, we can see that:

1. The town is elongated, with its longest axis orientated north/south.
2. There is a bypass road running from north to south to the west of the town.
3. A railway line runs through the town, more or less parallel to the bypass.
4. There is a railway station towards the south edge of the town.
5. A river runs through the south of the town.

If all of these ground features match, I would be happy to accept that this town is the one I am looking at on the map. Conversely, if one feature on the ground does not match what you see on the map, you would be unwise to conclude that you have a definite fix on your position.

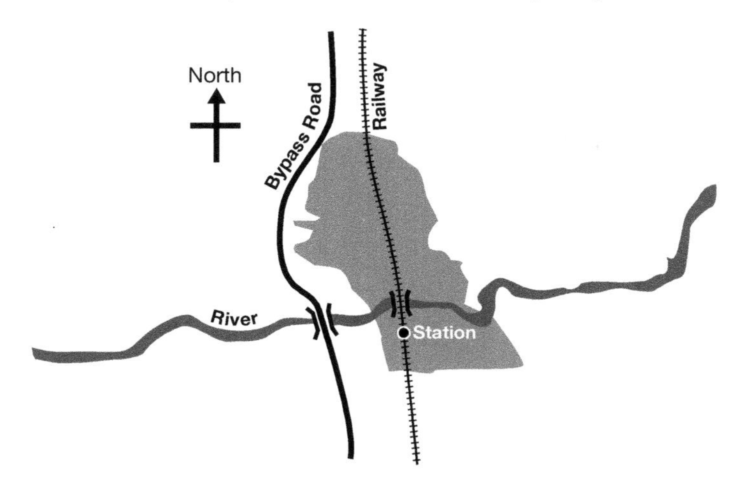

Fig 16.6 Confirming your position. Avoid jumping to conclusions when you are trying to establish your position from ground features. Use several features in conjunction with each other to identify a place.

Airspace

The chances are that if task planning has received even the smallest amount of attention, the route will not take the glider towards any airspace through which it is not allowed to fly. Occasionally, even although the ideal track cuts through the edge of such airspace, the glider will have to dog-leg around it.

Part of the problem with controlled airspace or restricted areas is that they tend to be centred on a point and to extend over a defined area of ground around it. Unfortunately, the ground where the surface boundaries of such airspace lie does not have thick blue or red lines drawn on it as is shown on the map. Ground features near such boundaries may be few, or orientated in such a way as to be less than helpful. If this is the case, all you can do is judge the distance of the boundaries from the central installation or airfield, or from whatever features you can see.

Controlled airspace that starts at a specific height above the ground, such as airways and terminal control areas, creates different problems and requires navigation in the vertical as well as horizontal planes.

To avoid infringing this type of airspace, you need to know two pieces of information. The first is how high you can climb before you reach the base of the airspace; the second is where, relative to a point on the ground below, the airspace starts. This second point is further confused by the fact that, when the airspace planners started planning, they found it necessary to step the base of airways and the like down as they approach major

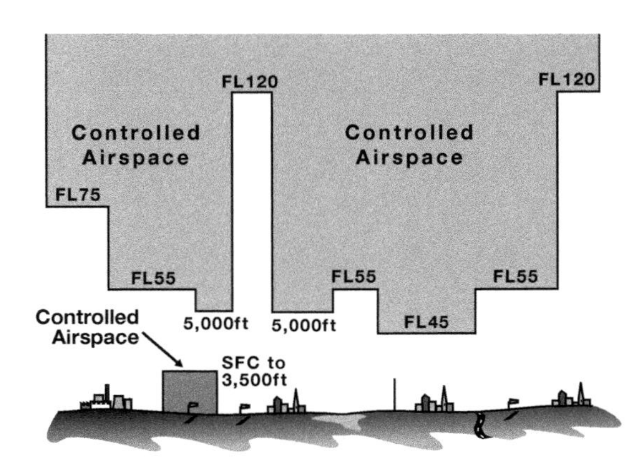

Fig 16.7 Controlled airspace. Relating controlled airspace boundaries to features on the ground adds to the workload of navigation.

airports. The result is controlled airspace that steps up and down in different parts of the country. The problem you have is recognising where, relative to some ground feature, it starts, steps up, steps down, or finishes. You can only know if you are approaching such a boundary if you keep a close eye on where you are.

Finding the height of the base of a particular piece of controlled airspace is further complicated by the fact that it is not defined relative to the height of the ground below it. (Not that this would be a great help anyway, as your altimeter almost certainly is not reading your exact height above the ground below the glider.) The base of controlled airspace is probably defined as either an ALTITUDE or a FLIGHT LEVEL.

ALTITUDE is defined as the height above mean sea-level. If your altimeter has been set to read zero when sitting on the ground at your home airfield, this setting is known as the QFE for your home airfield. When flying using this setting, the altimeter will display height above the airfield. If when sitting on the airfield, the altimeter is adjusted to show the height of the airfield above sea-level, it will now be set to QNH. All indications on the altimeter will now be heights above mean sea-level; that is, it will display ALTITUDE. (This method of finding the value of QNH will lead to slight inaccuracies, but none so great as to result in any airspace infringement problems.) The base of airspace defined by an altitude would appear on a map as a number of feet, e.g. 4,500 feet.

To further complicate the issue, the base of some airspace is defined as a flight level. A FLIGHT LEVEL is the reading on the altimeter, in hundreds of feet, when the altimeter sub-scale is set to read 1013.2 millibars (or hectopascals). This pressure setting is just one property of a mythical beast known as the standard atmosphere and, as a result, is known as the STANDARD ALTIMETER SETTING, or simply STANDARD SETTING. The base of controlled airspace defined as a flight level would appear as a number of hundreds of feet prefixed by the letters 'FL', e.g. 'FL45', which equals 4,500 feet with the altimeter set to 1013.2 millibars.

As atmospheric pressure changes daily (or even hourly) the base of an airway defined as a flight level will move up and down relative to the ground (through no fault of the airspace planners), depending on whether the local atmospheric pressure is high or low. The only accurate way to check whether you are approaching the base of airspace defined as a flight level is to set 1013.2 millibars on your altimeter's sub-scale.

If you choose to fly with your altimeter set to either QFE or QNH for the whole of the flight, then it is possible to work out the difference in feet between your chosen setting and the other two. The differences should be noted on a card or a kneepad to which you can refer in flight. To do this, set the altimeter to all three settings in turn while the glider is sitting on the airfield before the flight. Note the difference in feet between the setting you wish to use and the other two, along with whether the other settings are lower or higher than your chosen setting. Another option is to calculate the differences between the settings (1 millibar = approximately 30 feet).

Personally, I cannot be bothered with mental arithmetic while map reading and trying to stay in the air, and prefer to set the altimeter sub-scale to the appropriate setting for the particular airspace I am near. If you are going to use this technique, make sure that you note the other pressure settings before take-off, so that you can reset the altimeter when required. Final gliding is unrewarding if it ends 600 feet below the level of your goal airfield!

It is essential, if you are to utilise soaring conditions to the maximum, that you know exactly how high you can climb over any given place or in any particular area. There will be days when you can cruise at quite high airspeeds when the restriction on going higher will not be cloud or the top of the thermals, but controlled airspace. In these situations, it may be necessary to dolphin fly at high MacCready settings to stay within the law, but this can only be done efficiently if you know the extent of the airspace both horizontally and vertically.

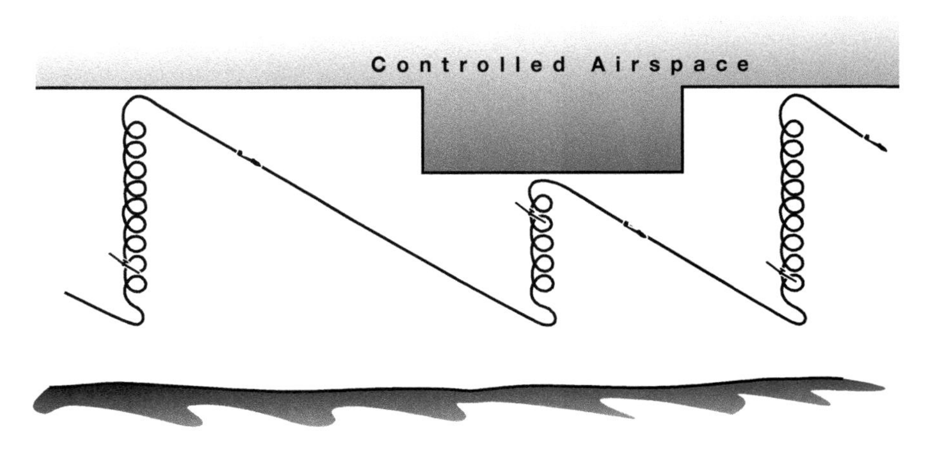

Fig 16.8 Knowing your operating ceiling.

Not all airspace restrictions are permanent and marked on aeronautical maps. Events such as air displays, balloon rallies, parachuting displays and military exercises, along with (in the United Kingdom) Royal Flights, crop up throughout the year and are commonplace during summer months. These events are notified to pilots through a system of bulletins and NOTAMs (an abbreviation of Notices to Airmen). In order to avoid becoming part of a jet formation team or a target in a military exercise, it is necessary to check such NOTAMs, note the position of the activity, and plan or navigate your way around it.

A sound knowledge of the rules regarding controlled and other airspace is of benefit, not only to allow you to keep clear of airspace from which you are prohibited, but also to allow you to use all of the airspace in which you have the right to fly your glider without the worry of prosecution. Any doubts, as mentioned earlier, can distract you to the extent that your soaring suffers. The better you know the rules, the more you will enjoy your cross-country flying. This is a key element in navigating without worries.

Global Positioning Systems

Global Positioning Systems (GPS) have had a major impact on gliding. Officially called Global Navigation Satellite Systems (GNSS), such systems, as the name suggests, use satellites to determine the position of the GPS receiver and hence the glider. They are very accurate, with the manufacturers claiming accuracy to within 50 feet or less.

Their compactness, lightness, low power consumption and relative cheapness make them the first practical electronic navigation aid usable in gliders.

GPS receivers display information either as numerical data, consisting of position (given in latitude and longitude), bearings, speed, track-made-good, course to steer and estimated time of arrival at a chosen point, or as a position on a moving map display. Some GPS receivers do not have the map facility.

There are many manufacturers and models of GPS of differing size, ease of operation and cost. As they all tend to be supported by comprehensive operating manuals, it is not necessary (or possible) to describe how to use a GPS in this text. What I will do is look at the uses and limitations of GPS as they affect the glider pilot.

Most GPS units have the facility to allow the user to input the latitude and longitude of a large number of positions on the earth's surface. Such a position is known as a WAYPOINT. In addition, some GPS units have a huge database containing the positions of all of the airfields in the world. With this information, the GPS can show the glider's position and all the navigation information necessary to steer the glider to whatever waypoint has been chosen. You can even enter a complete task or route, and the GPS will automatically update the track requirements as each waypoint (turning point in gliding parlance) is reached.

You would think that with such an aid at your disposal, you could throw away your maps. Well, this is certainly not the case. The GPS is, after all, only an aid to navigation. It still has its limitations, and as with all technology, reversion to the old methods occasionally has to be made. This is why so much of this chapter has been devoted to traditional glider navigation methods.

Limitations of GPS

As with all electrically powered equipment, a GPS will suffer failure from depleted batteries, blown fuses, tripped circuit breakers or faulty wiring connections. It is even possible for the unit or its antenna to suffer internal failure.

For a GPS to work, it has to be able to 'see' at least three satellites to give a two-dimensional (horizontal) position fix, and four satellites for a three-dimensional position fix (with a height indication). Normally it is not a problem for the unit to acquire these satellites, and indeed, it is not uncommon for a GPS receiver to be scanning eight or more satellites at any one time. If, however, the satellite positions are temporarily unfavourable, the trigonometry on which the GPS relies will not work, resulting in a POOR COVERAGE message and a loss of accurate position information. In this event, the best the GPS can do is give you data based on your last known position, speed and track. This is not an ideal situation, considering how often gliders change track and airspeed.

The same situation may arise if the antenna of the unit becomes masked from the satellites by part of the glider's structure – a less likely but possible scenario.

All of the above events may send you scurrying after your discarded maps, or even going into 'where am I?' mode if you had been relying totally on your GPS.

Then there is the possibility of the common failing of any equipment – human error. Operator error can manifest itself in many ways, from accidentally switching the GPS off to inputting the wrong latitude and longitude of a waypoint.

If the GPS is switched off accidentally, even after the power is reinstated, some time will elapse while the GPS receiver re-acquires the necessary satellite signals. The route information may have to be re-entered, depending on the mode in which the GPS was being operated. The result of inadvertently switching the GPS off is likely to be the loss of only a few minutes' worth of navigation information. However, if this happens at a critical phase of flight (such as an approach to a turning point or on a final glide) it is likely to cause more than a little inconvenience.

On the other hand, inputting the wrong latitude or longitude of a waypoint, or of your goal airfield, can have disastrous results.

Inputting the wrong co-ordinates could result in your displacing a waypoint by 100 feet or 1,000 miles, depending on how much you jumble the numbers. It is more likely that you will make only small errors when entering latitudes or longitudes. Even so, this could still mean that a turning point does not appear where you thought it would be, or that you do not fly through the turning point sector. It could result in your flying completely the wrong track, adding to the possibility of airspace infringements and your being totally lost and confused. As if this were not enough, one wrong input could result in the GPS being useless for a major part of the flight because of your having to ignore erroneous waypoint information.

The only way to recover the situation may be to input the correct co-ordinates whilst in flight – a task which is bound to lose you some time which should have been spent concentrating on soaring efficiently, and even more worrying, maintaining a good lookout.

If you are flying in England or in any other country that has the Greenwich Meridian (0°E/W) running through it, one easy way to get a waypoint's co-ordinates wrong is to forget to set 'east' or 'west' as appropriate. (Most GPS units default to either east or west when data is being input, and require the operator to switch from the defaulted hemisphere.) The resulting position error will depend on how far the waypoint is from the meridian. If it is only a mile or so from the meridian, then some confusion may result, but if it is ten miles or more then your whole track will probably cover unintended ground and require some shrewd deductions and clever map reading. (A similar problem will occur if you are flying in the Southern Hemisphere and your GPS defaults to northern latitudes when setting waypoints.)

When using a GPS, it is necessary to safeguard your flight navigation as follows.

1. Take extra care when inputting waypoint co-ordinates.
2. Do not forget the west or east (or north and south) sign.

3. Check and double-check what you have input.
4. Check that the length and track of each leg of the task, as given by the GPS, matches that measured from the map.

Even once the correct co-ordinates of a waypoint have been stored, the pilot may still have to select the correct waypoint in flight. To do this, it is necessary to remember where you are going and what name or code you have given the waypoint. Marking the GPS name or co-ordinates either on a card displayed in the cockpit or on a task sheet is a necessary safeguard against that other common human failing – memory lapse. (Note: If you have programmed the whole task into your GPS in the form of a route, then this may safeguard against this problem.)

Early GPS receivers presented information in an all-digital manner. It was not long before moving map displays were also available.

Surprisingly, the one piece of information that a GPS with an all-digital display does not give, in a format that is readily usable in a glider, is the glider's position. Admittedly, it does tell you the glider's position relative to the selected waypoint, but not to a number of ground features in the way that use of an ordinary map does.

Knowing where you are is essential if you are to avoid controlled airspace. As with navigation using a map, you need to know your position relative to ground features in order to deduce where airspace boundaries lie. In poor visibility, it is easy to rely too much on a GPS. If there is a lot of controlled airspace at varying altitudes above your track, your GPS may not give enough information to determine the airspace boundaries.

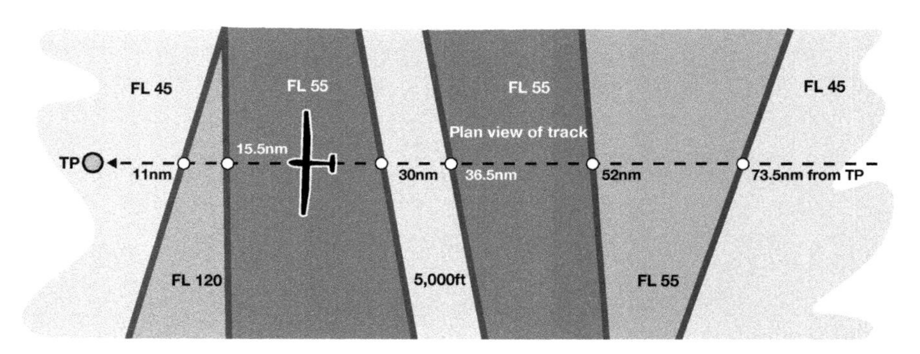

Fig 16.9 Use of GPS and a map to define airspace boundaries. By marking your map with the distances of airspace boundaries from the next turning point, GPS can be used to show where inadvertent airspace penetration is a danger.

One way to use the GPS so that you know when you are approaching an airspace boundary is to mark on the map the distance of the airspace boundary from the waypoint to which you are heading. You then know that when you are at that distance from the waypoint (as displayed on the GPS) you have reached the airspace boundary (figure 16.9).

This technique may be suitable if your track only flies under (or over) a few areas of controlled airspace, but is probably unmanageable if you are on a task that takes you under a lot of controlled airspace, the base of which has varying heights.

A GPS with a map display solves some of this problem, but the tendency is for the map screen to be too small and too cluttered, with only airfields, user-defined waypoints and VORs displayed.

Satellite navigation systems are improving all the time, as is the type of information that they can store and display. Many give a choice of displays, offering different map styles, data or a 'compass rose' presentation which can assist greatly with orientation relative to the waypoint selected. They often have extensive databases included as part of their specifications. It is even possible to have a GPS programmed to show all of the controlled airspace on your task route. However, the changing nature, horizontal boundaries, and lower and upper limits of controlled airspace require that such a database be updated regularly.

Even more useful is the introduction of 'handheld' computers into the cockpit. Receiving inputs from a GPS and a flight director, and programmed with dedicated soaring software, these provide information on a reasonable-sized screen. The screen itself can be 'touch-sensitive' enabling data to be presented quickly. Information on navigation, airspace, final glides and even the lift distribution in the thermal being worked can be displayed.

GPS is a fantastic navigation system. It is a very useful aid, especially to the glider pilot who wanders off track regularly to follow the best lift. If used as an extra source of navigation information, and not as the sole or prime source of navigation information, it will reduce the workload considerably during a cross-country flight.

The traditional map, the compass, and the pilot's eyes and brain still have to be used. In fact, leave the GPS at home on some days and you will probably manage to navigate quite well. Leave the map and compass at home and you may fare less well.

One last thought for you to ponder on the use of GPS: do not let the track line or bearing on your GPS display become a god to be obeyed. The GPS shows the ideal navigational track – not the best soaring track. I wonder how many pilots have sat in a field exactly on track when the lift was not on track.

Navigation, like so many other aspects of soaring, benefits from practice, which can best be gained while flying cross-country. Some pre-flight planning is essential – but my advice is to keep it practical and do not try to make navigation into a science. After all, you do not know where the lift will be relative to the ideal map track. Use your time before the flight to prepare for navigation during the flight, thus reducing your in-flight workload. Do not waste this time preparing maps to an accuracy that your flying will be unable to match.

Chapter 17

Turning Points

These days, most cross-country tasks involve flying the glider around one or more pre-declared turning points. Gone are the days of straight-line distance tasks with subsequent long, expensive road retrieves, unless the pilot is attempting a specific record flight in this category.

Proof of the glider having rounded a specific turning point is from evidence recorded on a datalogger which records information received from a GPS.

Selection of turning points

If you are flying in a competition, the competition director or task setter will decide what your task will be and therefore the turning points that you must fly around. In such situations, you will not normally have any say as to which turning point features you will use. (The exceptions to this are the so-called PILOT-SELECTED TASKS, such as the ASSIGNED AREA TASK.)

If yours is a non-competition flight, the turning points you choose to use will, to a large extent, be decided by the size, nature and direction of the task you set yourself.

Once airborne you will have to find and identify the turning points you have chosen. The ease with which you achieve this will, to some degree, be determined by how obvious the turning point feature is. An obvious, well-defined turning point feature will be easier to find, allowing you to fly straight to it and get on with flying the next leg. A poorly defined feature which does not stand out well from the surrounding landscape, or which can be confused with similar features in the area, will be more difficult to find, will take longer to identify, and can waste valuable time. It is therefore important to select obvious landmarks as turning points. Motorway junctions, places where roads and railways cross each other, prominent buildings (such as cathedrals and castles), reservoir dams and features on airfields, all make excellent turning point features.

In Australia, grain silos (usually next to railway lines) stand out well in what can be a comparatively sparse landscape by European standards.

Despite having a large number of easily identifiable ground features available, many pilots and competition directors still select features that are not well defined, or are difficult to find.

To help select reasonable turning point features, the British Gliding Association has compiled a comprehensive list of recognised turning point features covering the whole of the UK. Each feature is detailed with its latitude and longitude for GPS use, and is given a category depending on how easy the feature is to find. Other countries have similar catalogues.

Make life easier for yourself. Choose only well-defined, obvious ground features as turning points. Remember the challenge of cross-country flying is not to find some obscure turning point. It is completing the task or achieving a high average speed around the task.

Navigating to a turning point

In Chapter 16, Navigation, emphasis was placed on not trying to identify and use every small ground feature to check your position, but instead to use only major ground features. However, when approaching a turning point (say within 10 nautical miles of it) your general route finding must change to more detailed navigation.

You will now, instead of checking the glider's position, be trying to find a specific feature on the ground. To do this, you will have to check-off each ground feature and its relative position to the turning point. Doing so will not only help you to find the turning point feature, but will also allow you to orientate the turning point feature to the surrounding ground features, so that you can position the glider in the turning point zone. This is one area in which more detailed maps, such as the 1:250,000 scale map, are most useful.

In poor visibility, the relative position of local ground features can often be used to lead you to a turning point which may still be out of sight. In fact, should your turning point be on a line feature such as a motorway or a railway line, it is often useful to find the line feature and then fly along it until the turning point feature is found. However, the problem with this technique is deciding which way to fly along the line feature, once you have reached it. This problem can be solved by deliberately tracking to one side of the ideal track so that you know for sure in which direction to turn to look for the turning point feature. Such DELIBERATE ERROR NAVIGATION is well known to power pilots, and can be useful in poor visibility, especially when flying over terrain with few features.

If you are using GPS to aid navigation, it will guide you accurately to the turning point – assuming, of course, that you have correctly entered the turning point co-ordinates.

The turning point sector

Having found and positively identified the turning point feature, you will now have to fly the glider into the turning point sector. The turning point sector is a zone that is orientated relative to the turning point feature to make sure that the glider has been flown around the turning point, thus covering the total task distance.

The turning point sector is defined as 'a 90° degree sector on the ground with its apex at the turning point feature and orientated symmetrically to, and remote from, the two legs meeting at the turning point'. (See figure 17.1.)

In order to determine the turning point sector for a particular turning point, you must firstly bisect the smaller of the two angles created by the inbound and outbound tracks. Extend this line to the other side of

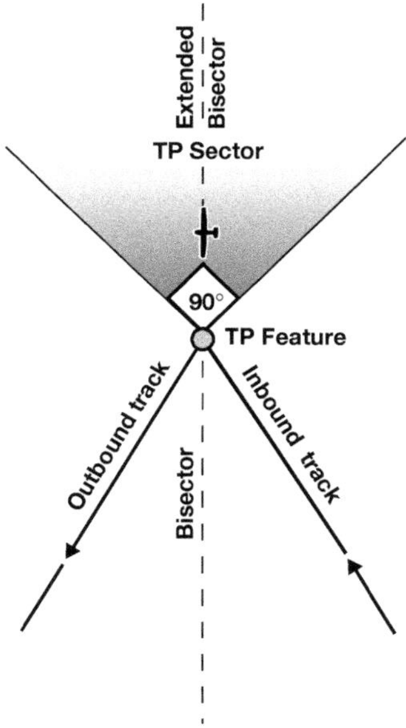

Fig 17.1 The turning point sector.

the turning point and draw a quadrant (90° sector) symmetrically about this extended bisector. GPS evidence must prove that the glider flew through this sector. If you cannot prove that you have flown through this sector, it will be assumed that you have never been to this turning point. Your flight will not be ratified. If the flight is for a badge or a record, it will not be awarded.

Positioning the glider in the turning point sector

As well as working out where the turning point sector lies, it pays either to mark it on the map or to make a sketch of the major ground features relative to the turning point sector. The most important part of the turning point sector is its bisector. It is therefore desirable to have some target feature close to the bisector of the inbound and outbound tracks to help you position the glider in the sector.

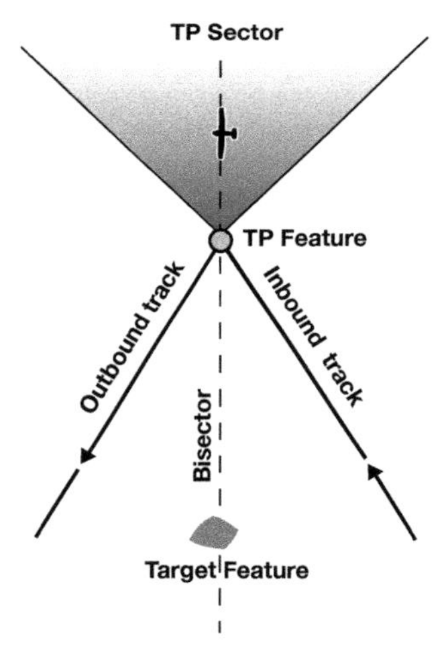

Fig 17.2 Turning point target feature beyond turning point feature.

Personally, I prefer to make a sketch of each turning point on a piece of card which will fit easily into the cockpit pocket.

Occasionally no target feature will lie beyond the turning point feature, and it may be necessary to use a feature within the turning point sector to aid alignment.

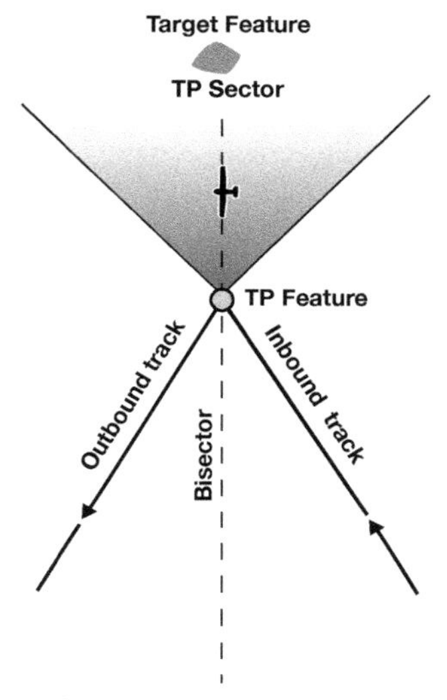

Fig 17.3 Turning point target feature on opposite side of glider.

Turning point features such as motorway junctions, railway stations or railway junctions can provide their own alignment aids by virtue of the line features they provide.

By using GPS, positioning the glider in the turning point sector can be incredibly simple. Before the flight, calculate the bearing of the turning point from the bisector of the turning point sector. When the GPS display shows that the bearing of the turning point feature is this number of degrees from the glider's position, then you are on the bisector of the turning point sector (figure 17.4).

With some GPS units, it will be necessary to use the GOTO mode, rather than the ROUTE mode, in order to use the GPS in this way. With the GPS in GOTO mode, the turning point feature's co-ordinates input, and the BEARING output selected, the GPS will continuously update the bearing of the turning point feature from the glider. As you turn through the turning point sector, the bearing indicated will change rapidly. The closer you are to the turning point feature, the quicker its bearing will change. Anticipate the correct bearing by monitoring the GPS display. As you cross the bisector your datalogger should have captured the evidence you need. Throughout this procedure, make sure you maintain a good lookout – turning points are often busy places.

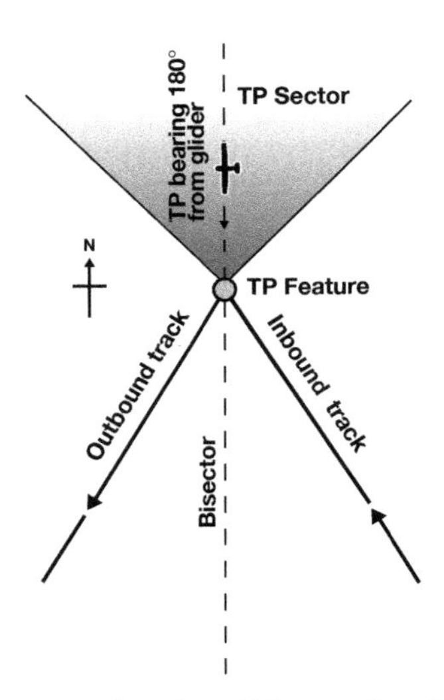

Fig 17.4 Turning point bisector bearing. When using GPS to locate the turning point zone, it helps to know the bearing of the turning point feature from any point on the extended bisector.

GPS turning point evidence

How the GPS evidence is assessed determines the way in which the glider has to be flown around the turning point.

The rules may vary. If they require at least one GPS position fix to be within the turning point sector, then care must be taken to keep the glider in the sector long enough to allow a position to be recorded in the datalogger. One way of doing this is to fly along the bisector of the turning point sector (away from the turning point feature) for a specific length of time before turning on the next track. This period of time will need to be almost as long as the interval between the position fixes being logged by the datalogger. (Some dataloggers increase the rate at which fixes are recorded when close to the turning point, while others even confirm to the pilot that a fix has been logged in sector by giving a 'bleep'.)

If the rules will accept a line drawn between two GPS position fixes, which crosses the bisector of the turning point sector, then care must be taken not to turn too close to the turning point feature in case the line between the position fixes falls on the wrong side of the turning point feature.

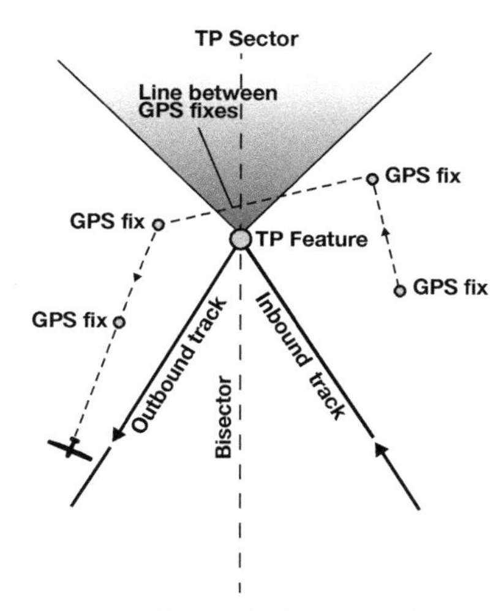

Fig 17.5 Line between two GPS fixes which crosses the turning point zone bisector. In some competitions, the rounding of a turning point may be accepted if datalogger evidence shows a line connecting two GPS fixes which crosses the turning point zone bisector.

Tactics at turning points

When approaching a turning point, there is probably only one thing more important than getting into the turning point sector and out again as efficiently as possible, having captured the evidence required to show that you have been there: that is, staying airborne.

There is no point in having evidence of rounding a turning point if you are sitting in a field next to the turning point a few minutes later. Unfortunately, too many pilots dive into turning points as if they were the finish line at the end of the task. It is necessary to think very carefully about where the lift will be found near the turning point, and how much height will be required to round it and reach the next thermal. Skyreading and sensible flying is every bit as important as getting the GPS evidence.

When flying a speed task, any appreciable wind will affect the optimum height at which you should reach the turning point. If the turning point is at the upwind end of the next leg, then it is acceptable to approach the turning point as low as is safe to do so and still find the next acceptable thermal after the turning point. This is because any lift found after the turning point will be accompanied by the bonus of being drifted on the desired track. Whereas taking excess height in a climb before an upwind turning point will waste time because of the tendency to drift away from the turning point.

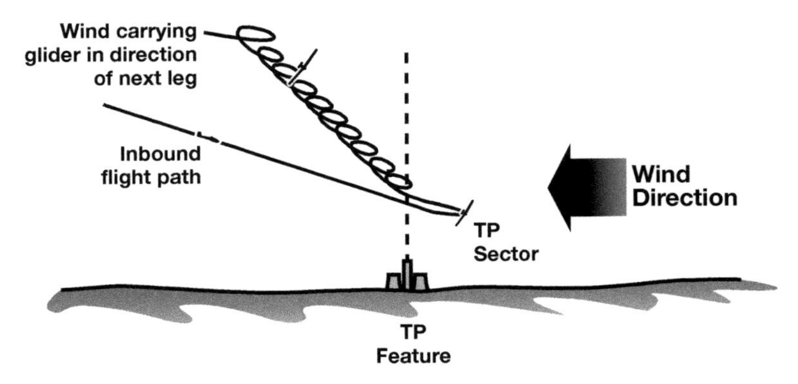

Fig 17.6 Tactics at upwind turning points. Approach upwind turning points as low as is safe.

On the other hand, when arriving at a turning point at the downwind end of a task, you should attempt to turn the turning point as high as possible, irrespective of where you think the next lift will be found.

Taking a climb some distance before reaching the turning point will result in the glider being drifted towards the turning point while climbing. Arriving low at a 'downwind' turning point will require a climb, possibly in weak lift, during which the glider will be drifting away from the direction of the next leg, thus effectively increasing the distance to be flown into wind.

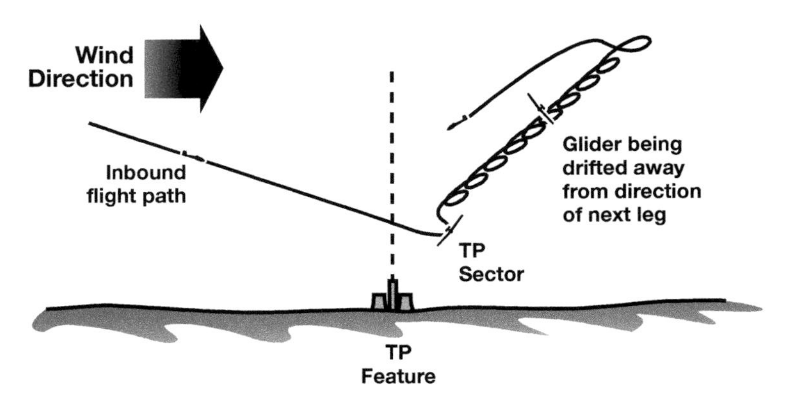

Fig 17.7 Tactics at downwind turning points. Approach downwind turning points as high as possible.

One situation which might override these rules is when a very strong thermal is encountered. In this case, you must weigh up factors such as drift rate, climb rate and the sky ahead to see whether you should use the thermal to climb. More often than not, it will be wise to climb in such strong thermals.

In strong winds, do not try to turn very close to an upwind turning point, as you may be drifted over the turning point feature and out of the turning point sector before you can get the evidence you require.

Your route around the turning point will determine how long you will take to log it and get on with flying the next leg.

If your task is for the award of a badge or qualification that does not require a high average speed, then you can afford to make sure that you get good turning point evidence by rounding the turning point feature at some distance, thus allowing time for the datalogger to record several fixes.

On the other hand, if it is a speed task, or a distance task requiring a high average speed for completion, then you will want to spend the minimum time flying around the turning point feature. In this case, you will want to turn close to the feature. Doing so will save vital seconds, but will result in your crossing the turning point sector at its narrower end. This means that you will be within the turning point sector for a shorter time,

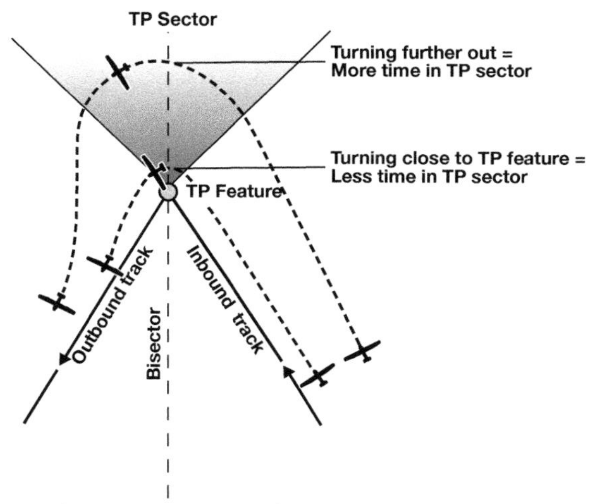

Fig 17.8 Distance of turn from turning point feature. Turning close to the turning point is quicker but allows less chance to get the evidence you need.

but this may not give a datalogger enough time to log a position fix within the turning point sector.

Chapter 18

The Final Glide

To achieve the highest possible average cross-country speed, it helps to fly the glider on the desired track without circling. Unfortunately, this is not normally possible for the whole of the flight. The best that can usually be achieved are periods of dolphin flight, glides between thermals, and the glide between the last thermal and the goal airfield, that is, the FINAL GLIDE.

A well-flown final glide can increase the average speed for a task considerably. A badly executed final glide can result in time wasted or even a landing short of the goal airfield.

To illustrate the effect on the average speed, let us look at a flight of 150 kilometres around a triangular task where each leg is the same length.

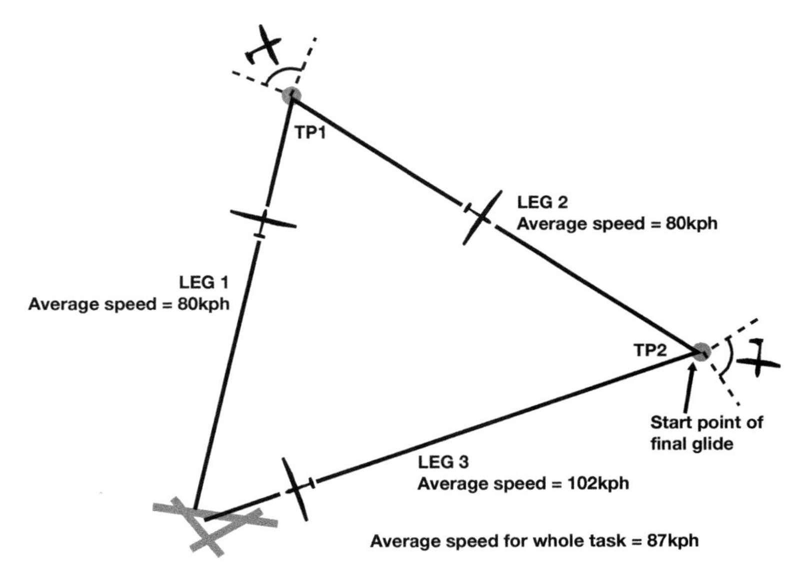

Fig 18.1 The effect of the final glide on the average speed for a task.

If the first two legs are flown in the classical 'climb and glide' style, perhaps with little sections of dolphin flying, then let us assume that an average speed of 80kph may be possible on these legs. If it were possible to leave the last turning point and glide the whole of the last leg at a constant airspeed, without stopping to circle, the average speed for this leg would be the same as the average speed of the glider on this leg. Even if we assume that this final glide is flown at the fairly slow airspeed of 55 knots or 102kph (approximately the airspeed for maximum glide angle for an average-performance glider), it will have a noticeable effect on the average speed for the whole task. Thanks to this final glide, the whole of the flight, which was averaging 80 kph, has become a flight with an average speed of almost 87kph. (An increase of almost 9 per cent over the whole task.) This simplified example, using fairly mediocre speeds, illustrates how important the final glide can be.

Flying an efficient final glide requires much more than setting off at the airspeed that gives the best glide angle and pointing the glider in the direction of the goal airfield. There are several factors which have to be taken into account if the glider is to get to the goal airfield as fast as possible. Let us deal with these factors in turn, and then look at how the theory works in practice.

From the initial discussion of the glider's polar curve, we saw that there is an optimum airspeed at which the glider will cover the greatest distance. If the air mass is neither ascending nor descending, and there is no wind, the optimum airspeed will be the airspeed that gives the best glide angle, that is, the best glide speed.

We also know that the most efficient performance will be achieved if we increase the glider's airspeed in sink and reduce it in lift. This rule applies just as much on the final glide as it does when gliding between thermals.

One aspect that was not covered when discussing speed-to-fly was the effect that any wind has on the selection of the optimum airspeed (although it was mentioned briefly with reference to crossing wave gaps). This has been deliberately left until now, as the wind only affects the glider's performance relative to the ground. In most cases, we are interested in the performance of the glider relative to the air mass within which it is flying, and not relative to the ground. As the final glide is to a point on the ground, the effect of the wind has to be taken into account.

Any head wind or tail wind will affect the glider's speed over the ground. As the rate of descent of the air mass in which the glider is flying will not change due to the wind, this means that the glider's glide angle relative to the ground will be different from that in calm conditions.

For instance, if a glider's best glide ratio of 43.3 to 1 is achieved at 52 knots (giving a still air rate of sink of 1.2 knots) then this glider can only achieve a glide angle of 43.3 to 1 over the ground in calm conditions.

If there were a 20 knot head wind, then the speed over the ground would be 32 knots if the same airspeed of 52 knots were selected. In this case, the glide angle *relative to the ground* would be just 26.7 to 1 (32/1.2 = 26.7 to 1).

By using the glider's polar curve, we can find a more efficient airspeed at which to fly when final gliding in a head wind.

To discover the optimum airspeed to use in a head wind, we move the origin of the airspeed axis to the right by the value of the head wind, in this case by 20 knots. (Again, for convenience, we are moving the origin of the axis to the right, rather than re-drawing the whole curve further to the left.) We now draw the tangent to the curve from this new origin. Vertically above the point where the tangent touches the curve, we have the new optimum airspeed for the best glide angle *over the ground* in a 20 knot head wind (59 knots). Notice that at this higher airspeed the sink rate of the glider has increased to 1.41 knots, but that the resulting glide angle

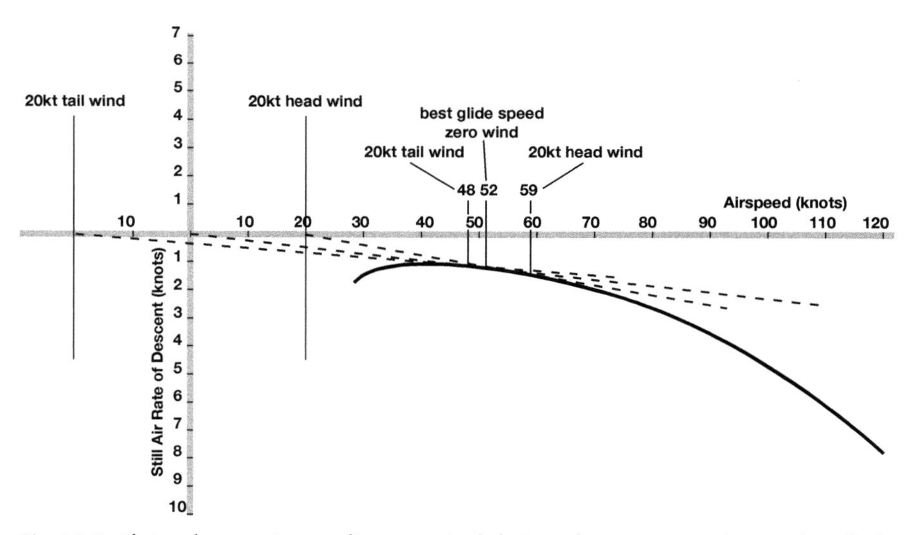

Fig 18.2 Flying for maximum distance. A glider's polar curve can be used to find the best airspeed to cover the maximum distance in any wind.

of 27.7 to 1 is still better than that achieved by flying at the best glide speed for calm conditions (26.7 to 1).

In the case of a tail wind the glider will be covering more ground, and it will pay to fly at a lower airspeed to reduce the still air rate of descent, thus achieving a better glide angle over the ground than would result from flying at the best glide speed for calm conditions. Using the polar curve for the tail wind case, the tangent to the curve is drawn from an origin which is moved to the left along the airspeed axis by the value of the tail wind component (in this case 20 knots). This results in an optimum airspeed that is less than the airspeed for best glide angle in calm conditions (48 knots). In this example, the glide angle over the ground is 68/1.12 = 60.7 to 1 (as opposed to 60 to 1, if the airspeed were not reduced). Note, however, that once the glider's airspeed is reduced below the speed for minimum sink, then it is pointless reducing the airspeed further.

Overall, the increase in the glide angle due to adjusting airspeed to match the wind component is generally very small. The most noticeable effect of the wind is that the reduction or increase it causes in glide angle *over the ground* will mean that a greater height will be required to reach your goal when there is a head wind component, and less height when there is a tail wind component.

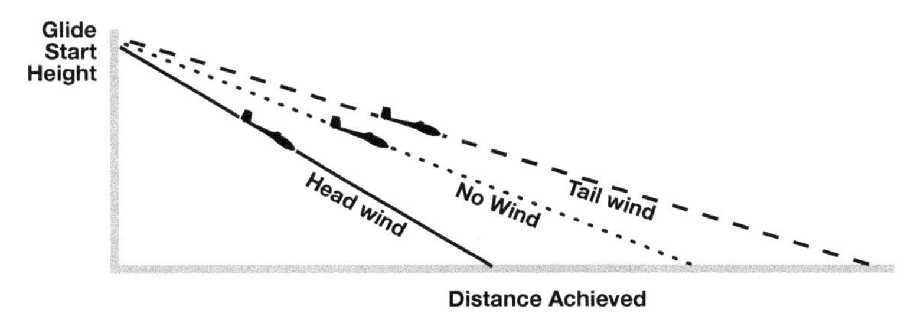

Fig 18.3 Effect of the wind on distance achieved. The distance a glider can fly from a given height will depend on the wind.

You will seldom find yourself setting up a final glide to cover the maximum distance in still air (no lift or sink, as opposed to no head wind or tail wind), except perhaps at the end of the day when no more lift exists. It is much more likely that you will be able to climb to a height that will allow you to glide home at a greater airspeed than that for maximum distance.

If you are trying to achieve a high average cross-country speed, the height to which you climb will depend not only on the wind and the distance to be covered, but also on the rate of climb being achieved immediately before leaving the last thermal. (Note that in this situation we do not use the average rate of climb for the whole thermal.) For instance, if you are climbing quickly, it will pay to climb higher than if you were climbing slowly. The extra height gained will allow a higher airspeed on the final glide and a shorter elapsed time for the combined climb and final glide. On the other hand, time spent gaining extra height in a weak thermal will not be regained during the final glide.

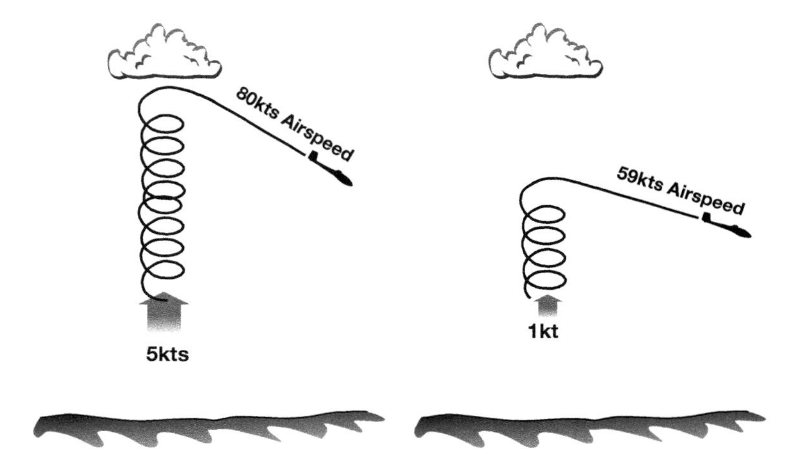

Fig 18.4 Rate of climb affects optimum final glide speed. If you are climbing quickly, it will pay to climb higher and fly the final glide at a higher average airspeed.

Whatever height you climb to should include a safety margin, in case you encounter an abnormal amount of sinking air on the final glide.

All of these considerations, along with the performance of modern gliders, make it difficult to calculate and set up an optimum final glide without using some form of calculator or computer. The old rule of thumb of 1,000 feet of height for every 4 nautical miles to be covered will invariably result in an embarrassment of height on arrival at the goal airfield when flying a modern glider. There are two types of aids to compute your final glide. The first is a form of slide-rule, and the second is a computer incorporated in the glider's flight director system.

JSW final glide calculator

Probably the easiest to use of the slide-rule type of final glide calculators is the JSW FINAL GLIDE CALCULATOR. These calculators come in a variety of models covering the whole range of glider performances. Therefore, you should use only the model designed for the type of glider that you are flying.

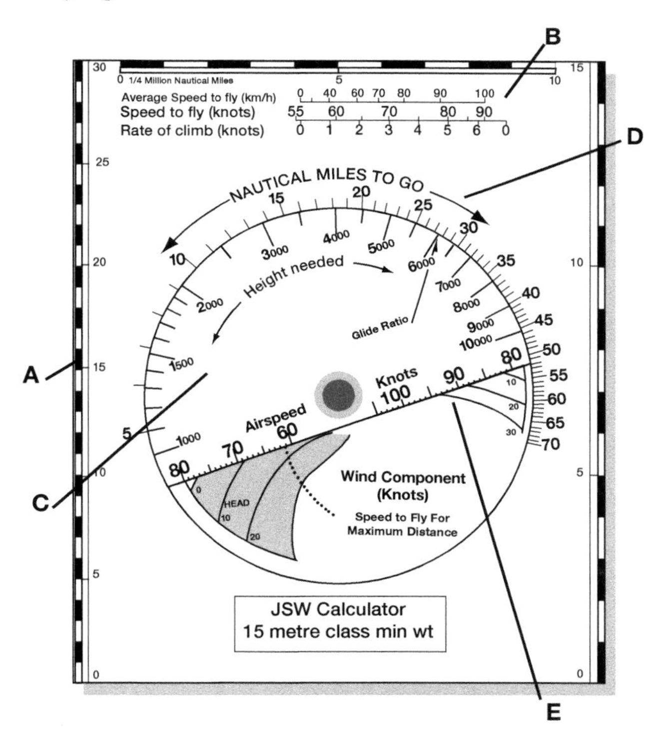

Fig 18.5 The JSW final glide calculator.

The JSW final glide calculator is pocket-sized and made of plastic. Along each side edge is a rule that is calibrated to allow easy measurement of distances directly from a map (A). Each rule has a different scale for use with different scale maps.

At the top of the calculator is a 'reckoner', which allows you to decide the best airspeed at which to fly the final glide, depending on the rate of climb being achieved in the last thermal (B). This is labelled the 'SPEED TO FLY' scale.

The main part of the final glide calculator is in the centre of the calculator. It consists of a semicircular slide that can be rotated over the body of the calculator (C).

Along the curved top of this slide is a scale which gives the 'HEIGHT NEEDED' to fly the appropriate distance shown on an adjacent scale on the body of the calculator, labelled 'NAUTICAL MILES TO GO' (D).

Along the bottom edge of the semicircular slide there is an 'AIRSPEED' scale which starts in the centre of the calculator and increases as the scale goes to the left. When the scale reaches the extreme left of the slide, it continues on the far right of the slide, again increasing in value towards the centre (E).

Beneath the rotating slide are a series of wind component curves, which are coloured red for head winds and green for tail winds.

Using the JSW final glide calculator

The best way to describe the use of the JSW final glide calculator is to go through an example of a final glide. (If you possess one of these calculators, it will be of benefit to use it to follow the example given. However, bear in mind that the figures you derive may differ from the text if your calculator is for a glider with a different performance.)

Let us assume that we are 14 nautical miles from our goal and that there is a 10 knot head wind. We encounter a thermal that gives us a rate of climb of 3 knots.

STEP 1. From the 'Speed-to-Fly' scale at the top of the calculator, find the speed-to-fly corresponding to a rate of climb of 3 knots. We find that the airspeed required is 70 knots.

STEP 2. Turn the slide until the 70 knots airspeed mark is touching the 10 knot head wind curve.

STEP 3. Read off the height required for the final glide (at this airspeed) against the 14 nautical mile mark on the 'Miles-To-Go' scale – in this case 3,100 feet.

STEP 4. Lastly, add to this height a suitable safety height margin, say 500 feet. *There is no safety margin built into such calculators.*

JSW final glide calculators not only come in a variety of models which cover the performance range of most modern gliders, they are also available in double-sided models. Some of these utilise the other side of the calculator to give the same final glide information, but with scales that are calibrated for use when the glider is being operated at the higher wing loadings consistent with the carrying of water ballast. On other versions, the reverse side of the calculator provides a simple wind component resolver.

There are other types of 'slide-rule' final glide calculators. Some involve a transparent plastic disc placed over a map in such a way that the goal airfield lies at the centre of the disc. This gives the pilot the advantage of being able to read off heights required without having to measure distances. However, unlike the JSW calculator, which can be used to rapidly calculate the height required to reach any point, be it a turning point or another airfield, these calculators cannot be quickly reconstructed to change the goal of the calculation.

Electronic final glide computers

Many modern variometer/flight director systems incorporate a final glide computer. This requires more or less the same data inputs as the JSW or other slide-rule types of calculator, although the final glide information presented may be different, often simply showing the height excess or deficit. Normally information such as the safety height margin can be input before (or at any time during) the flight.

When climbing in the last thermal, the wind component for the final glide, the rate of climb and the distance to the goal airfield are entered in the computer. When the final glide is started, the flight director system will continually update the height required and the distance to run. These calculations use an input from the glider's airspeed system to update the distance and hence the height required. If, after setting off on the final glide, the distance to run does not tie in with your position on the map, then it will normally be necessary to input a revised wind component. This will result in a new height requirement and possibly a re-setting of the MacCready value being used for the final glide.

If the system is coupled to GPS, then the only inputs necessary from the pilot are the safety height margin and the climb rate being achieved in the last thermal. Everything else will automatically be updated by the GPS.

Such instruments are excellent, although expensive. However, no matter how well-designed and programmed they are, they will always give accurate data based on the inputs made by the pilot. Any wrong inputs may result in the glider having too little or too much height – and if the wrong co-ordinates have been entered into the GPS, then you may not be going home! One drawback of relying on GPS-linked final glide computers occurs when enough height can be gained to start the final glide on the leg of the task before the final turning point. In this case, unless the whole route has been entered into the GPS (instead of using the GOTO function), the calculations will require measurements from the map and probably a reversion to the faithful JSW calculator.

Setting up and flying the final glide

Now let us look at the last section of a flight (including the final glide) and some of the considerations involved.

By the time you are close enough to home to be thinking of climbing to final glide height, you should have some idea of what the wind component will be on the final glide, albeit a rough estimate. This can be assessed by your drift while thermalling, the movement of cloud shadows, or possibly direct from your GPS.

Once you have reached a point from which you can climb high enough to final glide to your goal, do not be lured into using any old thermal to gain the necessary height to get you home – unless the sky ahead looks unpromising as far as thermic activity is concerned.

You should keep scanning ahead for the likely existence of thermals. If the probability of finding a good thermal seems high, then continue to fly on track at the airspeeds dictated by the appropriate MacCready setting until you encounter such a thermal, or until the height remaining dictates a more cautious approach. Should the sky ahead look dead, then reduce the MacCready setting and continue cautiously until lift is encountered.

Once you have found an acceptable thermal, start climbing. Only after you are established in the best lift should you begin to calculate the final glide.

At this point, it is essential that you remember that the height required is height above the goal airfield. If you have been flying on another altimeter setting, such as the sea-level setting (QNH) or the standard setting (1013.2 millibars) then it will be necessary to reset the altimeter sub-scale so that the altimeter will read zero on arrival at the goal airfield (QFE). If your climb or glide is likely to take you close to controlled air-space, you may either have to delay this altimeter setting change or do some mental calculations to allow for a different altimeter setting. (One millibar of pressure = approximately 30 feet of height.)

When calculating or setting your computer to show the height and optimum airspeed required for the final glide, remember that it is the present rate of climb being achieved at that moment which is important here – not the average rate of climb for the whole thermal. This means that should the rate of climb increase or decrease before you reach the height calculated, you should re-calculate the final glide, either climbing higher or leaving the thermal earlier and beginning the glide to your goal.

When you reach the optimum height and set off on the final glide, you should still fly as directed by the speed-to-fly ring or your speed director.

Think carefully about the best track home. In most of life, the shortest distance between two points is a straight line. In gliding, the second of these points may end up being a field short of your goal, unless you vary your track to stay under any cumulus. When cloud streets are present, running along under one of these until you are closer to the goal airfield may significantly increase your average speed or even your chances of getting home. Avoid flying for any length of time between clouds, especially between cloud streets. If there is any significant crosswind, then you will have to make allowance for this by flying a heading more into wind than one which points the glider straight at your goal. It may even benefit you to track under clouds that are on the upwind side of the track required, so as to have more of a tail wind for the last part of the final glide. Use this tactic only if the clouds to that side of track look promising or as good as those on or downwind of track, and there are no good-looking clouds on the direct track home.

In blue conditions it is easy to find oneself in a 'sink street' which can quickly rob you of any excess height. Watch out for prolonged periods of sink, and alter track (say by 30 degrees) if such persistent sink is encountered. Once in better air, resume your heading to the airfield. With luck, you will fly into a lift street and increase your average speed or at least improve your chances of making it home.

During the final glide, select prominent features on or close to track from the map, and measure their distance from your goal. As you approach each feature, calculate the height required from that point. On passing overhead or abeam the feature, check that the height remaining is close to the height required. If a GPS is being used, such distance checks can be continuous even if there are few prominent ground features. Even if you have a GPS, it is still a wise caution to confirm your track and distance regularly using a map.

Should the height be less than you expected at a checkpoint, then it may be necessary to adjust your MacCready setting or re-assess the strength of the wind component.

Suppose, for instance, that you pass overhead a checkpoint and discover that you have lost 300 feet from the required height. There has either been more sinking air on the glide to this point or the head wind component is stronger. If the reason is assumed to be sinking air, then simply re-calculate the new airspeed required and reset the corresponding MacCready value. If you believe that there is a stronger head wind, then re-calculate the final glide from this new height using a greater head wind component.

On the other hand, if you pass a checkpoint and are higher than you expected, you can either assume that you have flown through favourable air, or that the head wind component is less than you expected. Again, re-calculate your final glide, this time either at a higher MacCready setting (and subsequently higher airspeed) or for the new wind value.

This constant assessment is continued all the way home. Drawing concentric rings on your map, centred on the goal airfield at 5 nautical mile intervals, assists in speedy distance estimates. For the 5 nautical miles closest to the goal airfield, these rings should be drawn every one or two nautical miles. Obviously, this needs to be done before the flight – and as the goal airfield will often be the home base, this should be done when a map is first obtained. (This is also a good use for last year's map, which can be cut down to final glide size – perhaps making final glide maps for several airfields.) Even if your glider is equipped with GPS, a map prepared in this way is still a useful back-up to have.

If the final glide is a marginal one, then the last 5 miles can be regarded as a critical time. It is essential to know where you will land if 'it doesn't work out'. Picking fields ahead is probably acceptable down to 1,000 feet if you are experienced, but below this height the fields may have to be picked as you pass abeam them, with the view to turning in to one of them for a landing. This is a less than ideal situation. However, it is better to have options available one or two miles out than to 'travel hopefully' into the airfield fence. The important thing is to make a decision! If the last few miles are devoid of suitable fields or landing areas, then should you be there at a marginal height? After all, Murphy's Law says that the worst sink is bound to be half a mile from the airfield boundary!

If you still have water ballast on board, then start to dump it in good time for the landing. Depending on the amount you are carrying, this may need to be four or five minutes before touchdown. If there is any danger of having to make a hurried field landing, then get rid of your water ballast early. This will make the field landing easier and reduce the risk of damage should the field be rough. When selecting an approach airspeed, always remember that the heavier the glider, the higher its stalling speed.

On some gliders, the dumping of water ballast can lead to temporary airspeed indicator errors as a result of water droplets covering the static vents. With these gliders, this is another good reason for releasing the water ballast well before you start the pull-up at the finish line or you join the landing pattern.

Often, a final glide may involve flying towards an airfield which may not be visible until the last mile or so. This is particularly so if the final glide is marginal, visibility is poor, or you are flying into a setting sun. If the airfield is a grass strip orientated at an angle to the approach track (for instance, approaching an east–west strip from the north), it may be difficult to see, especially if there are trees along the airfield boundary nearest to you. In such cases, trying to visualise how the airfield will appear from the direction from which you are approaching may help save the embarrassment of missing the finish line, or having to manoeuvre at low level for a landing.

Once you are sure that you can make it safely home, any extra height can be converted to airspeed. This last attempt to gain a few extra seconds can be quite important in competition flying. However, care should be taken not to infringe local rules when arriving at an airfield where ordinary club flying is in progress. Even in competitions, organisers take a dim view of what they term 'dangerous flying', and have threatened to penalise pilots for poor airmanship during finishes. Often the pilot who makes a fast, low finish at the end of a final glide, when thermic conditions are not great, is the pilot who has not flown an efficient final glide. Never mind, though, he has probably impressed one person – himself!

Lastly, do not forget that the flight is not over after you cross the finish line or arrive over the airfield boundary. The glider is still airborne and you still have to perform a landing, possibly in a crowded environment.

The height at which you find yourself after the pull-up will depend on the performance of your glider and the airspeed at which you cross the finish line or airfield boundary. From a high-speed final glide, the pull-up may result in enough height to fly most of a circuit pattern. From a marginal final glide, it may just be enough to turn in and land – or even land straight ahead. It is important that you anticipate the height at which you will end up and plan ahead as to what you will do with it. If your height is going to be marginal over the finish line, then it is foolish to convert what little excess height you have to airspeed, only to run out of height and ideas for the landing pattern.

In windy conditions, watch out for the effects of the wind gradient. If the pull-up is in a downwind direction, then energy will be lost during the pull-up. On turning back to land, you will have to descend through the wind gradient, losing even more energy. The result may be a dangerously low airspeed for the round out.

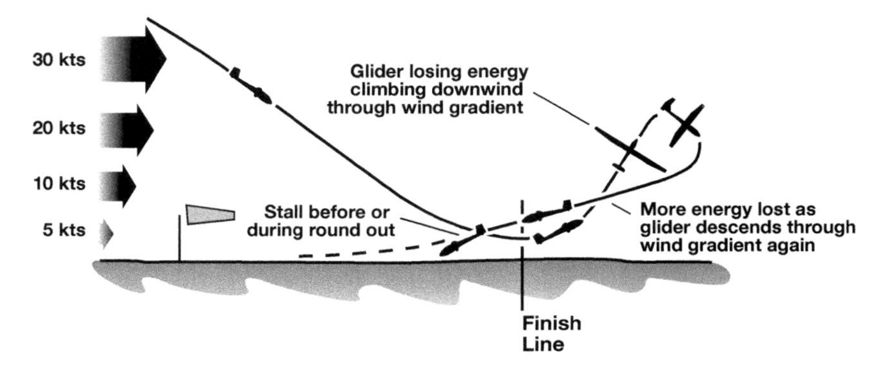

Fig 18.6 Double wind gradient. The glider will lose energy more quickly as it climbs with a tail wind through a wind gradient. More energy will be lost when it turns and descends into wind for a landing.

When there is a marked wind gradient, turning close to the ground can result in difficulty controlling the glider – or even the risk of spinning, as the lower wing will be travelling at a lower airspeed than the upper wing. This effect is increased when the glider has a large wingspan and is banking steeply.

It pays to think well ahead, and by the way – do not forget to lower your wheel!

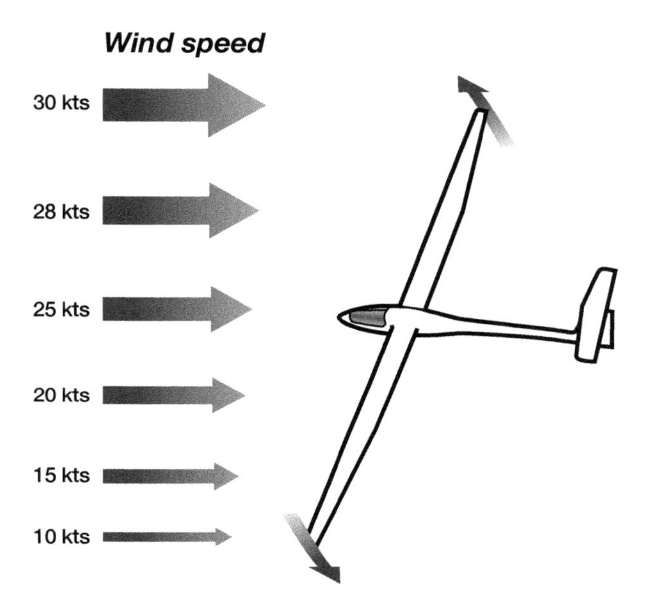

Fig 18.7 Turning in a wind gradient. A wind gradient can make it difficult to control a steeply banked glider.

Factors affecting the safety height margin

In the earlier example, I used an arbitrary figure of 500 feet for the safety height margin. Most pilots would consider this excessive, especially in competitions, as to gain this amount of height in a weak thermal may take several minutes.

As experience is gained in cross-country flying and especially at final gliding, you will be able to estimate the height required as a safety height margin. This margin will depend on various factors: the sky ahead; whether there is any evidence of lee waves; whether large areas of sinking air have been encountered previously; the airspeed at which the glide is being flown; the amount of contamination from dead insects on the wings; and sometimes the terrain over which you need to glide. These are items which no calculator or computer can take into account. Such devices assume a reasonable balance of rising and sinking air *en route*. Any allowances have to come from the pilot.

If the sky ahead looks as if thermals will be plentiful – and experience has shown that, even low down, such thermals will provide reasonable assistance in gaining or maintaining height – then it would be reasonable to reduce the safety height margin. However, if it is late in the day or when flying in blue conditions, then it is wise to start the final glide with a sensible margin of height in hand.

Should the latter part of the flight have shown any sign of lee wave activity, extra height allowance may be necessary in case wave sink is encountered during the final glide.

If the climb rate before leaving the last thermal was high, then the airspeeds used for the final glide will also be high. This extra airspeed may be enough of a safety margin in itself without adding any extra height in the form of a safety height margin, as airspeed can be converted to height at the last minute, if necessary. (This assumes that the glider's airspeed is not dwindling to maintain the required glide angle during most of the final glide.) If the airspeed is being bled off in order to make the distance, then there is a danger that the last few miles may have to be flown without any safety height or any extra airspeed. This is generally unacceptable and creates a risk of a landing short of the airfield, perhaps in a less than safe fashion.

One detrimental influence on glide performance that is easily overlooked or underestimated is the effect of 'bugs' or insect debris on the leading edges of the glider's wings. This is particularly noticeable on modern gliders, where the wing profile is critical, and any contamination will cause drag and disrupt laminar airflow.

For example, a glider with a nominal best glide ratio of 43 to 1 could easily have its performance reduced to 35 to 1 or less if the wings have accrued a large quantity of bugs on their leading edge. When this is the case, either the glider performance used for the final glide calculations must be down-rated, or a larger safety height margin must be used. Should the allowance made be that of a larger safety height margin, then this must be expected to decrease to a more normal figure as the goal airfield is approached. The secret of this technique is recognising what is an acceptable or an excessive rate of height loss as the final glide progresses.

If the terrain over which the final glide has to be flown, especially the last few miles, is unlandable, then this is a great incentive for not risking a marginal final glide. In these conditions, it is wise to allow a generous safety height margin.

Constant versus increasing MacCready setting

Theoretically, starting a final glide and flying it at a constant MacCready setting is the efficient way to fly, providing conditions do not change *en route*. However, in weak conditions, setting off at a lower MacCready setting than calculated and letting the glider get above the glide slope may be a safer option, and if this allows you to reduce your safety height margin, then this 'curved' final glide may even save some time. Moderate or strong thermic conditions or tail winds do not lend themselves to this technique.

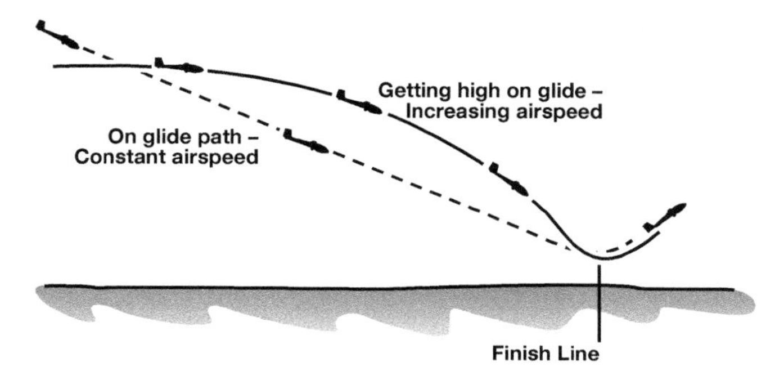

Fig 18.8 Increasing airspeed on final glide. In some conditions, it may be beneficial to start the final glide lower and at a slower airspeed than calculated and intercept the optimum glide path closer to your goal.

On final glides over short distances, be careful not to fly at too low a MacCready setting for too long – or else you may not be able to convert excess height into airspeed without the danger of exceeding Vne.

Badge and goal distance flights

Much of the above deals with flights where achieving a high average cross-country speed is important. If the task is a distance qualifying flight for the award of a badge or distance record, as opposed to a speed task, then getting home may be more important than achieving a high average speed over the final part of the flight. In fact, on such flights, the importance of achieving a high average speed may only be to reach a position from which the final glide can be commenced before thermal activity ends. In such circumstances, any pilot can be excused for climbing as high as possible before setting off on the final glide.

Flying an efficient final glide can be very satisfying. Not only can it mean getting home, but if a high average speed is important, as in competitions, it can save you time.

The final glide can be exhilarating, nerve-wracking or rewarding. Well flown, it is probably the most satisfying part of any task, as it is always accompanied by some sense of achievement.

Chapter 19

Landing Out

When you first start flying cross-country, it may well be your intention to land away from base in order to achieve the distance requirement to qualify for the FAI Silver Badge. Even for this early step in cross-country gliding, the rules allow closed circuit flights using turning points. Despite this, if you continue cross-country flying, and certainly if you intend flying in competitions, you will almost certainly end up landing out sooner or later.

No pilot enjoys landing away from base when the intention was to get home. If the landing has to be made in a farmer's field, not only will there be the inconvenience of the retrieve, but also a certain amount of risk to the glider. This is inevitable, as any field may have an unknown surface, with possible concealed obstacles or ditches. It is essential to reduce these risks as far as possible. This chapter, by looking at techniques, as well as field types and their hazards, aims to show you how to do this.

Landing on airfields and airstrips

It would be ideal, you may think, if every outlanding were on an airfield or airstrip. Surprisingly, this is not always the case.

If the airfield is large and active, you may be contravening some regulations by landing there. If you are able to make contact with the air traffic control unit at the airfield and permission is granted, then landing on a runway will normally prove the safest option. (The controller may offer you an alternative grass area but you will need to inspect this before committing yourself to land on it.) However, some airfields will not welcome gliders and it would be wise to have an alternative field in mind should the controller deny the airfield's facilities to you.

Disused airfields often look inviting but can offer different problems. The surface of old runways is often broken or has sections that are stepped because of subsidence. Fences may cross runways where the land is now owned by more than one landowner. Power wires or cables may be strung across the runways or the approach, and may be difficult to see. Weeds and scrub bushes may stem from the cracked surface or be perilously close to the runway edges. All of these hazards will require a long, critical inspection of the proposed landing area before committing yourself to a landing.

Private airstrips are often a good place to land, but with caution. The width of an airstrip can be hard to judge from the air and one that is comfortably wide enough for a small powered aircraft or a microlight may not be wide enough for a glider. Once safely on the ground, move the glider off the landing area. By doing so, you will not be blocking the airstrip should the owner's aeroplane return.

If you are lucky, your landing may be among fellow enthusiasts at another gliding site, but all too often it will be in a farmer's field.

When to select a field

When flying cross-country, the chances are that the fields below you are at a different height above sea-level from the airfield from which you launched. To add to this problem your altimeter sub-scale may be set to sea-level (QNH), or the pressure setting of your home or destination airfield (QFE). (The choice of pressure setting will vary upon the phase of flight, or often simply upon the pilot's preference.)

This means that, at lower levels, any attempt to judge heights above the ground will be estimations using your eyes rather than the altimeter. With this in mind, the heights that follow are only approximations of the glider's height above the ground. Like your in-flight estimates, these should become more accurate as the glider gets lower. The heights given apply in typical 'flatland' European soaring conditions, but may need to be increased in areas where suitable fields are few.

Above 2,000 feet above ground level, there will normally be no need to be thinking about fields. You should be concentrating on staying airborne and making progress cross-country. With 2,000 feet, a medium-performance glider will travel around eight miles in still air, so unless you are determined to land, press on and find another thermal. The only proviso to this course of action is to make sure that you are not using your height to fly over unlandable terrain. So, although you do not need consciously to be studying fields, do keep an eye on the type of terrain you are flying over and are approaching.

At around 2,000 feet, start keeping an eye on fields ahead and those you are passing, but do not become obsessed with field watching at the expense of soaring.

By 1,500 feet, your chances of finding a thermal are reducing. If no lift is found, you have about eight minutes' flying time left, some of which will be required for flying a circuit pattern to land. Therefore, you must start selecting a field. Fly the glider at a sensible airspeed so as not to 'dive off' height unnecessarily. Do not rush. You still have plenty of time to select the best field available rather than the first one that catches your eye. The visibility downwards from most gliders is limited to the area on both sides of the cockpit in front of the wings. A 360 degree turn will bring more fields into view. If you cannot see any fields that you fancy in the immediate area, then fly downwind. In this way, you will cover more ground and therefore be able to inspect more fields. This assumes, of course, that the terrain downwind looks acceptable and will offer reasonable fields.

In windy conditions, be careful not to overestimate how far your glider will go into wind to reach a chosen field. To do so may result, at best, in failing to reach the field with enough time to inspect it properly. At worst, you may not reach the field at all.

Should you encounter lift during this search, you should use it. If there is any significant wind, take care not to drift out of range of any fields you have earmarked, unless you have other fields in mind which are within reach as you drift downwind. Switching off the radio will also aid your concentration, so that you are not distracted by irrelevant chatter. *Do not be tempted into trying to call your crew; workload management is very important from this point onward.*

By 1,000 feet, you should have selected the field in which you intend landing. Indecision below this height can be disastrous. Stick to your decision and do your best.

Between 1,000 feet and the height at which you will begin flying the circuit pattern to land in the field, you should be positioning yourself for the circuit pattern and checking out the field's surface, its approaches, and whether there are any obstacles which might dictate which part of the field should be used for the landing. Do not allow yourself to get too close to or overhead your field, as this will reduce your view of the field and limit your inspection of it. Doing so may also cause difficulty in positioning the glider when you begin to fly the circuit pattern.

Using lift at this stage is still acceptable, but only if it does not place you in an awkward position for flying the circuit pattern or even out of reach of the field. If you do start thermalling at this height, take care not to become disorientated or to lose the selected field. This is best avoided by noting the field's position relative to some prominent landmark such as a group of farm buildings, a lake or wooded area.

At 800 to 700 feet, begin flying a circuit pattern to land in the field. Should you hit a thermal, do not be tempted away from flying the circuit pattern, unless you are very experienced. Even if you are an experienced pilot, decide at which point you will ignore any lift and commit yourself to the landing. Make this decision *before* you begin flying the circuit pattern. The more difficult the field looks for a landing, then the higher this point should be. The circuit pattern you fly should be as similar as possible to that which you would fly at your home airfield.

The heights given will vary with experience and practice. Very experienced pilots may select fields 'on the move' and will be able to assess fields more quickly than those inexperienced at field landings. The availability of fields will depend on the state of the crops or the terrain. This in turn will determine the caution necessary when deciding the height at which searching for a field should begin. Whatever heights are chosen for the above, adequate time must be left for assessment of the chosen field and to fly some semblance of a circuit pattern.

Selecting a field

The old adage of '**S**ize, **S**lope, **S**urface, **S**tock' still holds good when it comes to selecting a field. However, before you get down to these details, you will probably have rejected a large number of fields, almost subconsciously. Why?

Some areas may be hillier than others, consequently having smaller and more sloping fields. The predominant colours of others may be indicative of standing crop. Others may be too close to a power station or substation and are therefore likely to be strewn with power cables.

After you have done this initial rejection, you can concentrate on the advantages and disadvantages of the fields with which you are left. Even this will not take long, and soon your mind will be set on possibly as few as two or three fields. In some instances, you may not have much problem choosing a field; if an area has a large amount of crop, there may only be a couple of fields even remotely suitable. (A similar problem will occur on hazy days, when the poor visibility may limit the number of fields in view. Not only will this give you fewer fields from which to chose, it will also make it difficult to compare the surfaces, partly because of the limited number of fields in view for comparison, and partly because the lack of direct sunlight striking the fields may eliminate shadows, making it more difficult to judge the type of surface.)

Now you can start to look at the remaining fields in detail, comparing them and looking at their relative merits.

Size

Obviously, a field has to be large enough to allow the glider to round out, hold off, and have a ground run without ending up in a fence. Size in this context mainly means length in the proposed direction of landing, although the field must be comfortably wide relative to the glider's wingspan.

A field length of around 250 metres would be the minimum for a landing in a modern sailplane, assuming no wind. However, experienced pilots could probably land in shorter fields – although one hopes that their experience would prevent them from having to do so. Any obstructions on the downwind boundary of the field will effectively reduce the length of the field and require it to be longer. (For instance, approaching over a line of trees which are 15 feet (approximately 5 metres) tall will prevent the glider landing in the first 150 feet (approximately 50 metres) of the field. Thus, a distance of around ten times the height of the obstruction to be overflown will be used up before the glider is at ground level.

When considering the length of a field, any wind should be taken into account. As the wind affects groundspeed, a landing into wind will reduce the glider's groundspeed and therefore the length of field required (effectively making the field longer). Landing downwind will increase the glider's groundspeed and require more distance before the glider comes to rest. Therefore, if an approach is to be made with a tail wind, a longer field will be required. Providing other factors do not prevent it (such as slope), you should attempt, if possible, to make any approach into a field into wind.

In light winds, the wind direction can normally be assessed by the drift of smoke from local chimneys or fires; or if the wind is stronger, from the drift being experienced while the glider is circling. Crops rippling in the wind will give some indication of the wind strength and direction.

In the absence of any ground indicators, cloud shadows moving across the ground will show the direction of the wind at cloud base. However, the strength and direction of the upper wind will not necessarily be the same as the surface wind. The upper wind strength will often be greater than that of the surface wind. Nevertheless, its direction will usually be within 30 degrees of the wind at ground level and can be taken as the same for the purposes of assessing a field's suitability.

Any lee wave effects, sea breezes or cumulonimbus clouds in the area may affect the wind's strength and direction. Such effects may often result in a reversal of the local wind direction from the general flow. Such effects are hard to predict and, in the absence of any visual wind indication, the only safeguard is to choose a field long enough to offer a safety margin. In any event, it is wise to land well away from any large storms.

Occasionally you may hear an instructor or a pundit saying that a good field is one with good fields before and after it, in case an undershoot or an overshoot is necessary. While it is ideal if the local terrain does not adversely affect the approach, if a pilot cannot get a glider into a reasonable-sized field without undershooting or overshooting its boundaries, then that pilot should not be flying cross-country. Such considerations are thus superfluous.

Slope

Slope in a field can present considerable problems for the pilot of a landing glider. Any slope of a field's surface will greatly determine the direction of landing. Even a gentle slope will necessitate a landing up the slope. Any attempt to land downhill will result in difficulty in putting the glider on the ground. Even if this is achieved, there is a great danger that the glider will continue rolling (perhaps despite attempts to stop it, using the wheel brake) into the boundary fence or hedge. Landing across a slope is normally difficult to achieve and if the glider inadvertently turns on the ground run, it may roll down the slope.

Therefore, landings should always be made up any slope. On some occasions, this may mean that a landing will have to be made with a tail wind. If the wind is light, then this in itself should present no problems as the uphill gradient will soon stop the glider and reduce any risk of overrunning the field's boundary.

Often the problem is one of identifying that a field has a slope. To do this, it is necessary to inspect the field from at least two sides to give different perspectives. Here we have another good reason for not flying overhead the field when inspecting it, as looking down on a field from directly overhead will not assist in identifying any slope.

The neighbouring terrain will usually give a clue to the likely presence of any slope. If the field is near a line of hills then the chances are that it may slope downwards away from the hills, even although compared to the hills it looks perfectly flat. The slope of a field situated in a hilly area can often be very difficult to judge, as the neighbouring slopes deny the pilot a horizontal surface with which to compare the field's surface. Often, the lie of a hedge or a field's boundary will give a clue to the presence of a slope.

Fields next to streams, rivers or lakes will tend to slope towards the water. However, where a large river meanders through a wide flood plain, the fields within the curves of the river are often flat. This is not the case if the river is meandering through some form of gorge or ravine.

As a general rule, any slope visible from around 1,500 feet will be a reasonably steep slope, and in most cases, too steep for a landing.

Having established that any landing in a sloping field should be made uphill and never downhill, one must be cautious when making such a landing. Several hazards await the unwary pilot when landing up a slope.

The airspeed used on the approach will have to be increased above the normal approach speed to allow for the round out, which will be through a larger number of degrees than when landing on a level field. If the airspeed selected is too low, there is a danger that the glider will stall before the round out is completed, resulting in at best a heavy landing.

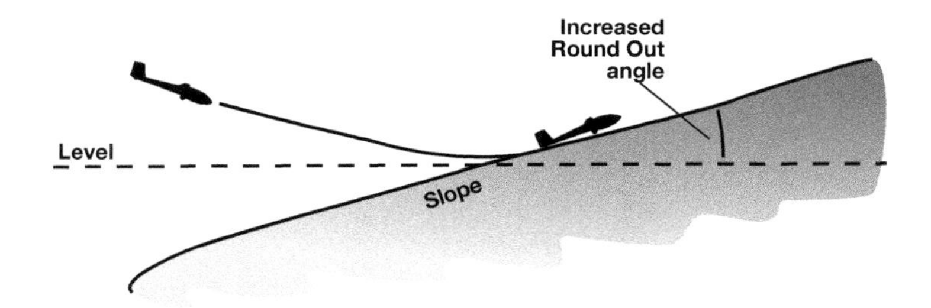

Fig 19.1 Rounding out uphill. An uphill landing will require a higher airspeed on the approach to allow for the greater angle of round out.

This extra airspeed, possibly as much as 15 knots, even in a modern glass fibre glider, must be achieved at the beginning of the approach before any airbrakes are used. The airspeed indicator must be monitored more rigorously than on a normal approach, and if necessary the airspeed must be adjusted immediately. This is because the visual clues which the brain is receiving when approaching an uphill landing area will give the impression that the approach angle is too steep. The subconscious tendency will be to lower the glider onto the more familiar shallower approach angle using the airbrakes. If this is done, the glider may be flying a flatter approach path than is possible. If this is the case, the airspeed will decay, resulting in the glider not having enough airspeed for the round out and subsequently landing heavily. If the slope is moderately steep, then the glider may even stall before it reaches the field.

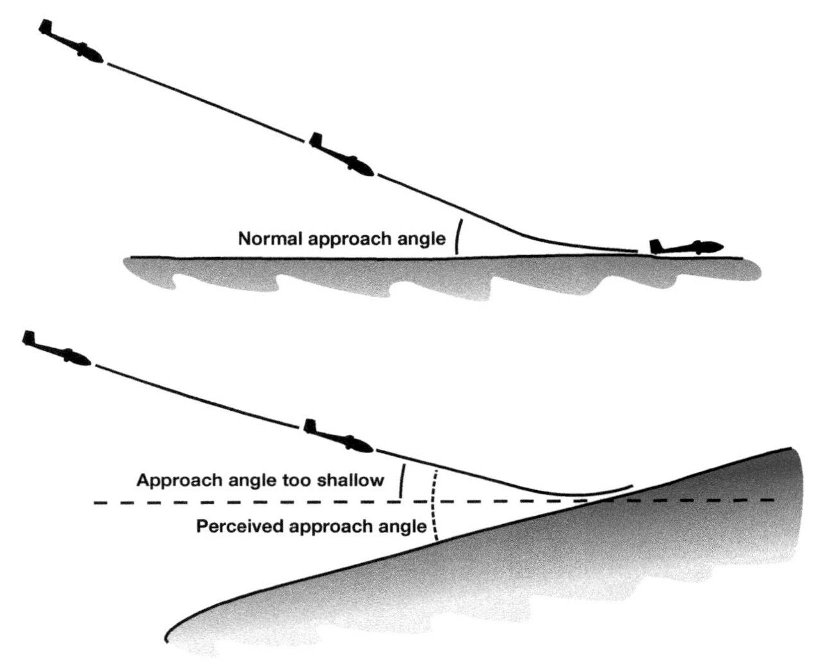

Fig 19.2 Misleading visual clues when landing uphill. On an approach to land uphill, there will be a tendency to flatten the approach angle, leading to insufficient airspeed.

As if these hazards are not enough for one day, if there is any significant head wind on the approach, there will also be the problems posed by descending through a possible wind gradient – and possibly even curl over and sink caused by the hill which is being approached. One technique that you may find useful when faced with a difficult field and a greater need than normal to monitor airspeed, is to read aloud the indicated airspeed. This may aid your concentration and ensure you take the airspeed indicator into your scan every two or three seconds. Obviously there is no point in monitoring the airspeed this closely if you do not react and correct it immediately if it varies from the selected airspeed.

Surface

It is essential that, as far as possible, the surface of the field you choose is suitable for a landing. Otherwise, either the glider may be damaged or, if the field contains crop, expensive damage may be caused to the crop. There follows a summary of field types and their characteristics, which should help you understand the problems posed by some types of field surfaces.

Pasture fields

Pasture fields are usually acceptable for a landing glider, subject to there being no livestock in them. If the area is a crop-growing one, then the question has to be asked, 'Why is the farmer not growing crop in that field?' It may be simply that the field is too small to make it worth the effort, in which case it may be too small for your needs also. It may be that there is too much slope or the surface is too rough or badly drained to risk taking combine harvesters or other machinery into it. Although a pasture field that is devoid of livestock may seem ideal, most require careful consideration as to whether or not there is a better alternative within reach.

In the UK, many pasture fields display historic undulations known as RIDGE AND FURROW. These parallel ridges, which are often a metre or more from trough to crest, haunt fields, particularly in England. Sometimes it can be hard to identify this hazard, depending on sun angle and visibility. Avoid these fields, as the ridges can be too narrow to land along safely. Landing across these ridges will wreck a glider and possibly its pilot.

'Set-aside' fields

'Set-aside' is a European system where farmers are encouraged (by government policy) to leave land uncultivated for a year or more. Set-aside fields can be treated in a similar way to pasture fields, except that they have at least been cultivated in the recent past.

Set-aside is still a bit of an unknown quantity. Some of these fields have proved to be safe landing places, while others have resulted in undercarriage damage due to the roughness of the surface. The surface of some set-aside fields is unpredictable, and very tall weeds and thistles may create the risk of a ground loop. If the weather during the last cropping season was wet, then such fields may be badly rutted.

Many farmers seem to be spraying these fields with some sort of defoliant or weed killer, making them more acceptable as a landing place.

Those that have been chemically treated appear as golden, untended stubble from the air. (So far, none of my toes, feet or anything else has dropped off from walking through these treated fields – fingers crossed!) I would advise that you think twice before landing in this type of field.

(Whether set-aside fields continue to be a feature of the European landscape will depend on the agricultural policies of the European Union. The practice may well cease as quickly as it appeared.)

Brown fields

Brown fields usually fall into three categories: ploughed fields, harrowed fields and fields which have crop so thin that the soil shows through when viewed from the air.

Ploughed fields are usually identified easily from the air and are best avoided. The furrows are easily visible from a height of 1,500 feet. They are often quite deep and will cause damage to the surface of a landing glider. Undercarriage doors will also be at risk of being torn off. If landing in a ploughed field is unavoidable (and in some countries this may be the only safe option) then land along the furrows – **NEVER** across them.

Harrowed fields make somewhat safer landing fields than ploughed fields. If the soil is very soft, then there is a danger that the glider's main wheel will sink into it, damaging the underside of the fuselage and undercarriage doors. In this situation, there is also a risk of the glider decelerating suddenly and the potential for a ground loop if the tail tries to overtake the rest of the glider.

Brown fields with thin crop will contain either young cereal crops such as wheat or barley, or small broad-leafed crops such as peas, beans or potatoes.

Landing in young cereal crop will be similar to landing in a harrowed field, except that the crop will tend to bind the soil together and the chance of the glider sinking in will be reduced. At this age the crop will not sustain any lasting damage and the farmer will not normally show much concern.

Broad-leafed crops usually appear darker and sparser than cereal crops when seen from the air. Avoid landing in them if they seem particularly lush, as they are probably reasonably mature and you may cause damage to the crop. Potato fields should be avoided. Potatoes grow in furrows, which are quite deep, presenting the same dangers as ploughed fields.

The chances of the glider's wheel digging in when landing in a brown field will vary with the type of soil and, more so, the wetness of the field.

If the soil is baked dry then the risk might be small. On the other hand, with a soaked or waterlogged surface, even soil bound by crop will be soft enough to bog down any glider. In the event that you do land in a wet, brown field, you had best request lots of crewmembers as it is almost certain that you will have to carry the de-rigged glider to the trailer.

NOTE: *If a severe swing or a ground loop has occurred on landing, the glider must be examined by a qualified inspector, as whiplash of the tail areas can cause serious damage that only an experienced eye will spot.*

In some countries (for instance, South Africa) cattle-grazing land may be no more than open, uncultivated country. In such cases, your only chance of choosing a reasonably smooth area in which to land, which also avoids tall crops, may be to choose a harrowed or cultivated field which has a surface that appears brown from the air.

Cereal crops

Cereal crops, such as wheat, barley or oats, present no problems to a landing glider when they are young. At this stage, the stalks of the crop are very short and the considerations of landing in a brown field with young crop apply.

It is as these crops grow that the problems arise, and these problems increase as the height and density of the crop increases. As the growing season continues, the colour of many cereal crops will change from green to a golden brown. Once the crop has grown above a height of about 24 inches (60 centimetres), any glider landing in these crops risks being substantially damaged or possibly written-off!

The problem lies in the density as well as the height of the crop. On landing, the stalks of the crop will drag against the glider's wings and tail areas. The compressed crop will form a strong barrier. In fact, it will usually be stronger than the tailplane fittings of most gliders. If the glider has a low tailplane, then this, or large parts of it, might be ripped away from the glider. At any rate, the tailplane fittings will risk damage. The rear fuselage may also suffer from this attack, possibly breaking just forward of the tailplane.

A 'T' tail may fare somewhat better than a cruciform tail in such a landing, as the tailplane will be above all but the higher crops. However, 'T' tail gliders are not immune when landings are made in tall crops. The other great danger (apart from the farmer) is that a wing will either touch the crop just before touchdown or will decelerate quicker on contact with the crop than the other wing. If either of these events occurs then there is a very real danger that the glider will ground loop. Such a ground loop will be sudden and fierce. The stress placed on the rear fuselage and tail area during such a ground loop is very high, and after such an event, it is fairly common to find that the rear fuselage has been badly damaged or broken off completely. Having the mass of the tailplane at the top of the fin in such situations adds to the whiplash effect, thus increasing the risk of damage.

The only certain way to avoid such damage is to avoid fields with high, standing crop. If a landing has to be made in such a field, then landing on the crop at as low an airspeed as possible with the airbrakes closed might reduce the risk of damage.

When wheat or barley is harvested, the combine harvesters leave rows of straw lying on the ground. This is easily identified from the air. Unlike hay or silage, these rows of straw may lie quite deep (two feet high or more). If you land your glider in such a field then there is a very definite chance of building a 'straw-stack' on one wing, should one wing touch the straw before the other. A ground loop will almost certainly follow.

Within a day or so after harvesting the crop, the farmer will bale up the straw and the field will be cleared, leaving only short stubble (unless the bales are left lying in the field). Such stubble fields are ideal landing areas. Even before all of the straw is baled, there may still be cleared areas easily large enough for a safe landing. Watch out for stray bales which have been missed. The work in progress in the field and the straw rows make it easy to spot these fields.

Once a field has been completely cleared, it can sometimes be difficult to determine whether you are looking at a stubble field or a field of golden crop, especially in hazy conditions. The two can be distinguished by the following points:

* Crops will ripple if there is any significant wind.
* Flocks of birds 'sitting' on fields will normally be an indication of stubble, not crop.
* Random tractor marks all over the field will be evidence that straw bales have been lifted and therefore the field has a stubble surface.

Oil-seed rape

From the air, in its early stages of growth, oil-seed rape is difficult to distinguish from wheat or barley. Even at this stage, it is somewhat denser than these cereal crops. As it grows, it very quickly becomes tall and tangled, presenting a serious hazard to any landing glider.

When it starts to flower, it is unmistakable with a profusion of vivid yellow flowers. When it has reached this stage the crop will be so tall and dense that even walking through it would be difficult.

Some weeks later, as the flowers are shed, the rape will return to a greenish colour, and the only difference visible from the air between it and other crops may be its duller colour and its very tangled and chaotic appearance.

Unlike other crops, which tend to be collected when, or soon after, they are cut, oil-seed rape is cut and left to dry on top of its stalks. From the air it might be obvious that you are looking at a cut crop, but this crop still presents hazards. The crop is cut high up on the stalks and these stalks are both strong and fairly tall. The small oil-seeds are held in pods which, as they dry, become brittle, making it possible to harvest them. Any disturbance to the drying crop will shake the seeds out of these pods and that part of the crop will be lost. So if you land in rape after it has been cut, you will not only risk severe damage to your glider but also the wrath of the farmer who will not be enamoured by the idea of your crew stomping through the field carrying bits of glider.

Oil-seed rape is a high-value crop. Avoid landing in such fields if at all possible.

Tractor wheelings

All of the crop fields mentioned above hide another hazard for a landing glider. When the fields are sown, the tractors doing the work will make wheel tracks (known as 'wheelings') in the field. These wheelings will be used each time the crop has to be treated with fertiliser or insecticide. With each subsequent use, these ruts risk getting deeper and more sharp-edged, especially if the earth is soft or wet. As the field dries out in summer, the wheelings will bake hard. Wheelings will be apparent throughout the life of the crop and, to a lesser extent, after it has been harvested.

Should a glider's wheel run into one of these ruts, there is every chance that damage will be done to the undercarriage and possibly elsewhere as the structure of the glider flexes.

When landing in a field that displays wheelings, always land parallel to the direction of the wheelings – **NEVER** across them.

Often the presence of wheelings makes identification of a deep crop easier.

Silage and hay

Silage and hay, when it is growing, is difficult to differentiate from cereal crops that are still green. In this state, these grass crops offer the same problems as cereal crops such as wheat and barley. The difference becomes obvious once farmers begin cutting hay or silage fields – and in doing so, provide us with some excellent landing fields.

These cut fields are usually safe for landings, whether the silage or hay is lying or has been collected. They are easily identifiable by the strips of cut grass lying awaiting collection or, if the hay or silage has already been collected, by the very pale colour of the grass. Often there will be groups of such fields. All you have to do is select the most suitable.

The surface of cut silage or hay fields is usually smooth and firm. If rows of cut crop are lying on the ground, try to land parallel to these rows, avoiding the cut crop if possible. Landing across these rows will probably not cause a problem, but the glider's deceleration may be rapid and under-carriage doors may be ripped off.

After the crop is lifted, these fields will often have bales or rolls lying in them, or stacked at the edge. This helps identification. Provided there is plenty of space between these bales or elsewhere in the field, these are good landing fields – but watch out for the occasional bale which has not been lifted or has fallen off the cart.

Maize

Maize (often known as corn on the cob, or mealies, depending in which part of the world you are) grows to heights of over 6 or 7 feet (over 1.8 metres) and has very thick stalks. In its very early stages of growth, it is not tall or dense enough to damage a glider. At this stage, the soil can be seen through the crop when viewed from the air. As the crop matures, it soon becomes a hazard to landing gliders. Avoid landing in this crop when it has started to grow above knee height or when more leaf than soil can be seen from the air.

Supported crops

Crops such as strawberries and raspberries are often grown in rows with 'stick' supports. Landing in such crops will probably cause damage to a glider and should therefore be avoided.

When selecting a field, try to visualise what the farmer is doing with the field (e.g. why he has left it fallow, has ploughed it in that direction or not harvested it). This might give you a clue to the surface. A basic knowledge of the farming calendar will help you understand the state of the crops and which ones will be a danger to a landing glider at a particular time of year. In certain conditions, removing your sunglasses may allow a better assessment of a field's surface, especially if your sunglasses are of a type that distorts colours.

If you are flying in certain hot countries (such as some parts of Australia), crops may not be a problem. Because of the high temperatures during summer, crops are harvested before they are scorched and thus before the best part of the soaring season really gets under way.

Other possible landing areas

Sports fields

Sports fields can frequently offer large enough areas in which to land, providing several football pitches or similar are adjacent. These areas should be avoided if there are any players or other persons on them. Apart from the more obvious goal posts, sports fields often have poles to mark pitch boundaries, and the grounds are often tiered into different levels. These features and the risks to third parties make such areas less than ideal for landings. Over the years, many gliders have landed on golf courses. While these provide many large areas where it is possible to land, you cannot expect a warm welcome if you have interrupted someone else's sporting passion by adding to the difficulty of the course by a factor of one glider.

Heath

Landing in areas of open, uncultivated country cannot be recommended. The surface of heath or moor is almost certain to be rough and undulating, perhaps with ditches or streams. If covered with heather or bracken, this in itself will pose a hazard to the glider, endangering the underside of wings, fuselage and low tailplanes. Such vegetation may also conceal sizeable boulders and rock outcrops.

Sandy areas

Barren, sandy areas, as in semi-desert or desert, will present a totally unknown surface to the pilot. Unfortunately, if you find yourself flying in a country where such surfaces cover large areas, you may have little choice but to land where you can. Using common sense is the only hope, avoiding any obvious obstructions or rough ground.

Beaches

Occasionally, when distance flying or cliff soaring, you may find it necessary to land on a coastal beach. Land well away from people or animals. Watch out for obstacles and rifts in the surface. The firmest surface will be near (but not in) the water. The glider will probably sink into the drier, softer sand further away from the water's edge. Watch out for the tide coming in, or you might have a seaplane on your return from telephoning the crew! (Enlisting volunteers to move the glider up the beach away from the sea would be a wise, early priority.)

Roads

In some countries, landing on roads may be a safe alternative, but this seems to be a practice only used in wild, deserted areas, where the roads are straight and traffic occasional.

Stock

Avoid fields containing any livestock. Horses stampede and risk injuring themselves. Cows trample gliders' wings, and destroy fabric and glass fibre finish by licking and rubbing against it! Sheep simply get in the way and have been known to leap up in fright into the tailplanes of gliders overflying them.

If you have to land in a field with cows, do not leave the glider. They are curious beasts and will be around it by the time you return. If possible, try to enlist a helper to protect the glider, as most cattle seem highly trained in flanking manoeuvres. Horses are less of a problem in this sense but their owner will not be amused.

Given no choice but to land in a field containing animals, a landing with sheep is preferable to a landing with horses or cattle.

Obstacles

The one type of hazard not covered in our mnemonic is 'Obstacles' (or if you prefer, 'Obstructions'). They come in a variety of forms, some more visible and therefore easier to avoid than others.

Power wires or cables

Large power cables, such as those used for national electricity grids, tend to be more easily spotted than smaller cables. The main problem with these is that they are supported by pylons, which may be many fields apart. They may also branch off to other pylons. It is essential, therefore, to scan the whole area around your selected field and not just the field itself. Discovering such cables on a base leg or long final approach is not conducive to a perfect field landing.

Smaller power cables and telephone cables supply farms and houses and are supported on poles. These cables may cross fields or follow the line of roads. They can also have branch lines, which create a veritable web above some fields. If there are such poles in a field, often any tractor wheelings present will be seen to make a short detour before continuing their normal parallel pattern. This is where the tractor has had to go around the obstruction. Such a sign can show where an otherwise hidden pole may be and reveal the possibility of cables across the field.

The poles supporting a cable that follows a road can be difficult to see if some of the poles are concealed in hedges or by trees. Watch out for these cables if you are approaching a field through, or just above a gap in trees – especially if the field is next to a road or track.

In Australia, power cables, locally called 'SWERs' (Single Wire – Earth Return), are strung from trees as well as poles. These are difficult to see. The only way to reduce the risk of colliding with one of these (or any other cable) is to assume that every building is fed by one and try to deduce its position.

Fences

Fences come in two varieties, either constructed of wooden, metal or concrete posts with barbed or plain wire stretched between them, or electric fences. The former are sturdier, more permanent features and as such are more visible, but also more dangerous. A glider running into a post and wire fence will be damaged by the posts and the wire may cut through the glass fibre or canopy, injuring the pilot.

Electric fences are hard to see from the air. Fortunately, being of much lighter construction, they will normally break or the stays will collapse, if a glider runs into one. However, if the electric fence wire is anchored at its ends to a traditional post fence and is strong enough, this type of fence can cause damage to a glider.

Sometimes both of these types of fence are difficult to see from the air. As they are normally used where cattle are or have been grazing, the easiest way to spot their presence is by looking for differing shades of green in the field's surface, caused by uneven grazing. If any adjacent grass areas are of different colour, assume that a fence exists where the colour change occurs and avoid landing across that part of the field.

Another indicator of a possible fence line is a drinking trough or troughs in the middle of a field. They have probably been placed in that

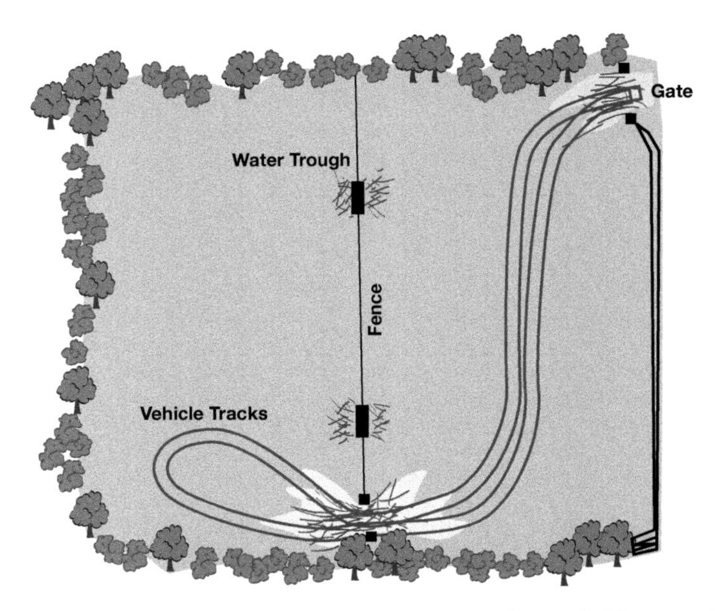

Fig 19.3 Water troughs and tracks. A trough in the middle of a field may indicate an unseen fence across the field. Animal or vehicle tracks converging on or changing direction at one point may indicate a gateway or gap in an unseen fence.

position so that animals can use them from either side of a fence. The field is probably not one, but two fields.

If multiple tracks made by animals or tractors cross a field towards a specific point and then change direction towards a gate, the point at which the track changes direction is almost certainly where the track goes through a gate or a gap in a fence. Again, if suspicious, assume the field is divided by a fence and plan your landing accordingly.

Vehicle tracks

Any vehicle tracks or paths running across a field should be assumed to have edges which are rutted or stepped. Plan your landing run to avoid running across such tracks, otherwise you may damage or collapse the glider's undercarriage.

Vehicles

Tractors, combine harvesters, carts and other vehicles, whether working the field or 'abandoned', all add to the problems by cutting down the useful field area, and some may even inadvertently move across your approach or landing run. Give them a wide berth if possible.

Careful selection and assessment of a field is the only way to reduce the chance of damage to your glider. Most field landings are carried out without incident because pilots take time to inspect their chosen fields. When doing so, do not let convenience, such as a nearby farmhouse or a road, influence you. Neither should another glider already in the field affect your decision. Do your own field selection and assessment.

When selecting a field, your priorities in order of importance must be:

1. Your safety and that of others.
2. The safety of the glider.
3. Minimum damage to crop or stock.

and least important

4. Convenience.

We met the Vega pilot on the road near the field in which he had landed. He replied to our greeting with the words, 'I am OK but the Kestrel is not'.

The Kestrel 19 had landed in a 'field' which was actually two fields separated by a post and wire fence. It had rolled into the fence at speed and had been badly damaged by the impact. Fortunately, the pilot was unhurt.

The Vega pilot had inspected the 'field' for himself and had not merely been content that another, apparently undamaged glider had already landed in it. He had spotted the fence from the air and chosen his landing point accordingly, making a safe landing.

Getting into the chosen field

By the time you begin cross-country flying, you will have flown many circuit patterns at your home airfield. Your skill at flying such circuit patterns will not be in doubt, but your versatility at positioning a glider, which comes from experience, often means that you may take flying the circuit pattern less seriously than when you were less experienced. It is when faced with landing the glider in a field that all the nagging from instructors about tidying up your circuit flying will be remembered and appreciated.

Having selected a suitable field, there is only one safe way to get into it – in an organised fashion. This means flying some semblance of a circuit pattern.

Ideally, the circuit pattern you fly will not be unlike that which you would fly at your home airfield, and so I am not going to dwell on the basics of circuit planning. What I will do is point out the differences between flying a circuit pattern into a field as opposed to flying one at your home airfield.

The first reference that will be lacking when landing away from base will be an accurate height reference. As mentioned previously, the altimeter will be of little use, as the indicated height will probably bear little relation to the height of the ground over which the glider is flying. Fortunately, most glider pilots are used to 'eye-balling' at least the lower parts of the circuit pattern – and providing that the altimeter readings are not allowed to distract you, then you should be able to manage your height accurately enough to be within the scope of the airbrakes on the final approach. When judging height in this way, always refer to the height of the landing point and not the terrain around you. In this way, should you fly over a hill or valley on the downwind leg or base leg, you will not find yourself manoeuvring unnecessarily to adjust height.

The next problem is one of judging distances. Intentionally or not, at your home airfield you will use local features on the ground to check your distances from the airfield or landing area. You may even start your circuit pattern habitually at the same position each time – perhaps opposite the upwind end of the airfield. Neither of these references will be present when landing out. There will be no familiar features.

The second habit (starting the downwind leg at the upwind end of the airfield) can lead to some problems if translated to a field landing. Assuming that your home airfield is between 3,000 and 4,000 feet long (1,000 to 1,200 metres), you will be lucky to find a suitable farmer's field this long. If the field chosen is much smaller and you start your circuit pattern opposite the upwind end of the field, you will have a very short downwind leg. This in turn will give you insufficient time to get rid of your excess height and the whole affair will be cramped and rushed. The only answer is to begin the circuit pattern well upwind of the chosen field (perhaps three or four field lengths upwind if the field is small). Once you have started flying the downwind leg, your only reference point must be the intended touchdown point.

The importance of flying a circuit pattern when landing in a strange field goes beyond the need simply to position the glider for the final approach. The chances are that you will not have seen this field before

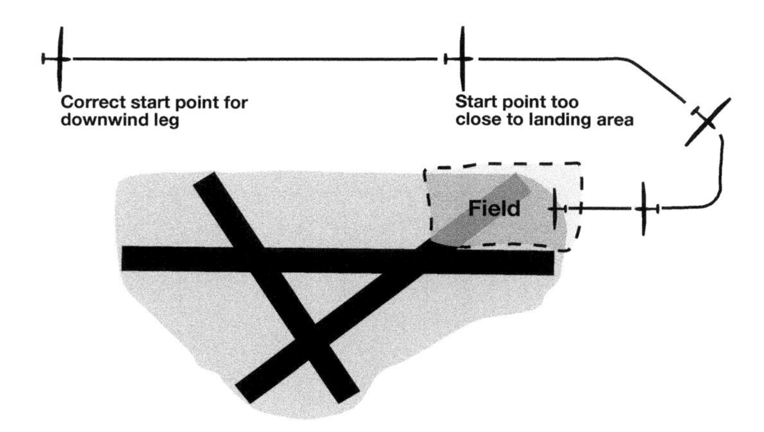

Fig 19.4 Comparative size of an airfield to a farmer's field. Few fields will be as large as your home airfield. This must be borne in mind when planning your circuit pattern for a field landing.

and therefore the surface and the approaches will be unknown to you. Flying a circuit pattern allows you to check out the field, looking for power wires, ditches, rough ground or fences from differing perspectives. Although it may be a bit late to select another field, should some problem be spotted from the lower heights of the circuit, you will still have time to select a different approach line to land in a different part of the field. Flying a circuit pattern will also give you another chance to assess and allow for any slope of the field's surface.

When flying the circuit pattern, continuously scan ahead looking for any cables or power lines, which may be a problem on the crosswind leg or approach. Anticipate the effect of neighbouring terrain. (Will that nearby hill give lift on the downwind leg or curl over and sink on the approach?) Complete the same pre-landing checks that you would normally do when landing at your home airfield. This will ensure that you will not forget anything essential, such as lowering the undercarriage, when the workload is high.

The base leg and final turn should be carried out well back from the boundary of the field. This will give you the maximum opportunity to adjust your height with the airbrakes. A final turn made over or close to the boundary is always disastrous when an accurate short landing is required. (Next time you turn onto the final approach at your own airfield on a windless day, note how far back you are compared with the size of the local farmer's fields.) Do not hesitate to use airbrakes on the base leg or during the final turn if you think that you are too high. However, use them with caution so as not to end up in an irretrievable undershoot situation.

As you straighten the glider on to the final approach, again think 'cables'. Keep an eye out for power cables or telephone wires along the boundary of the field or in the field itself. Often the poles that support these wires will be concealed by trees or hedgerows, leaving only the wire drooping across a gap in the trees. The only safe answer, if the field is large enough, is to cross the field's boundary above the height of the trees, opening the airbrakes to let the glider down immediately it is safely past any possible obstruction.

Should you misjudge your approach and end up low, or should you fly through an area of sink before you reach the field, do not try to 'stretch the glide' by raising the glider's nose. Such action will reduce the airspeed, creating the risk of a stall as well as increasing the time spent in any descending air. Instead, 'dive' the glider at the top of any obstacles on the field's boundary. If you can make good this glide path, you will be able to pull up over the obstacles to reach the field. (This assumes that you have closed the airbrakes if they were being used.) Such a technique will only be possible if you identify the potential undershoot early enough.

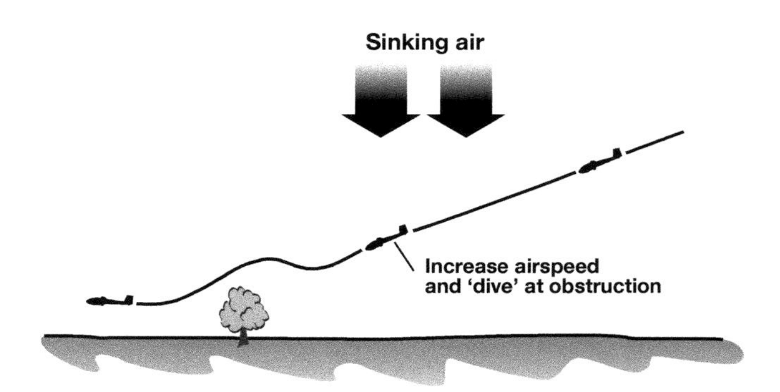

Fig 19.5 Undershooting a field boundary. If an undershoot looks likely, use any extra airspeed to dive at the obstruction rather than try to stretch the glide by reducing airspeed.

Your aim should be to get the glider close to the ground in the first safe part of the field, thus giving you the maximum distance of field ahead within which to land and bring the glider to a halt.

When close to the ground, **do not** be tempted to 'put' the glider on the ground while it still has flying speed. Round out fully and hold off as you would for a normal landing. This will ensure that the glider has as slow an airspeed on landing as possible, reducing the stresses on it (and on you) should the surface be rough. (The exception to this rule is if you are in danger of 'running out of field' and into the upwind boundary. In this case, forcing the glider onto the ground and using the wheel brake – or even inducing a ground loop by placing a wing tip on the ground – may literally save your neck. This assumes that you do not manage to bounce the glider during this hurried attempt at a landing.)

Once on the ground, apply the wheel brake. Some purists might say that the wheel brake on gliders is for emergencies only. As you do not know where in the surface ahead the next rabbit hole or ditch is, it is better to stop the glider as soon as possible. For this reason a glider that is fitted with a wheel brake should not be flown cross-country if the wheel brake is unserviceable.

I had just landed my glider in a field of lush grass. No problems were encountered with the landing or the ground run. On walking back down the landing run towards some farm buildings, I almost fell over a chain harrow which lay concealed in the long grass about twenty feet from where the glider had touched down!

Sideslipping is a useful skill to have when excess height needs to be dumped. However, unless you are practised at sideslipping, there is a danger that the glider will accelerate and end up with excess airspeed, making landing difficult as well as inaccurate. Such sideslipping practice must include sideslipping while adjusting the airbrakes throughout their full range. Despite the usefulness of sideslipping, it is far better to fly an accurate circuit pattern and approach at a sensible, controlled airspeed than to rely on your ability to sideslip.

Once on the ground

If the conditions are windy, secure the glider. This is best done with pickets. If the glider's parachute is used to hold the wing down, place it in its carrying bag first. It is better to avoid using this important piece of safety equipment for this task, if at all possible. Never use a parachute for this purpose if it is raining or is likely to do so. If necessary, wait or signal for help to move the glider into the shelter of buildings or trees. Remove valuable items, such as portable GPS receivers and dataloggers, and lock the canopy, ideally hiding the remaining contents by using the canopy cover.

If the landing has been made close to a village, town or school, then it is likely that sightseers will start towards the glider. Discourage them politely, as the farmer may not be happy if you encourage members of the public into his field. It may be necessary to enrol the assistance of one or two of the adults to control the enthusiasm of younger sightseers while you are away finding the farmer.

The next task is to find the farmer or landowner to inform him of your arrival and to telephone your crew. If any damage has been done to a crop (or if the farmer insists there has been) supply him with the name and address of the glider's insurer. Do not part with any money (except to cover the cost of your telephone call) and do not admit liability for any damage. It is not your place to pre-empt the insurance company's assessors.

Before you telephone for your crew, establish your position accurately and obtain precise details of how your crew can find you. A quick call on a mobile telephone giving an assumed landing position has led to many an unnecessarily long wait for the arrival of a retrieve crew.

Leave any gates as you found them. If in doubt, leave them closed or check with the farmer.

Even a well-planned and well-executed field landing can go wrong. On a practice day for a European championship competition in France, one of the British team members was about to land in the only available field in the area. On the approach, to the pilot's surprise, the field began to fill with water. His only option was to continue and land in the field, which was now deep in water. Soon the glider was also full of water.

The young crop in this particular field required regular watering, and to do this, the farmer had opened a sluice-gate from an irrigation system. Fortunately, no serious damage was done and the glider was dried out and ready to fly when the competition began.

Section 3

Personal Improvement

Chapter 20

Personal Improvement

Personal improvement in soaring could more accurately be called *self*-improvement. After you have reached that early stage in your flying when you are allowed to fly solo without first having check flights with an instructor, you will most likely be left to your own devices – without much, if any, further instruction. Most of the improvements to your flying will come from your own efforts and the experience you gain, often as a result of disappointing events. This chapter aims to give you some ideas and exercises that, ideally, would be given by instructors but, sadly, seldom are.

Self-criticism

The first personal requirement you need to advance your standards is the ability to be critical about your performance. Such self-criticism must be controlled. It must be as objective as possible. It must not be destructive and undermine your confidence.

When you get something wrong or do something badly, analyse what went wrong, assess how well you corrected the problem and decide how you will avoid making the same mistake again. On some occasions, such as when a mistake in co-ordination or thermal centring is made, the corrective action can be taken immediately. Other errors will require deeper analysis once the flight is over.

The important point is that when a mistake is made, you do not gloss over it or make excuses to yourself for your poor performance. The message is simple, 'Learn from your mistakes'.

Self-improvement

Improvement in ability is often difficult to measure. In soaring, badges, awards and competitions all offer some measure of achievement, but the degree of overall improvement can only be judged over many flights and perhaps years of soaring.

In the early stages of your gliding, when the number of skills being learned is large, the acquisition of new skills and improvement in the ability to make decisions is more obvious. Later, when soaring is the main aim, measuring improvement is less easy. For instance, achieving a qualifying flight for the award of a badge is always accompanied by a feeling of achievement, but given the soaring conditions on the day, could you have done better? Completing a closed circuit task may give you a good feeling, but are you ignoring the fact that you could have completed it at a greater average speed or flown a larger task, if only you had flown better?

Short of flying in a competition, there is no way of knowing how well you have really done. One thing is certain. There is always room for improvement.

The award system, which gives badges for certain achievements, is still a useful system to help measure improvement. However, once the celebrations are over, take a long look at what you have achieved and where you go next. If, for instance, you have managed your Silver Distance in a glider with a 50 to 1 glide angle, when other club members are achieving the same distance in gliders which have performances below 30 to 1, you probably should have gone further. Many pilots buy performance and think they are buying skill.

The need for practice

Improvement at any skill normally comes from practice. Improving one's judgement comes from experience. Practice can be specific to a particular skill (e.g. thermalling) or may occur while chasing another goal (e.g. thermalling while flying cross-country). Improved judgement can be gained in many ways: from practice (e.g. skyreading), from personal, often bitter, experience (e.g. the wrong assessment of a final glide which ends in an outlanding), from instructors, other pilots or books.

Exposure to different situations is essential to improvement, and therefore you need to fly as much as possible and set yourself tasks and exercises designed to expand your ability. This is especially important when a pilot is still relatively inexperienced, as there are many new things to learn and weather conditions to encounter. Once you are more experienced, you can afford to be more selective about the days on which you fly, but you must keep in practice.

Never hesitate to fly with someone who can add to your knowledge. In the early stages of solo or cross-country flying, this may be most instructors. Once you are more experienced, it will probably be with pilots who are renowned for their soaring ability. A more experienced pilot can often show you tricks that you might take years to discover on your own. Possibly, you may never learn them otherwise.

At one club where I ran a soaring course there was a pilot called 'Al'. Al had a glider of reasonable performance that he never flew. When asked why, he said he was waiting for a 500 kilometre day to complete the distance requirement for his Diamond Badge.

I wish him the best of luck, but think he might have a better chance, when that day arrives, if he stays in practice.

The right type of practice is very important. As your achievements and ambitions grow, you will be attempting larger tasks. These will demand concentration and decision making for longer periods. Despite the fact that soaring is fun, it is surprising how fatiguing a long flight can be. Flying for long periods and attempting to keep up determined, accurate flying will help build up your ability to perform well when faced with a long task which may well end in difficult soaring conditions.

To gain the most from a flight, always take off with at least one aim or purpose. The aim of a practice flight may be to improve a weak point in your soaring or to correct mistakes made during a previous flight or flights. Of course, it can be the completion of a task or the achievement of a high average speed.

After a flight, list the low points in your performance and the mistakes you think you made. Analyse your flight. Discuss the task and the soaring conditions with other pilots who also flew the same or similar tasks, and use this to aid your analysis and to judge your performance. Make sure your interaction with other pilots is meaningful, and avoid bragging or ego massaging. Exchanging information means giving as well as taking.

General handling practice

One of the main reasons for flying regularly is to keep in practice. Practising regularly keeps your handling skills at a reasonable standard. Skills such as aileron and rudder co-ordination and accurate airspeed control are essential to both soaring and safety. It is these skills which are the first to suffer if you have not flown a glider for some time.

Not only will poor co-ordination result in losing thermals or failing to climb in weak lift, it can also result in dangerous situations such as a stall or a spin. Poor airspeed control can have detrimental effects on both soaring and landing accuracy, which is crucial when landing in small fields.

Use any spare flight time you have to explore your glider's handling. Experiment with the whole airspeed range. See how stable the glider is at high airspeeds and how much height you can expect to gain when converting airspeed to height, as you might do from a final glide. Observe the handling at lower airspeeds. Find out what warning signs are present near the stall and what the stall and recovery are like. All of these exercises will help you operate the glider with confidence.

Flying frequently will also allow you to become familiar with your glider, its handling and its instruments. Being aware of a glider's vices or shortcomings increases your chances of flying it safely and allows you to utilise its full potential. Getting used to its instruments (especially the variometer system), and what they are telling you, should enhance your understanding of what the air is doing and thus improve your soaring. Some flight director systems have so many functions that much in-flight practice is required, firstly to understand how to use them and then to remain proficient in their operation. (While flying with unfamiliar instruments, do not let them distract you from keeping a good lookout.)

Thermalling practice

There are many aspects of thermalling which can be improved with practice. Everything from entering thermals to exiting them needs to be practised. Learning to centre quickly in thermals takes time – and once a technique is established, it must be used regularly if it is to be carried out efficiently on every occasion. As each day will produce thermals with different characteristics, it is necessary to work on thermalling in many different weather conditions.

Being comfortable in thermals in which a large number of gliders are climbing is an aspect of thermalling which can only be achieved by practice in such situations. To remain safe in such circumstances, your airmanship will have to be of a very high standard. If you fly at a gliding club which has only a few gliders, you may think that your airmanship is of an adequate standard to deal with your local environment. Watch out – a national championship race with 50 gliders may be heading your way or using your airfield as a turning point.

Although most aspects of thermalling can be practised while local soaring, the real test of your soaring efficiency will come when thermals are weak and you are still far from home on a cross-country flight, or when you are desperate to achieve the maximum rate of climb and the highest average cross-country speed in order to win a race. It is therefore important that you start working on your thermalling technique from the earliest days of solo flying and use every opportunity to improve it, even when local soaring.

Occasionally, an attempt to centre in a thermal is so poor that it is blatantly obvious that you could have done better. A short, highly personal jab of mental abuse and a promise of better concentration usually improves the performance in the next thermal. Sadly, it is not always that easy to measure how well you have centred the glider in a thermal.

As the process of centring and re-centring is a continual one throughout the whole climb in the thermal, it is worth comparing climb rates in different thermals during the same hour of the day. Admittedly, each different thermal will ascend at a different rate, but you should gain a feel for how well you are doing and see some improvement with each new thermal. For instance, decide on a minimum height at which to enter thermals (say, about 1,500 feet) and climb approximately 1,500 feet before leaving each thermal. Using a stopwatch, time how long the chosen gain of height takes. Descend to the entry height again, enter another thermal and try to better your previous best performance. This exercise can also be fun when flying in a two-seat glider with another pilot of similar ability and attempting to outdo each other's climb performance.

Probably the best measure of your centring and overall thermalling ability comes when you try to outclimb another glider in the same thermal. Never miss such an opportunity to practise your skills. While local soaring, if you see another glider in a thermal, go and join it. Establish your glider in the lift (ideally below or at the same height as the other glider) and climb for all you are worth. The challenge should not start until any extra airspeed above that used for thermalling has been dissipated. (Do not fool yourself into counting a pull-up to get rid of excess airspeed as part of your climb in such a situation.) This exercise is all the more educational when the other glider is of a similar performance and is flown by a pilot whose ability is respected.

Cross-country flights will often end in weak soaring conditions when you are left struggling to find sufficient thermals to get you close enough and high enough to commence a final glide. Such 'end of the day' thermals may offer such weak lift that any inaccuracies in your flying may change what should have been a slow (sometimes painfully slow) climb into a gradual loss of height. Take every opportunity to practise in such weak conditions while local soaring.

When trying to climb in such a feeble thermal, set yourself a realistic target height to achieve (say, 1,000 feet above the height at which you found the thermal). Imagine that this is the last thermal of a 1,000 kilometre flight, and that the target height is needed to get you home with no height to spare. Work the thermal with this degree of determination.

Even when conditions are good, assuming that you are restricted to local soaring or have returned from a cross-country flight, local thermalling practice can be useful. Setting yourself artificial restrictions, such as putting a low ceiling on your climbs (say 1,500 or 2,000 feet) and then leaving every thermal at this height, will increase your workload and force you to try harder or be faced with a landing back on the airfield earlier than you had wished. Not only will such an exercise force you to work hard at your thermalling, it will also require you to identify the position of your next thermal efficiently. If conditions are good and thermals are plentiful and predictable, then you should be very selective about which thermals you are willing to accept for a climb. This, together with the selection of a low height band, will provide you with a good workout – even if it turns out to be a relatively short one.

Taking a wire launch (winch or car launch) as opposed to an aerotow launch is also an effective way of increasing the workload in the early part of a flight. Unlike an aerotow launch, which can deliver the glider to a thermal and to a height where you have many options, a wire launch normally demands that you find lift for yourself and usually from a lesser height.

It may even be that a pilot trained totally using aerotow launching will take longer to feel confident when searching for or using thermals when low than colleagues who have trained totally using wire launching. If this is the case, and assuming you have been trained totally using aerotow launching, the low-level thermalling exercises given may be worth trying before you commence cross-country flying.

Every thermal climb, whether to cloud base, to a preselected height or to a point where the rate of climb falls below a pre-determined level, offers a chance to practise exiting a thermal. Decide where you are going to go to find the next thermal *before* you reach the exit height. Practise accelerating *before* you leave the lift. Note the height loss as you accelerate and note the total height lost in clearing the sink zone around the thermal. Did you accelerate too early, too late, too little or too much? Experiment and practise to see if you can improve on your technique in subsequent thermals. Try other techniques and see which you prefer.

Being able to thermal well in all conditions is an essential part of soaring, and as such, it is worth practising in a determined manner.

Skyreading practice

Although skyreading could be encompassed in the section on thermalling practice, it is such an important aspect of soaring that it warrants further discussion in its own right.

Gaining the ability to interpret what the clouds are indicating is a major acquisition to any soaring pilot. For this reason, the more time you dedicate to skyreading (whether while airborne or on the ground), the better your chances of appreciating where the next good thermal will be found.

Unlike many aspects of soaring, skyreading practice can be free from launch fees. You can observe the changing sky and the formation and dissipation of clouds from your garden, from a train, while on the beach, while walking the dog – even from your office window when your boss is not watching. Any time you spend studying the clouds and trying to work out which ones are growing and which are evaporating, where new cumulus clouds are likely to appear, and which areas have remained devoid of cumulus for some time, will add to your understanding of what the air is up to.

Do not leave it at that. Try to imagine that you are up there in your glider and try to decide where you would go to find the best lift and to avoid sink. Give yourself an imaginary task across the sky, firstly from east to west and then north to south. By doing this you make sure you do not always pick the easiest route. Try the same route ten minutes or so later. What does the sky look like now? Are the clouds under which you thought thermals might have been found still there or have they broken up, spread out or vanished?

When airborne, you not only have the chance to visualise which clouds are being fed by thermals but also the chance to prove if your theories are correct. When local soaring at a safe height, cruise around the area, visiting cumulus clouds which you expect to be indicators of thermals. Do not stop and climb in every thermal found. Instead, look for another likely cloud, test your judgement again and move on. Explore some clouds which you would normally reject and see if they do indeed prove useless. When you get to a height at which you need a climb, pick the best thermal you can and climb in it. When high enough, you can continue your research. This exercise should be practised when you are close to your home airfield so that you can land there and take a further launch if you cannot make contact with a thermal when you need to.

During your skyreading exercises, keep an open mind and do not simply think of isolated thermals. Constantly ask yourself if other types of lift may be present and indicated by the shape or behaviour of specific clouds. Look for thermal streeting, wave influence or sea air effects and try to work out the effect these would have on your flight if found to be present.

Ground reading practice

Identifying ground features which may be the source of thermals is much more difficult than practising skyreading. Any wind will make linking a thermal to a definite ground feature an imperfect science. If the glider is to be flown low enough to be reasonably certain which feature has given rise to the thermal, it will probably be operating at a height where an imminent landing is a distinct possibility. For this reason, only ground features close to the home airfield can be investigated.

Identifying thermal triggers, especially man-made ones, is a somewhat more definite way to identify the ground feature which allowed sufficient surface heating for a thermal to form. The main point of such an exercise is to try to work out where a thermal may be and test your theory. Store the results in your memory for future reference. If you encounter a thermal unexpectedly, your first action should be to look upwards to see if there is a cloud forming, and your second to look down to see where it came from. If no cloud has yet formed, you have found a newly triggered thermal, assuming, that is, that cumulus clouds are evident elsewhere. Remember the type of surface which spawned this thermal; it may save you some day when you are low and far from home.

Weather assessment

If you are to be able to judge what sorts of conditions are going to be experienced further along your track or later in the day, then much more general weather assessment than skyreading for thermals is necessary. Although much of this will be gained from the weather forecast, you will still need to be able to estimate, firstly, whether the forecast is accurate, and secondly, whether other elements, not included in the forecast, are having an affect on soaring conditions.

For instance, is the thickening upper cloud which is appearing tonight's front coming in early or is it another feature that was not forecast. If it is the former, it may adversely affect the quality of thermals for the rest of the day. If it is the latter, it may be a more short-lived phenomenon, which, after its passage or dissipation, may allow the return of good soaring conditions.

Like skyreading, weather assessment can be practised every day for the rest of your life. Watch and listen to forecasts. Observe the sky, the barometer (or altimeter), the wind direction and the actual local weather. Listen to VOLMET and the actual weather reports for the airfields around the country. Try to forecast the weather for the hours ahead and then check whether you are correct. Did that front, whose cirrus cloud you saw this morning, arrive when you expected it? Did convection continue under the altostratus cloud for as long as you predicted? How much of the day looked soarable? What sort of task would you have set?

There are many excellent books on meteorology. Read as many as you can. Try to gain an understanding of general weather mechanisms, the clouds and the wind changes associated with them. Such knowledge can help you in flight or when task planning. It may even help you decide on which day to stay at home and paint the spare room.

Navigation practice

Undoubtedly, the best navigation practice is gained while airborne and actually having to find your way around a task route. The more cross-country flights you do, the better you will be able to handle the workload which comes from flying the glider, staying airborne and navigating at the same time. The more 'new' ground you cover, the more experience you will gain. Flying cross-country on days with poor visibility will add extra difficulties to the task of navigating and in turn will widen your experience.

Even during the early days of solo flying, before you are allowed to venture cross-country, you can work on navigation exercises which will help you to orientate, interpret the compass and appreciate its limitations, distinguish ground features and estimate distances. To facilitate this learning process, always carry a map in the cockpit when local soaring, and set yourself exercises such as those that follow.

Before you launch, select several (about ten) easily distinguishable ground features within local soaring range of the airfield. These features can be road junctions, public houses, local villages, churches, bridges, etc. They do not have to be features as obvious as the ones you would select as turning point features on a real cross-country flight. Measure their track from the airfield and the heading you think you will need to fly to reach each of them, bearing in mind the effect of any wind. Once airborne, attempt to reach each of them in turn, positively identify the feature and fly back to the airfield before setting off to find the next feature.

The above exercise can be varied by attempting to miss out the return leg to the airfield, but instead flying direct to another ground feature. This will require that you either estimate the track and heading in flight (a useful orientation exercise), or draw tracks and calculate headings before you leave the ground. Whichever method you intend to use will be much more disciplined and beneficial if you number the ground features in the order that you intend to visit them.

The exercise can be modified further, and incorporate a soaring judgement element, by attempting to cover all of the ground features within a set time or simply as fast as possible. If this variation is used, then you can either stick to the pre-declared sequence of visiting and identifying the ground features (in which case you will have to assess when it is safe to fly towards a feature without encountering too much sink), or allow yourself the freedom to change the sequence of visiting the ground features, depending on how you interpret the sky in the direction of each of the ground features. Again, it makes the exercise more challenging if you decide before you launch which of these methods you will use.

Of course, the expression 'ground feature' could be substituted for 'turning point' and an element of turning point practice included. This would make a more complete exercise; however, in the early stages of soaring, you may have a high enough workload without the accurate orientation involved with turning points. As your ability increases, such a combined exercise is well worthwhile.

One great aspect of an exercise involving local landmarks is that you can set off to identify these and compare them with the map legend, even when it is not particularly soarable. All you need is a high aerotow (say 3,000 feet) and features which are close enough to the airfield to reach. If there is any significant wind, take care not to select any features which are too far downwind, or you may not be able to reach them and still return to the airfield.

By using several variations, exercises such as this can include an element of competition involving a group of pilots. Such exercises add purpose to a flight as well as improving a pilot's ability to use maps and orientate.

When flying near the home airfield, you will usually have some idea of its position relative to yours. This orientation is often subconscious, but works on the same basis as the orientation skills which are an essential element of successful navigation.

Learning to be constantly aware of the sun's position will make it a useful aid to be used in conjunction with the compass.

Practise assessing your position by referring to two or more positively identified ground features (not the home airfield) and then pinpoint this on the map. When doing this, check the relative bearing of the features from each other and compare this with the expected bearing estimated from the map. Finally, confirm the accuracy of your position estimate by judging your distance and bearing from your home airfield. Start using such ground feature/map orientation, rather than always referring to your home airfield, to judge your position when local soaring.

Practise using maps to judge distances and identify distant ground features (more distant than those you can reach and still be local soaring). This will prepare you for cross-country flights to come. The distance at which you can clearly see features such as towns, power stations and hills, etc. will vary depending on the visibility, the height of your glider, the sun angle and, to some extent, the presence of cloud shadows. This means that there are no hard and fast rules about the distance at which a feature can be seen. On days with excellent visibility, you may be able to see such a long way that distances are foreshortened, and as a result underestimated. On others, it may be difficult to see features less than a mile or even a kilometre away. Learn to estimate distances and check your estimations with the distance on your map. (Distance rings on maps have uses other than setting up a final glide.) The day will come when you will use such distance estimates to determine your location and perhaps prove to yourself that you are not lost after all.

Use every opportunity to study the map which you will use in flight. Get to know what its symbols indicate and how it depicts ground features, especially those essential to navigation, such as towns, railway tracks, roads, rivers, canals and airfields. Work out how it portrays the position and heights of controlled airspace. Investigate this aspect by studying an area where there is a lot of controlled airspace with varying bases and ceilings.

When at home, set yourself tasks which you can draw on your map or maps. Make these tasks practical ones which start from your home airfield and use turning points that you may use some day. Be ambitious but realistic in the tasks you set. Prepare your maps exactly as you would for a flight. Input a hypothetical wind into the exercise (perhaps the wind on the day you try the exercise). Determine the track for each leg and the heading you expect you would have to fly to make good each track.

Now work along each track on the map, checking each symbol you come across. If there are any you do not know or cannot remember, then refer to the map's legend in order to clarify their meaning. Be specific with the meanings. For instance, the symbols used on many maps may distinguish between active, disused, military and civil airfields. This may seem academic, but when it comes to pinpointing or re-discovering your position, or deciding whether it is wise to land on a particular airfield, such detail is important.

Having gone over the task on the map looking at ground features, start again at the beginning and study the airspace symbols and heights affecting each track. Note which of these pose a problem to your hypothetical flight. Look carefully to see which heights are defined as altitudes and which as flight levels. The difference in altimeter setting may be considerable and, on a real flight, could cause an airspace infringement. If the meaning of any of the airspace symbols is not clear, refer to the map legend. If the restrictions or conditions of flight in any of the airspace along your route are unknown to you, or if you have any doubts about the legality of glider flight in such airspace, check it out. You can do this by reference either to the airspace publications which are held at your club, or to your national gliding authority's publication on the rules for glider pilots. The latter will probably make easier reading and be more specific to gliding. (Each year in the UK, the British Gliding Association publishes an article in its magazine which gives the latest airspace situation, and explains in which airspace a glider pilot is permitted to fly and under what circumstances. A copy of this article is a useful inclusion in your flight bag if you fly in the UK.)

Some early airborne navigation experience can be gained from flying around a route in a motor glider or light aircraft. Although this may appear to be a useful exercise, in practice, without all of the circling, climbing, diversions to find lift and the workload of a normal gliding cross-country, such exercises are unrealistic. Another problem is that they are expensive, as it takes some time to get away from familiar landmarks; if the home airfield is in a prominent place (e.g. on or close to a hill) then you may never be out of sight of it. The value of such an exercise is doubtful. A far better orientation exercise arises when a student has to re-establish the motor glider's position and find the way home after the completion of a high workload exercise such as field landing training.

Once you have completed some actual cross-country flights, your navigation skills will improve rapidly because of the experience gained. However, always flying tasks which are in the same direction or from the same airfield will eventually result in familiarity with the landmarks on these routes. To increase your experience, set yourself tasks which take you in different directions or else go and fly from an airfield in a different part of the country. Not only will this enhance your ability to navigate, it may also improve your soaring skills by presenting you with different types of soaring or unfamiliar weather conditions. The ultimate in this extension of experience will come from soaring in different countries, which may have different terrain and visibility, and also require the use of maps which have a different scale, accuracy and legends from those you are used to.

Of course, you can always take the easy way out and use GPS for all your navigation. However, navigation skill involves more than watching a GPS display, as you will soon discover if your GPS ever fails.

Learn the elements of basic navigation first, and then enjoy the benefits of modern technology later with the knowledge that you are able to revert to map and compass navigation, if necessary. My advice to all pilots who have a GPS fitted to their glider, and who wish to improve their ability to navigate, is to leave it switched off during some of your flights, and practise basic navigation skills. You never know when you might need them.

Final glide practice

The best way to practise final gliding is with caution and due consideration of your level of experience. This does not mean that you should only start practising final glides after you are allowed to fly cross-country. On the contrary, useful final glide exercises can be practised even when you are restricted to local soaring.

Gaining an appreciation of final gliding involves understanding how far your glider can go from a given height. This, in turn, will not only help you judge what is a safe distance from the home airfield when local soaring, but also give you more confidence when deciding whether a cumulus cloud (and potentially its thermal) can be reached with the height available.

The first requirement is to mark your map with concentric circles centred on your home airfield. On such a practice map, these should be spaced at a distance representing two nautical miles. Mark the distance from the airfield next to the appropriate circle in such a way that you do not obliterate any important map detail. This 'final glide map' will be used throughout your final glide practice exercises, and can be useful any time you are local soaring as a quick reference to the distance of a ground feature or your distance from the airfield.

The other useful tool to have is a final glide calculator (such as the JSW final glide calculator described in Chapter 18).

When at home or whenever the opportunity arises, practise using the calculator and run through some hypothetical final glides. Note the distances that can be covered from the sort of heights normally possible with climbs in thermals. Note the effect of different winds on the distances possible. See how helpful a tail wind can be and how much more height is required to penetrate into a head wind. Check to see how much extra airspeed a fast climb can allow, and how much extra height such a climb and glide will require.

Combine your exercises with the calculator and with your final glide map. Start your final glide scenario from different points on the map and mentally go through the eventualities of encountering lift or sink on the final glide. Try the same final glide scenario with different wind conditions. Generally, get used to using the map and the final glide calculator.

Once airborne on your local soaring flights, choose thermals which are at an increasing distance from the home airfield, while maintaining a sufficient height to return to the airfield with a safe height margin, even if you fly through some sink. When at a reasonable distance, say 5 or 6 miles, calculate the height required to glide back to and still arrive overhead the airfield with 2,000 feet of height remaining. Start this final glide and see if your calculation of height loss was accurate.

Find a thermal, climb away again and repeat the exercise from a series of different distances and at different airspeeds. As you become more confident, you can reduce your safety height margin. (Most of the 2,000 feet suggested was to increase your chances of finding another thermal so that you could continue the exercise.)

Eventually, you will be able to select a MacCready setting based on your last climb and fly the final glide according to speed-to-fly theory, rather than simply at or close to a preselected airspeed.

Even when the weather conditions are not conducive to soaring, a tow-plane pilot can be instructed to deliver the glider to a specified position at a pre-calculated height, thus allowing the glider to practise a final glide back to the airfield. Such an exercise should be supervised by an instructor.

The following exercise, demonstrating the rewards of speed-to-fly technique, involves two gliders and requires a greater amount of organisation if it is to work.

Take two gliders of similar performance. Select a point about 15 miles from the airfield. Calculate the height required to glide this distance at the airspeed at which the glider's best glide angle is achieved: that is, best glide speed. To this height add a safety margin, say 500 feet.

Re-calculate the height required for the same final glide if the climb rate in the last thermal is 4 knots, and add the same safety height margin.

Now the fun part! Aerotow both gliders simultaneously (either as two separate aerotows or, even better, a dual aerotow) to reach the predetermined position at the height which was calculated as sufficient for the final glide at the best glide speed. When the gliders arrive at this position and height, the first glider releases and sets off on the final glide at its best glide speed.

The second glider stays on aerotow and the tow-plane continues climbing in the same position to the height which will allow the glider to fly the final glide at an airspeed (or MacCready setting) appropriate to a rate of climb of 4 knots in the last thermal. (This is about the rate of climb achievable by most tow-planes with a single-seat glider attached. If the combination is capable of climbing faster than this, the tow-plane pilot should adjust the rate of climb to about 4 knots.) When the glider reaches this height, its pilot releases and starts the final glide at the appropriate airspeed.

Which glider will reach the airfield first? Will the extra time spent climbing have been wasted or will it be regained in the faster glide? Ask your duty instructor to set up this exercise and you will find out for yourself.

In practice, the distance to the start point of the final glide may have to be varied, depending on glider performance and weather conditions. During the final glide, it will be necessary to adjust the glider's airspeed depending on the height and distance still to run to the airfield. (The safety height margin selected may have to be increased depending on pilot experience.)

Once you have gained some experience at cross-country flying, you can start using what you have learned from your final glide practice for real. Hopefully, the experience which you have gained from these exercises will allow you to make your first real final glides with confidence. The more you do, the more efficient your final glides should become and the better you will be at judging how much safety height margin you will require.

A final glide, whether for practice or for real, normally ends with some excess airspeed. This is usually converted to height by 'pulling up' after arriving home or crossing the finish line, if the final glide is at the end of a speed task.

Judging such a pull-up takes practice, and therefore you should end your practice final glides in this way, in order to learn to estimate the likely height gain and rate of airspeed loss, as well as acquire a feel for the glider's handling as it decelerates. As with practising final glides, begin by practising these pull-ups at height and reduce this as your experience, judgement and skill increase.

Finally, when practising final glides, always remember that other pilots, often a lot less experienced than you, are operating from the same airfield. For this reason, let the duty instructor know your intentions and conduct your flight with the highest degree of airmanship and consideration for other pilots, some of who may be learning to fly more traditional circuit patterns.

Training for field landings

Before you are permitted to fly cross-country, you must be capable of safely landing a glider in a field. There are several exercises which can be practised to prepare you for this event.

Undoubtedly, the most effective of these exercises involves using a self-launching motor glider. With such an aircraft (and a suitably qualified instructor) not only can you select and assess several fields in one training session, but you can also demonstrate your ability to fly a circuit pattern and approach 'into' each of the fields chosen. Each approach can be terminated and the motor glider flown to a different area, where another field selection and approach can be made.

Other exercises which will enhance your judgement and skill, although not that of field selection, include practice landings in a pre-arranged field (with the farmer's permission, of course), landings on a little used part of your own airfield, and flying from other gliding sites where you are not familiar with local ground reference points and where circuit flying may be more demanding.

On every approach into your home airfield, you should attempt to rigorously maintain a preselected airspeed (say within 1 or 2 knots) and every effort should be made to 'spot' land the glider, stopping before an imaginary boundary. An approach made with the intention of 'just getting the glider on the ground' is a practice opportunity lost.

At club level, spot landing competitions can be organised to encourage skill and judgement improvement. Such exercises also impart purpose to flying on unsoarable days or during evening flying sessions. Certain rules have to be incorporated into such a competition, but these are fairly basic and are designed to encourage fairness and promote a high standard of flying.

Firstly, all landings must be 'normal' and fully 'held-off'. This prevents pilots touching down at high airspeed simply to 'arrive' on the chosen spot.

Secondly, as well as having a well-defined, easily seen marker at the touchdown point, there should be a limit to the length of the landing run allowed (in the form of another set of markers simulating the position of a fence or ditch at the end of a field). This second limitation will go a long way to preventing touchdowns being made at high airspeed.

The only other ingredient is a judge or two to adjudicate and measure landing attempts for accuracy. Bad airmanship or unsafe flying should result in disqualification.

Such an exercise can involve any pilot capable of carrying out a landing. Even student pilots who are not yet cleared to fly solo can participate with an instructor on board as a safety pilot. Inexpensive (fun) prizes can be provided as 'carrots', possibly awarded for the best performance by pilots in particular experience bands.

The first approaches and landings at a strange airfield will require most of the same judgements as landing in a farmer's field, with the exception of judging the nature of the surface. An airfield with which you are unaccustomed will lack all of the local features that you probably subconsciously use to judge your circuit pattern and approach when flying at your own airfield. This is another good reason for flying from gliding sites other than your own.

Identifying and determining the depth and density of crops is an ability which comes with experience. In order to gain this experience, you must keep your eyes open and observe the state of the crops as they change throughout the year.

Such crop watching can be carried out both from the air and on the ground. While local soaring, look at the fields in the area and try to identify the crops and judge how high and how dense each crop is. Decide in which fields it would be safe to land and which to avoid. If you are unsure about any crop, note the position of the field relative to the airfield or some obvious ground feature, such as a road or building, and try to visit the field by car or on foot to see if you can clear up your confusion about the field's suitability for a landing.

When travelling in a car, study the crops you pass, their colour, height and density. Decide which are too tall or too dense to land in safely. Note those which are short or sparse enough to allow a landing to be made without damage to the glider.

When you do land in a field during a cross-country flight, make use of the time you spend waiting for your retrieve crew to walk the local tracks or lanes and study the neighbouring crops. Pay particular attention to any crops about which you were unsure and, for this reason, may have caused you to reject a field during your field selection. See if your assessment was correct.

When you talk to the farmer or the farm employees, take the chance to discuss the state of the various crops and to increase your knowledge of what is happening in the fields over which you fly. Gaining an understanding of what farmers are up to will help you work out which crops are which, and where it is safe to land.

Flying from new sites

We have already discussed some of the advantages and experience that can be gained from flying at gliding sites which are unfamiliar to you. However, to emphasise the value of such visits, here is a précis of these advantages.

Each new gliding site may:

* Have different circuit requirements, dictated by airfield size, terrain, airspace considerations or traffic.
* Require an awareness of unfamiliar local rules.
* Offer soaring or even weather conditions, the nature of which you have little or no experience or have not experienced for some time.
* Present new navigation considerations, because of the lack of familiar landmarks and possibly the presence of controlled airspace.
* Offer you the chance to experience a new method of launch, along with the need to consider launch failure recovery drills in a new environment.
* Offer the chance to fly types of glider not available at your own club site.

Experience of any or all of these unfamiliar aspects will add to your knowledge, enhance your judgement and possibly improve your skills.

Many clubs organise expeditions for their members to other gliding sites. Joining such an expedition is worthwhile, as you can work as a team with familiar faces and benefit from the knowledge of the local pilots who are your hosts.

Even if you have to organise your own visits to other gliding clubs, these can be just as rewarding, especially if you choose an expedition site which offers the maximum number of new experiences.

If you are relatively inexperienced, research your choice of expedition site well. Unless you are intent on gaining the experience on offer by flying a two-seat glider with an instructor (not necessarily a bad thing), avoid going to a gliding site where a pilot of your experience will not readily be allowed to fly solo.

Speed flying training

Once you are capable of staying airborne to the extent that most of your flights are of a reasonable duration, it is time to start thinking in terms of speed flying. Practising, from this early stage, the techniques that will be used when flying cross-country will help you prepare for the tasks to come.

Initially, this training can begin with being selective about which thermals you choose to use, and climbing in those you do select as efficiently as possible. Add to this some attempt at using speed-to-fly technique when gliding between thermals, and you have taken the first steps towards becoming a cross-country soaring pilot.

Next, start setting yourself tasks which you can attempt while local soaring. It is surprising how large a task you can achieve while still remaining within gliding range of the home airfield.

For instance, let us assume that your glider is capable of covering 4 nautical miles for every 1,000 feet of height loss. If the tops of thermals are 3,000 feet, you can afford to be 4 nautical miles from the airfield and will still have 2,000 feet in hand. Of this 2,000 feet, you will need around 1,000 feet for your circuit pattern. (Even this gives some safety margin.) This leaves a height band of between 2,000 and 3,000 feet in which to operate.

Given that there is no significant wind, you can afford to set a task with turning points at any point up to 4 nautical miles away from the airfield. As long as you are at or above 2,000 feet at any of these turning points, you should be able to return safely to the airfield. This may not sound like a large task, even for a pilot confined to local soaring, but if this task were a triangle, with three turning points, none of which are the airfield itself, the distance around it would be approximately 21 nautical miles (39 kilometres). Such a task is almost 80 per cent of your first cross-country target – Silver Distance! (Figure 20.2)

Of course, there is no need to set a task this large. The great thing about tasks involving local soaring is that you can complete the task as many times as you like during the day. One advantage of a smaller task is that you can attempt to increase your average speed for the task each time you attempt it.

If there is any significant wind, you may have to select turning points which are upwind. This will allow you to choose turning points that are

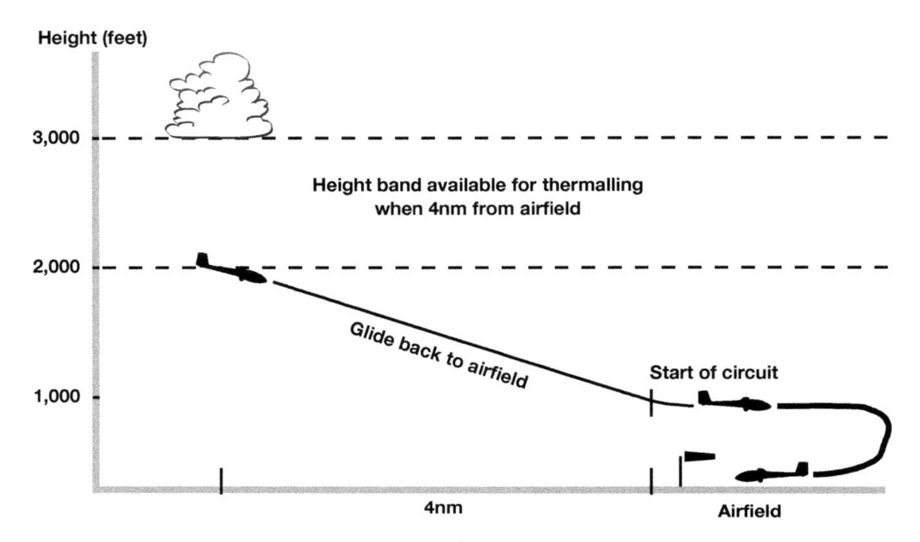

Fig 20.1 Local soaring. Assuming your glider can cover 4 nautical miles for each 1,000 feet of height, a 3,000 feet thermal depth will give adequate margins for 'cross-country' training exercises while local soaring.

further away from the airfield and still be within range of it. If the wind at flying levels is very strong, then you might be better selecting just one turning point which is well upwind and practising thermal street flying on a goal-and-return task.

If the airfield is chosen as the combined start/finish point, then final gliding can become one of the exercises practised during each attempt at the task (providing you finish with enough height to continue soaring). If you choose to include this aspect of cross-country soaring, do so with the caution described in the section on practising final glides.

Should you decide to select a combined start/finish point which is remote from the airfield, soar to it and start the task from a pre-determined height (usually as high as you can climb on your first attempt at the task).

If you are interested in monitoring the improvement to your average speed for the task, then on subsequent attempts use the same start point AND the same start height. Similarly, the finish height chosen should be specified and should be adequate to return to the airfield if you fail to soar after finishing.

One way of increasing the accuracy of your estimations of average speed is to time your flight up to the point where you not only cross the start/finish point, but also climb to the height from which you started. One word of caution needs to be injected here. If you find that your average speed to complete the task is improving by a large amount as the day progresses, it may be the soaring conditions which are improving. Your best performances will probably be achieved at the time of day when the thermals are reaching their greatest height and strength. Do not let this fact dampen your enthusiasm too much, as your average speeds remain an indication of what you are capable of during the best part of the day.

The great aspect about flying such TINY TRIANGLES, as they are commonly known, is that they can encompass many aspects of cross-country soaring. You can work on your thermalling, skyreading, speed-to-fly and dolphin techniques, map reading, turning point technique and even final glides. You can measure your performance and see which areas of your flying need improving. You can cram in a lot of practice without the risk of landing out.

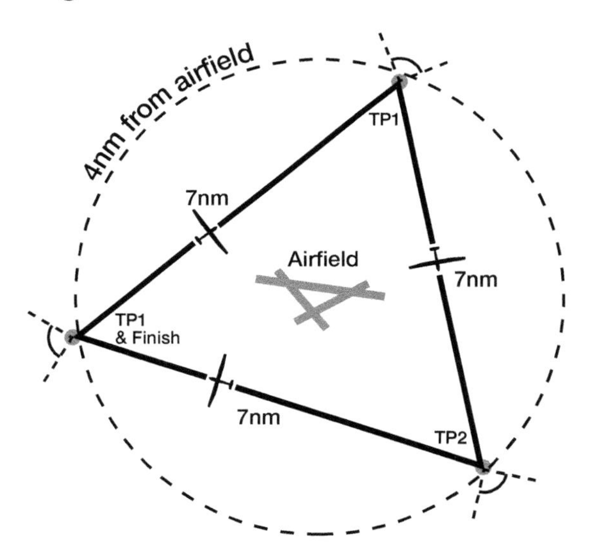

Fig 20.2 Tiny Triangle. Sizeable tasks can be carried out without leaving the local area.

There is, of course, no reason why your task has to be a triangular one or a goal-and-return flight. It can have as many turning points as you wish. (There is something not right about the name Tiny Quad – maybe Quick Quad would catch on better?)

If your airfield is close to another where glider landings are welcome, then it may be that, with your instructor's permission, you could count being within range of this second airfield as 'local soaring', in which case you could enlarge your tasks, once you have tired of the smaller ones.

Tiny tasks need not be limited to early solo pilots who are not yet allowed to fly cross-country. More experienced pilots can use them to improve their techniques or to get the bugs out of a particular part of their flying. Being local tasks, they are also ideal for two-seat glider instructional flying on the techniques involved in cross-country. In poor soaring conditions, they could even form the basis of club-level competition.

Practice during actual cross-country flights

Undoubtedly, flying cross-country is the best way to practise cross-country flying. Once you are cleared to leave the local soaring area and start flying cross-country, the fun (and to some extent the hard work which comes with any challenge) begins.

Every task you fly, whether completed successfully or not, will bring new experiences and increase your knowledge and ability. No longer will you be able to make a run for the airfield every time you get low. You will have to find a thermal or land out. There is nothing quite like the threat of spending what is left of a good soaring day in a field to make a pilot concentrate on finding and successfully using a thermal.

Once you start flying longer tasks or tasks in difficult soaring conditions, you will, on occasions, fail to complete a task and end up landing out. Do not despair. Firstly, you are gaining experience and confidence at landing in fields or at unfamiliar airfields. The knowledge that you can do this safely will relieve some of the concerns which many pilots have when they begin flying cross-country. Secondly, every failed attempt at a cross-country task is still more kilometres for your logbook and more experience gained.

Most pilots have a target they wish to achieve in their cross-country flying. Perhaps it is a 300 kilometre or 500 kilometre flight. On many days, soaring conditions will simply not be good enough for such flights. On such days, it may be worth flying tasks which cover part of the larger task that you are hoping to fly when the ideal day arrives.

Maybe a suitable task would be a goal-and-return flight to, or part of the way to, one of the turning points of a larger triangle. In this way, you will experience some of the navigational problems and get to know some of the landmarks on at least part of the proposed larger task. This will make it easier when you do attempt it. Relieved of some of this workload, you may be able to concentrate on your soaring, fly more efficiently and stand a better chance of success when you do attempt the larger task.

Cross-country flights in a two-seat glider with an instructor should not be regarded as a backward step in your progress or as a test. Instead, they are well worth embarking on, especially if your instructor (or more generally P1) is an accomplished cross-country pilot. One flight with a pilot with ability may teach you more than you could teach yourself in 100 cross-country flights. Do not miss such opportunities; they can also be great fun. One word of caution is to be selective about with whom you choose to fly. Make sure you are not simply paying to be taken for a ride. On a good soaring day, you should expect to do most of the flying, with the exception of low thermalling and demonstrations. At worst, you should share equally the total amount of flying.

Another method of gaining from a better pilot's experience is to take part in an 'escorted cross-country' or 'lead-and-follow' exercise. Such exercises involve a leader (an experienced cross-country pilot or instructor) flying in a single-seat glider, and a couple of followers, flying their own gliders. The leader literally leads the gaggle of gliders around the task.

At first, this may sound a haphazard method of teaching cross-country soaring, but provided the leader has the ability and patience necessary for such lead-and-follow flying, the followers can learn a great deal.

If such training is to work, each of these sessions must be preceded by a thorough briefing. In order to achieve the maximum benefit from such a flight, it is important that you do not simply follow the leader thoughtlessly. Try to work out where the leader will go next. Why has he or she gone to that cloud and not the one you had in mind? Why has the leader stopped to climb when the gaggle is not low, or passed up a particular thermal?

Taking part in escorted cross-country flights can teach a pilot almost every aspect of cross-country soaring, except perhaps navigation. Doing so can give you the chance to compare your ability in both climbing and gliding with that of other pilots – including the leader, whose ability you have accepted as your present target.

Cross-country training flights need not consist of long tasks to be successful. In fact, two-seat glider and escorted cross-country exercises are often more useful if their distance is kept to between 100 and 200 kilometres. This is normally enough to allow most of the important techniques to be demonstrated and practised, and to show up any parts of your flying which need attention. Longer flights may result in your forgetting some of the important points before you have time to assimilate them.

Competitions

Competing with yourself is never easy. Unless you get the chance to measure your ability against that of others, you will never know how good or how poor your standard is. Competitions offer a direct and, as far as possible, fair comparison. Many pilots who thought they were good or even excellent soaring cross-country pilots have been brought down to earth (perhaps literally) when they have entered their first competition.

In many countries, the standard of pilot ability displayed in competitions is marvellously high. What this means is that there is a classroom in the sky, just waiting for any soaring pilot interested in improvement – but be prepared for a possible lesson in humility.

Whether you see yourself as competitive or not, competitions can improve your flying at an incredible rate. Just to fly in the company of top pilots for a small part of a task (they tend to lose you, quite quickly) can give you an insight into how they do it and what is possible.

Even regional competitions are used by some national championship pilots as warm-ups or practice sessions for bigger competitions. Therefore, even these lesser competitions can offer a lot of experience and much fun.

Rated competitions, such as national and regional championships, are not the only events which offer such comparisons and educational opportunities. Many clubs organise task weeks which have all the elements of larger competitions at club level. Being somewhat more informal and less cut-throat than rated competitions, task weeks are a good introduction to the competition experience.

One aspect of any competition or task week, which will amaze and encourage a first-time competition pilot, is that sizeable cross-country tasks can be completed on days when soaring conditions are so poor that you would not normally have bothered rigging your glider.

Task and competition weekends complement the experience which can be gained from larger events and can be entered at much less expense, although usually resulting in less flying.

On a daily basis, a common task flown by a number of club pilots offers comparison of personal ability and greatly adds to club spirit. Such group tasks permit elements such as pair flying, and mutual help with retrieving gliders which have landed out. After a common task has been flown, pilots have a chance to compare notes and thus increase their knowledge by discussing the task and soaring conditions with other club pilots who attempted the task.

Training courses

If you feel that your progress (or even your enthusiasm) has levelled off and needs a boost, what you probably need is a soaring training course. Such courses are run for pilots of all experience levels by clubs and national gliding associations.

These courses have a faint resemblance to competitions, in that everything is organised for you. However, as the emphasis is on training, you need not worry if you shy away from competition. The instructor will tailor the flying exercises, tasks, briefings and lectures to your needs and experience.

Although some courses may effectively give private tuition, most involve one or two instructors and a small group of course members. This situation allows comparison of standards, interchange of ideas and techniques, and generally enhances pilot motivation. Add to this the knowledge which can be gleaned from the standard of pilots who instruct on this type of course, and the progress you make can be noticeable.

The structured nature of soaring training courses allows common tasks to be flown by course members, as well as two-seat glider training and escorted cross-country flying. Local training exercises can be set on unsoarable days.

If you are considering attending a soaring training course, I suggest that, if possible, you enrol on one which takes place in the early part of the soaring season. By so doing, you can use the rest of the season to put what you have learned on the course into practice, while the enthusiasm which such a course can create is still at a peak.

Flight analysis

Analysing your performance, unless you have done incredibly badly on a task, can be difficult. Several obstacles combine to reduce the effectiveness of objective analysis.

Subconscious dishonesty in the form of a reluctance to accept that you could have done better, forgetfulness which leads to details being overlooked or remembered inaccurately, and genuine lack of recognition of areas where performance was poor, all prevent thorough self-debriefing. (So you do not suffer from any of these, then? – OK, see what I mean?)

Comparison with other pilots' performances may offer some clues to anything lacking in your performance, but may not highlight where improvement is necessary. What is needed is a systematic self-appraisal of your performance, not only for each flight but also for each part of every flight.

Much of the initial analysis of your performance during a flight can be carried out as the flight progresses. A mental note can be made of any mistakes and aspects of the flight which could have been conducted more efficiently. It may be possible to rectify some of these failings immediately or as the flight progresses, but do not dwell on them at the expense of your performance for the rest of the flight. Time lost due to such inefficiencies cannot be made up. It is history. Concentrate on the present and the future and try not to make the same mistake twice.

As soon as possible after you land, take some time to make a list of the parts of the flight which need improvement. This list will include the mental notes you made during the flight and any subsequent thoughts you have had.

After the adrenaline and the elation (or disappointment) of the flight have reduced to sensible levels, work your way through the flight. Look at each part of it in turn. Give yourself marks out of ten for your performance in each aspect of your flying and on each leg. How was your start? How good was your thermalling on each leg? How well did you cope with the weather conditions? Did you lose too much height or time at the turning points? Could the final glide have been more efficient? From this self-debriefing, add to the list of aspects needing attention which you made earlier.

If you have the facility to analyse the flight using barograph or GPS/datalogger information, this should be used to pinpoint any more problem areas of the flight or to clarify what went wrong. Such devices can be used to analyse thermal centring technique, confirm the rounding of turning points, check for airspace infringements, scrutinise the use of detours, analyse final glides and a whole lot more besides. This is another area where GPS has proved a useful tool.

Having created a 'critique list', study it and highlight the areas which need most attention. Think of what went wrong and how you can improve that part of your flying. Will a specific problem improve with practice during cross-country flights or do you need to spend time working on it while local soaring without the distractions and concerns that accompany cross-country flying? Do you need to seek advice from an expert? Will instruction in a two-seat glider help improve the situation?

Tackle each problem area in turn, beginning with the one you see as the most damaging to your overall performance. Work on it whenever the weather conditions allow. By improving each element of your flying in this way, the whole of your flying will become more efficient. By keeping a file containing your critique lists, you can check to see which areas are consistently a problem, and have more chance of recognising areas where improvements have been made.

No pilot can ever say that his or her flying is so good that it does not need improvement. The above analysis techniques should be used after every flight, whether the flight is a cross-country task or a local soaring improvement exercise.

Psychology

Some pilots do not even consider their psychological state as having an effect on their performance. Some have the correct mental attitude to their soaring and never doubt their ability. As a result, their soaring benefits. However, many pilots do have doubts about their ability to perform well in some aspect of soaring, and such self-doubt may well affect their decision making, reducing their performance and their enjoyment.

For instance, lack of confidence in your ability to land a glider safely in a field may lead to apprehension when you get low, or mean that you tend to cling on to height unnecessarily. This in turn may mean that you take too many climbs, possibly using weak thermals. The result will be low average cross-country speeds and a reduction in the size of the tasks achievable. Ironically, this inability to achieve reasonable average cross-country speeds will reduce your chances of completing larger tasks, and in turn increase the chance of having to land the glider in a field short of your goal. Such apprehensions are common after a pilot has had the bad experience of damaging a glider while attempting to land in a field. At the very least, such apprehension will reduce your concentration and soaring efficiency when you are low and conditions are weak. This is the time when you need to be most efficient.

Other pilots fear doing badly on tasks or in competitions. Landing out early on a task when everyone else completes it is a dread that most pilots have. Instructors often worry about doing badly when some of their students are watching.

Analyse your fears and try to master them. Do you find yourself continually making excuses to yourself as to why you are unable to fly cross-country on a particular day or why you cannot take part in a club task? 'No retrieve crew' or 'must be home early' may be fronts for deeper concerns.

During a competition task when the soaring conditions do not live up to expectations, many pilots feel that they will not make it home, and as a result, decide to land at a convenient airfield, only to find that other pilots who they were with on track, and with whom they were capable of keeping up, made it home. If you think you can do something, then you are in with a chance. If you set your mind on failure, it is hard to succeed.

Using the exercises mentioned previously, work on the areas which concern you. Seek advice from an instructor for whom you have respect. If you are concerned about doing badly in front of your fellow club members, attend a course, task week or competition at another gliding site where this pressure is not present.

If you have previously achieved a degree of success in the particular aspect of soaring which is causing concern, then you know that you are capable of the same success again. Ability does not change significantly from day to day. Frame of mind and mental attitude do. Control these and you have gone a long way to achieving success.

Before any flight, remember your good performances. Relish the great moments of personal glory which you have had. Do not think of the bad flights or the off days. Think positively. You CAN do it!

Knowledge of equipment

Possessing a basic knowledge of how your glider flies, and the importance and operation of its various parts, is essential if it is to be operated safely. A greater knowledge of the aerodynamics involved will allow you to use the glider more efficiently.

An understanding of how any items of ancillary equipment function, and of their limitations, will give you confidence in their use and will save confusion arising from operator error.

Spend time studying the glider you fly, its structure, peculiarities, control and flight systems, performance figures and handling notes.

When the Certificate of Airworthiness inspection is due, ask the inspector if you can assist with the inspection or any work required. This is an excellent way to discover more about your glider and might even reduce the cost of the work.

Practise preparing the barograph or datalogger and using the GPS and final glide calculator. Read their manuals or operating instructions and try to understand how they work.

Check the glider's trailer regularly to make sure that it is roadworthy and that all of its fittings are serviceable. Make sure that you are familiar with how they work. Discovering how to put a glider into an unfamiliar trailer at midnight in a thunderstorm is not conducive to good humour among crewmembers.

Try to gain as thorough a knowledge as possible of all the equipment you are likely to need.

Books, magazines, videos and lectures

There are many books on gliding and soaring. These range from training manuals for student glider pilots to books by world champions telling you how to win competitions. Some tell of the experiences (good and bad) of great soaring pilots, while others offer a history of the sport. There is even a novel that is about a fictitious, epic glider flight.

All of these have something to offer (even the novel). Each book is bound to contain some information, facts, techniques or ideas which may be new to you. The newer you are to the sport, the more you will discover.

Magazines on soaring usually contain pilot's tales, articles on meteorology, soaring techniques and descriptions of soaring conditions in your own or other parts of the world.

Videos and films on soaring are often interesting as well as amusing. Most give some insight into what is available in the sport. Some show the problems faced in championships. Others give insights into how the experts deal with various soaring conditions.

Lectures given by, or forums involving, experienced pilots are always worth attending. Not only are these usually informative and entertaining, but also you will get the chance to ask questions and seek expert opinion on the subjects being covered.

Any knowledge should be greedily grasped and any possible source of knowledge investigated. Never miss an opportunity to increase your knowledge. Every little snippet of information filed in your memory might be useful one day and you never know when such knowledge might be needed.

Chapter 21

Badge Flying

From the moment you take up the sport of gliding, you will be attempting to gain qualifications which will be proof of your ability and skill. The initial national qualifications are used by gliding clubs as the requirements to fly certain types of glider and to fly in more demanding weather conditions. Later qualifications are recognised internationally as a measure of a pilot's competence and standing within the sport.

Most of these qualifications are recognised by the award of a certificate, and entitle the holder of a particular certificate to wear a badge signifying the achievement of this standard. For this reason most awards are known by their badge name: for example, Silver Badge, Gold Badge, etc.

The various qualifications are graded to offer a progressive challenge to the pilot; therefore, each in turn is harder to achieve and will stretch the pilot's skills and demand greater judgement. Generally, the standards necessary for the gaining of these certificates are set by an international organisation, known as the FEDERATION AERONAUTIQUE INTERNATIONALE (FAI), although national gliding associations, such as the BRITISH GLIDING ASSOCIATION (BGA), often set their own initial and intermediate qualifications to provide more graded challenges and encourage pilot development. In the following text, British awards which are not recognised internationally are annotated by the letters 'BGA', whereas those which are, are annotated 'FAI'.

As the requirements for the various qualifications will occasionally change, the official rules and conditions for the grant of any particular award should be consulted before attempting any qualifying flight.

The following is a list of the various awards, their requirements and, where appropriate, advice on achieving them. Where useful exercises can be practised while attempting to gain a qualification, such exercises are discussed.

'A' Badge (BGA)

* One solo circuit.
* Demonstrate a knowledge of the basic Rules of the Air.

Your first solo flight is an achievement, the details of which you will always remember. This is the climax of all of the training in general handling and judgement since you took up the sport of gliding. It is also your first flight without some instructor nagging in your ear. After this milestone, you can start to learn to soar.

The test to demonstrate your knowledge of the basic Rules of the Air is designed mainly with collision avoidance and general airmanship in mind. National gliding authorities normally produce some form of booklet detailing the Rules of the Air as they apply specifically to glider pilots. This should be read, as it will contain all of the information required to pass this test and those which follow.

'B' Badge (BGA)

* A soaring flight of at least five minutes at or above the previous lowest height after launch.

Once upon a time, when gliders had a lot less performance than they have today, this flight would have been an achievement. These days, it is so easily achievable that it seems just another reason to give away money to gain a certificate.

Bronze Badge (BGA)

* 50 or more solo glider flights (or 20 flights in a glider and 10 hours pilot in command). Powered aeroplane flying experience will give some dispensation.
* Two soaring flights. Each of these flights must last for 30 minutes or more if launched by winch, car or bungee, or 60 minutes or more if launched by aerotow. The height of release from such an aerotow must not exceed 2,000 feet.
* A minimum of three 'check' flights with a fully rated instructor.
* Successful completion of a written test paper which will include Air Law, Principles of Flight, Navigation and Meteorology.

The Bronze Badge flying test requires the candidate to demonstrate basic soaring ability, a good standard of flying skill, good judgement, a reasonable knowledge and understanding of soaring-related subjects, and a high standard of airmanship.

It is important that, while attempting to achieve the required number of launches and the soaring flights, you work hard on your general handling skills. Special emphasis should be put on good control co-ordination, airspeed and airbrake control. Setting yourself high standards will be of benefit for future soaring, and will help raise your flying to a standard which will help you pass the flying tests associated with this qualification. Never hesitate to ask for an instructional flight or advice if you are unsure of where you should set your targets.

At this stage of your flying, local soaring is still an achievement to be relished. For this reason, it is best that you simply enjoy your early soaring flights. You will be experiencing and learning so much that part of the problem will be assimilating and storing this new knowledge. Every bump of the air, patch of sink or minor scare you experience will put you in good stead for the epic flights in the years to come.

Although much of the knowledge required to pass the written tests will be picked up from your instructor or from being around the gliding club, much of the theory will have to be learned from books. THE GLIDER PILOT'S MANUAL contains all of the theory required for the Principles of Flight examination. Other books are available which deal with meteorology and aspects of air law as it affects the glider pilot.

Although the flying tests require you to demonstrate a high standard of control and airmanship, they take into account your relative inexperience. They are therefore nothing to get nervous about. Great emphasis will be put on stall and spin avoidance, and recovery drills. Airmanship will carry very high marks. Failure to maintain a good lookout will certainly result in your failing the test.

Part of the test is designed to examine your ability to land in a field. Therefore, the examiner will be looking for a reasonable level of approach control. To test for this, you may be asked to land within a defined area of the airfield which is smaller than the normal landing area and has an approach direction not commonly used. Alternatively, this part of the test may be done in a motor glider, with approaches being made into actual fields.

Many pilots tend to stay at home when the weather is too windy or visibility is too poor to allow them to fly solo. Such pilots miss a lot of experience, which at this stage of their flying is invaluable. On such days, instead of mowing the lawn or painting the house, try to get an instructional flight. After all, once you get your Bronze Badge, you will probably be allowed to fly in such adverse weather. It is far better to have found out how exciting some conditions can be when you are with an experienced instructor, than to have your first encounters with the pitfalls of such days on your own.

Cross-country Endorsement (BGA)

* Two soaring flights, one having a minimum duration of one hour and the other a minimum duration of two hours.
* A flight in which the candidate demonstrates to an instructor the ability to select suitable landing fields.
* A flight in a motor glider (or light aircraft), demonstrating the ability to make safe approaches into at least two fields.
* The candidate must satisfactorily plan a triangular task of approximately 100 kilometres and demonstrate good navigation and airmanship while attempting the task in a glider, motor glider or a light aircraft.

Until 1996, the minimum qualification which had to be held by a glider pilot in the UK before he or she was permitted to fly cross-country was the Bronze Badge. Then it was decided that pilots would be better equipped for cross-country flying if they had completed longer soaring flights and received instruction and testing in navigation, field selection and simulated field landings using a motor glider. For some reason (unknown to the author) it was decided not simply to increase the requirements of the Bronze Badge, but instead to introduce the Cross-country Endorsement to the Bronze Badge.

Undoubtedly, motor-glider training is a superb way to practise field selection and 'field landings', but problems arise if your club does not have a suitable motor glider available. If this is the case, it will be necessary to arrange for this training and the subsequent tests to be conducted at a site which has a suitable motor glider and an instructor qualified to carry out these tests.

Some preparation can be done at home, starting well before you are ready to do the tests. Practise using maps and plotting courses, and read up on field landing procedures. Ask your instructors to set you short-landing exercises into different areas of your home airfield. Study local fields and crops both from the air and from the ground. If you have not flown in the particular type of motor glider to be used for the test, ask for a training session before the actual test. This will give you confidence, help you understand what the instructor conducting the test will require and probably save you money in the long run.

Silver Badge (FAI)

(This award has three parts, known as SILVER DURATION, SILVER DISTANCE AND SILVER HEIGHT.)

* SILVER DURATION	One flight lasting not less than 5 hours from release to landing.
* SILVER DISTANCE	One flight covering a distance of at least 50 kilometres. The flight can be made as an undeclared flight to a point at least 50 kilometres from the start point or as a pre-declared flight (such as a goal-and-return or a triangle) where at least one of the legs is 50 kilometres or more in length. In either case, the loss of height between the start point (release height or logged point in the start zone) and the finish point must not exceed 1% of the total distance flown.
* SILVER HEIGHT	One flight during which a gain of height of at least 1,000 metres (3,281 feet) is achieved, measured from the lowest previous point after launch.

This is the first of the internationally recognised awards.

Silver Duration

The Silver Duration flight is literally an endurance flight, especially if you do not make good use of the time. The priority has to be to stay airborne to gain the award and so you should not risk getting low.

Despite this requirement, you should attempt to use the time when you are at a safe height to carry out exercises which improve your skills and judgement. Such exercises include thermal centring (try to outclimb any other gliders you come across), map reading, field selection and general handling.

If soaring on a hill, a 5 hour flight can be particularly boring. Once a safe margin of height is gained, move away from the hill and use the excess height to search for thermals or wave lift, or simply to practise the above exercises before rejoining the hill lift to regain height.

Silver Distance

The flight attempting to cover Silver Distance is likely to be the first time you have gone out of gliding range of your home airfield while flying solo. The priority is that you have a safe and enjoyable cross-country flight and achieve the distance required. Often this flight will be a straight distance flight, ending at a goal airfield which is known to welcome visiting gliders. On the other hand, the flight may involve flying a goal-and-return task or a triangle, ending up at your home airfield. Whichever type of task you attempt, your aim should be to stay high and cover the distance. There is no requirement for a high average speed, so do not take chances and increase the risk of landing short of your goal.

Start this cautious approach to the flight from the moment of launch. Find a thermal, climb as high as you can in it and relax. If the day is good enough for such an early cross-country flight, then you will not need to set off immediately. The weather pattern will have to sustain reasonable soaring conditions for some time. Sample different thermals and get a feel for the day. Only when thermals are going to a reasonable height (or near to the maximum forecast for the day) and the sky on track looks good should you set off.

If there is any significant wind, it will be wise to attempt a straight-line flight and to select a goal airfield which is downwind. As an alternative, a goal-and-return task may be possible, making use of thermal streets.

When attempting a straight-line flight for your Silver Distance, you should be aware of the distance in excess of 50 kilometres which may have to be flown to qualify for the award (the so-called one per cent rule). For instance, if the goal airfield is 60 kilometres away from the airfield of launch and both airfields are at the same altitude (same height above sea-level) then the following example shows the height to which you can launch and still qualify for the award of Silver Distance.

EXAMPLE:
1 kilometre = 3,281 feet
Distance = 60 kilometres = 196,860 feet
1% of 196,860 feet = 1,968 feet
Maximum Launch Height = 1,968 feet

In this case, you must release from aerotow at a height not greater than 1,968 feet. (Allowing a margin of 200 feet for altimeter error is a wise precaution.)

If the same goal airfield is 500 feet lower than the airfield from which you will launch, then the calculation must take this into account. In this example the maximum launch height would have to be reduced by 500 feet: that is, 1,468 feet.

If a goal airfield is so distant that the difference-in-height rule (the one per cent rule) is not a problem, or if you are flying to an area where there is no suitable airfield, then it is helpful to mark arcs or circles on your map which relate to the distance that you will have to cover from various launch heights (table 8).

By doing this, you free yourself from any launch height restrictions and can easily check in flight how far you must go to complete the qualifying distance. Such estimates must allow for the difference in terrain elevation.

TABLE 8

EXAMPLE:
Height of start airfield = 500 feet above sea-level
Height of landing point = sea-level

LAUNCH HEIGHT (FEET ABOVE AIRFIELD)	DISTANCE REQUIRED (km)
1,100	50
1,250	54
1,500	61
1,750	69
2,000	77
2,250	84
2,500	92

Note: These figures assume that the landing point will be at sea-level, and therefore give the worst case and probably some margin for qualifying.

If you choose to fly a goal-and-return or a triangular task to qualify for your Silver Distance, you will have to prove that you have rounded the turning point(s). Prepare for your arrival at the turning point(s) and fly around each of them carefully. REMEMBER NO EVIDENCE – NO BADGE.

As with any distance task, you must return with evidence of its completion. This will not only mean satisfactory turning point evidence but also a barograph trace to prove that the flight was made without an intermediate landing. The carrying of a barograph is a requirement for all of the awards from Silver Badge onwards and so you should get into the habit of preparing and stowing a barograph/datalogger for every flight.

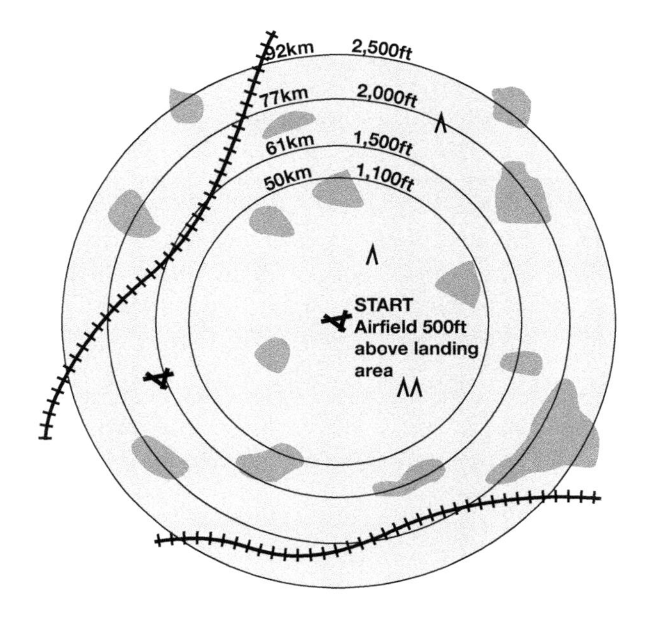

Fig 21.1 Silver Distance one per cent rule. A quick reference map.

Silver Height

The gain of height of 1,000 metres (3,281 feet) for Silver Height may not be a problem in many countries where thermals often go to great heights. However, in countries such as the UK, a glider pilot can wait for months for a cloud base that is high enough to allow this gain of height. In these situations, gaining the required height becomes an opportunity task and it therefore pays to carry a barograph/datalogger on every occasion on which you fly. The chances are that you will gain your Silver Height on a day when you were not specifically attempting it.

100 Km Diploma (BGA)

* Two flights, each covering a distance of 100 kilometres or more, around a pre-declared goal-and-return or triangular course, one of which must be achieved at a minimum handicapped speed of 60 kilometres/hour.

This award was introduced to fill the considerable gap between the 50 kilometre flight for the distance part of the Silver Badge and the

300 kilometre flight which is part of the Gold Badge. The speed requirement on one of the tasks is to give you some idea of the sort of speed you will have to achieve to complete a 300 kilometre flight. It is also the first flight which you will attempt where the achievement of a reasonable average speed becomes as important as staying airborne. The requirement for a closed circuit flight is to give you experience of turning points and final gliding.

As far as the first 100 kilometre task is concerned, your primary aim should be to complete the task and gain practice at soaring, navigation, rounding turning points and, hopefully, a final glide. A high enough average speed to qualify for the speed part of the diploma should only be a secondary aim. If the soaring conditions turn out to be very good, then by all means attempt to achieve a high average speed – but not at the risk of an outlanding. If you do achieve the required average speed on this attempt, then great, your task for the next 100 kilometre flight is not dependent on speed.

Assuming your first 100 kilometre task has been achieved at less than the qualifying speed of 60 kilometres per hour, then your second 100 kilometre task becomes a speed task. Now you must start making all the same decisions which you will have to make for longer tasks or competition flights. You will experience the dilemma every pilot has to face one day, that is, to take chances to increase your average speed and risk landing out, or fly conservatively and risk not completing the task (in this case at 60 kilometres per hour or faster). This is the type of decision that the Cross-Country Diploma is designed to make you take.

Gold Badge (FAI)

(This award has three parts, known as GOLD DURATION, GOLD DISTANCE AND GOLD HEIGHT.)

* GOLD DURATION	One flight lasting not less than 5 hours from release to landing.
* GOLD DISTANCE	One flight covering a distance of at least 300 kilometres.
* GOLD HEIGHT	One flight during which a gain of height of at least 3,000 metres (9,843 feet) is achieved, measured from the lowest previous point after launch.

Gold Duration

The good news is that the 5 hour duration flight, which you probably did to gain your Silver Badge, will count as the Gold Duration. Thus, you do not have to go through that ordeal again. The bad news is that your first attempts at the required 300 kilometre flight will probably take you 5 hours or more. However, time does pass more quickly when you are working hard.

Gold Distance

After the Silver Distance flight, or even your first 100 kilometre triangles, an attempt at a 300 kilometre flight is quite a challenge. Speed is an essential part of the task. (However, if the task selected is a straight-line task to a goal airfield which is downwind, then achieving the necessary distance will be less difficult if the wind is strong.)

It is now that you must utilise all of the skills and judgement (such as speed-to-fly technique and skyreading) which you have hopefully worked hard to learn. To fly slowly is to fail to complete the task and end up with an outlanding.

Gold Height

The gain of height required for the Gold Badge, unlike the Silver Badge height gain, is more of an opportunity award. In most countries, such a gain of height will not occur during normal thermal soaring. It will require specific weather conditions, and if it is to be achieved safely, a certain amount of organisation on the part of the pilot.

The weather pattern will have to be conducive to producing either thermals (and possibly very large cumulus or cumulonimbus clouds) or mountain lee waves which go high enough for the necessary gain of height. In most countries, your preparation will have to include oxygen equipment and warm clothing.

As climbing in cumulonimbus clouds seems to be less common these days and since cloud flying is illegal in some countries, achieving Gold Height is more usually the result of a climb in mountain wave. This will possibly require an expedition to an airfield where wave lift is common.

Diamond Badge (FAI)

(This award has three parts, known as DIAMOND GOAL, DIAMOND DISTANCE AND DIAMOND HEIGHT.)

* DIAMOND GOAL	One pre-declared flight covering a distance of at least 300 kilometres which ends at the start airfield.
* DIAMOND DISTANCE	One flight covering a distance of at least 500 kilometres.
* DIAMOND HEIGHT	One flight during which a gain of height of at least 5,000 metres (16,404 feet) is achieved, measured from the lowest previous point after launch.

Diamond Goal

Most pilots complete the requirements for this 300 kilometre closed circuit flight when they achieve the distance flight requirements for Gold Badge. This means that this need not be repeated for the award of the Diamond Badge. If, however, the 300 kilometre flight for the Gold Badge was achieved without a return to the home airfield then this will not qualify for the Diamond Goal award.

Diamond Distance

The 500 kilometre flight for the award of Diamond Distance is effectively a pure speed flight. A high average cross-country speed is essential to complete the task. Keeping up a good average speed throughout the whole flight means that you cannot afford many (or sometimes any) low scrapes, but, at the same time, neither do you have time to float aimlessly around at the top of thermals. You must fly with determination.

One reason why 500 kilometre flights are quite difficult in countries such as the UK, with its small land mass, is that it is relatively rare to have the good soaring conditions necessary for such a flight covering a large enough task area or lasting long enough. Since the rules for such awards now include the flying of quadrilateral tasks, it is easier to plan a task that avoids flying too near a coast with its cooling sea air. (Some pilots, usually those who flew a triangular task for this award, regard flying anything other than a triangle as an easy way out and are derisive of those who use three turning points. Ignore these snobs. The rules are there to be used, and if it is an award you are after then use the rules, as best you can, to your advantage.)

Hot countries, such as Australia and South Africa and some parts of the USA, regularly produce conditions favourable for such flights. Despite much of Europe's less favourable climate, many 500 kilometre flights are completed each year.

Diamond Height

Gaining almost 17,000 feet is certainly an enjoyable achievement. You will end up in a world of fantastic views. Unfortunately, if the climb is made in temperate latitudes, outside air temperatures at altitude will be low.

To achieve Diamond Height with the minimum risk will involve wave soaring. Again, if your own airfield is not a wave-soaring site, then an expedition to a gliding site renowned for its wave conditions is the best way to achieve your aim. Prepare well before the expedition and, if you are new to wave flying, seek advice from pilots who have flown from the airfield in question and who know the problems associated with wave flying at that venue.

It is best if you can spend some time at the wave site (at least a week, but preferably longer). The one factor that usually denies pilots the gain of Diamond Height is the weather. Even the best wave sites suffer from weather patterns which are unfavourable for the formation of usable wave some of the time. As no one can guarantee the weather, this is one aspect which is hard to solve by organisation. Occasionally, the problem will not be the lack of waves in the sky, but that the wind is of a strength and direction which makes it unsafe to launch. Alternatively, the area may be suffering from too much low cloud. In these situations, it may be worth arranging to take the glider to another site for the day. Mountain effects can be very localised, and a drive of 50 miles or so may offer you less cloud, a runway from which it is safe to launch, and possibly your Diamond Height.

Diplomas and Badges

Diplomas and badges are awarded for flights covering distances of 750 kilometres – in single or two-seat gliders (BGA), 1,000 kilometres (FAI) and 2,000 kilometres (FAI).

The diploma awarded for the completion of a flight of 1,000 kilometres or more was, for a long time, regarded as the greatest award a glider pilot could achieve. Even many good cross-country pilots who already hold the Diamond Badge look on the 1,000 kilometre diploma as being beyond their ability, and seldom attempt flights much greater than the 500 kilometres which they have already achieved.

It was for this reason, and to provide a less daunting target, that some national gliding organisations introduced 750 kilometre diplomas for both single and two-seat gliders.

However, as glider performance improved and, just as importantly, pilots' attitudes have changed, 750 kilometre flights are being achieved more often. (Even in Britain, with its small landmass, the 1,000 kilometre 'barrier' has been broken, and each year more pilots in the UK attempt 750 kilometre flights.) As a further challenge, there is now an FAI diploma for the successful completion of a 2,000 kilometre flight.

Not all that long ago, the 500 kilometre flight was the challenge for most glider pilots. It no longer is. In soaring, your targets are where you set them.

Appendix 1

Speed-to-Fly Ring Construction

This appendix shows you how to construct a MacCready speed-to-fly ring to fit onto your variometer. In order to calibrate a MacCready ring for your glider, you will need to obtain a performance curve (usually known as a polar curve) for your particular type of glider. This is often the first problem.

Many glider manufacturers provide a polar curve in their publicity material when their glider first appears on the market. The same polar curve will usually be included in the glider's operating manual. Sadly, such performance figures do not always equate accurately to the performance of the production sailplanes which are eventually produced. This means that if figures from such a polar curve are used, the MacCready ring that you produce may be somewhat optimistic as far as the airspeeds at which it suggests you fly are concerned. If such a polar curve is the only source of performance information available, then you will have to allow for such optimism when you come to draw the graphs which follow.

Fortunately, a large number of sailplanes have been test flown by independent assessors specifically to provide performance information. If you can obtain one of these reports on your type of glider, then you have a much better chance of constructing a realistic MacCready ring.

Firstly, take the polar curve and make a larger-scale copy of it on a piece of graph paper. If necessary, convert the units on the manufacturer's or test polar curve to those that you use on your glider. You will use these units throughout your construction of the MacCready ring.

Next, draw the tangent to the curve from the point on the vertical axis which corresponds to zero knots rate of climb. From the point where the tangent meets the curve, draw both a vertical line to the horizontal axis (showing the best indicated airspeed at which to fly for this expected rate of climb) and a horizontal line to the vertical axis (giving the still air rate of sink expected if the glider were to fly at that airspeed).

Draw tangents from other points on the vertical axis indicating various rates of climb, and construct the horizontal and vertical lines from the point at which each of these tangents touch the curve.

Make up a table similar to table 9, and note on it the figures pertaining to the rate of climb from which each tangent was drawn.

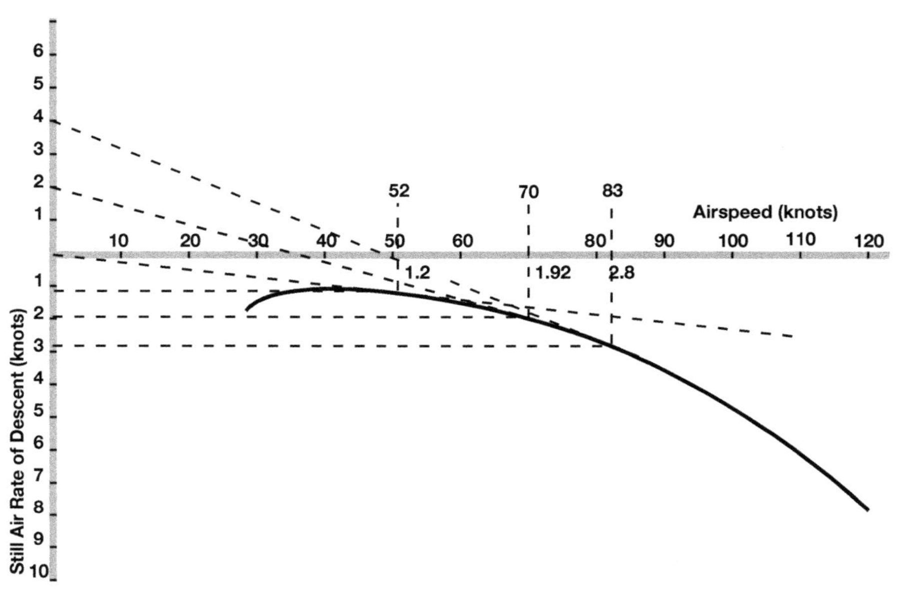

Fig A1.1 Using a polar curve to construct a speed-to-fly ring.

TABLE 9

Note: All figures are in knots.

RATE OF CLIMB	SINK RATE	RATE OF CLIMB + SINK RATE	INDICATED AIRSPEED
0	1.2	1.2	52
1	1.5	2.5	60
2	1.9	3.9	70
3	2.2	5.2	75
4	2.9	6.9	83
5	3.2	8.2	87
6	3.8	9.8	93
7	4.0	11.0	95
8	4.4	12.4	98

The third column of table 9 is simply the rate of climb value from which any one tangent was drawn added to the still air sink rate expected at the corresponding indicated airspeed.

You now have all the data you need to calibrate your MacCready ring. You will be using the figures in the third column (the sum of the rate of climb and the rate of sink) with the corresponding indicated airspeed to construct the ring. However, it is easier to read the airspeed off the MacCready ring if it is calibrated at 5 knot intervals, rather than at the odd intervals derived from the graph. It is therefore necessary to construct another graph in order to find the vertical speeds which apply for airspeeds at 5 knot intervals.

Plot the figures from the third column (rate of climb + sink rate) against those of the indicated airspeed column (using the horizontal axis for airspeeds). Select points on the airspeed axis at 5 knot increments, and draw a vertical line from each of these airspeeds to the plotted curve, and from there, horizontally to the vertical axis (rate of climb + sink rate axis). Note the value on the vertical axis relating to each airspeed on a table as follows (table 10).

TABLE 10

Note: All figures are in knots.

INDICATED AIRSPEED	RATE OF CLIMB + SINK RATE
50	1.1
55	1.75
60	2.5
65	3.25
70	3.9
75	5.2
80	6
85	7.5
90	9
95	11

Now you need to fit the blank MacCready ring to the variometer.

Mark an index arrow on the blank ring opposite the zero rate of climb/descent mark on the variometer scale. Keep the ring in this position throughout the whole calibration procedure.

Opposite where the point on the variometer scale which indicates the value in the right-hand column of table 10, mark the corresponding airspeed on the MacCready ring.

You now have a MacCready ring for your glider. When flying, the index arrow can be positioned opposite the average rate of climb expected, and the variometer needle will, as well as indicating the rate of descent of the glider, indicate the airspeed at which you should be flying.

If your variometer is fitted with a calibrated leak device to give air mass

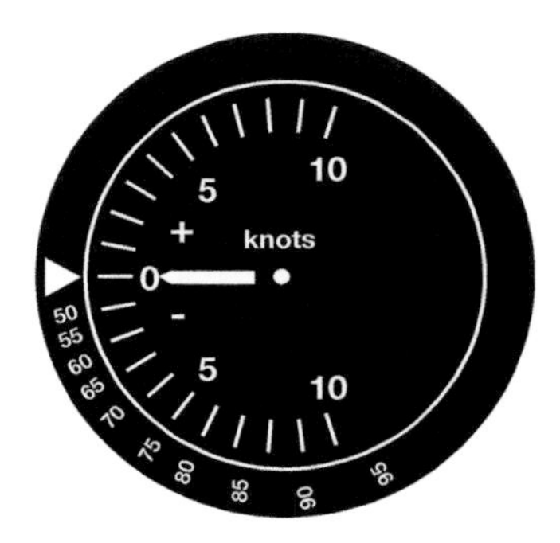

Fig A1.2 Marking up a speed-to-fly ring.

readings, then it is necessary to use the rate of climb column (column one) in table 9 throughout, rather than the combined rate of climb and sink value (column three). In other words, as the variometer will only indicate the rate of descent of the air, instead of the total rate of descent of the glider, table 9 need comprise only column one and column four. Table 10 would be derived from a graph plotted using these two columns.

If your system has a changeover valve which allows your variometer to display either total rate of descent or the rate of descent of the air mass, then you can use the different halves of the MacCready ring for these different calibrations.

Fig A1.3 Speed-to-fly ring – calibrated for air mass.

Alternatively, you may wish to have a second scale on your MacCready ring to give the higher airspeeds required when carrying water ballast. This will require you to recalculate the above using a polar curve which takes into account the increased weight of the glider.

Appendix 2

Compass Swinging

Compasses are much more complicated devices than they first appear. There is more than one reason why a compass can give erroneous readings. One problem is that DEVIATION (the error caused by ferrous parts of the glider, ferrous objects placed in the cockpit, or magnetic fields generated by the glider's electrical systems) may not be the same on all headings, and therefore a simple adjustment is not always possible.

If the compass error is constant on ALL headings, then you may be lucky. The chances are that the COMPASS CARD (the part of the compass which has the heading numbers painted on it) is out of alignment. In this case, all you have to do is adjust the heading scale by the amount of error. For instance, if the compass is over-reading by 10 degrees on all headings, simply realign the compass card by reducing the indicated heading by 10 degrees. If it is not possible to adjust the compass card, then you will have to make allowance for the error.

However, life is seldom that easy, and many compasses have a compass card which is permanently fixed to the magnet assembly. It is therefore rare that such misalignment of the heading scale is possible.

The chances are that the compass indication will be inaccurate by a differing amount on different headings. Now you are faced with the problem of reducing the deviation error. The following method is a simplified one, and would not satisfy the aviation authorities responsible for powered aircraft, who would expect errors to be corrected in a more mathematical fashion. However, as your glider contains much less metal and a lot fewer electrical devices than a wide-bodied airliner, and as you are unlikely to be flying thousands of miles non-stop, your compass will not need to work to the same accuracy.

The first thing you will have to do is ascertain what errors, if any, your compass has on various headings. To do this you will have to do a 'compass swing'. This should be done on a part of the airfield which is as 'magnetically free' as possible. In other words, do not do it next to trailers, a hangar composed of metal girders, or near an area where there are likely to be underground power cables. If there is a light aircraft maintenance base on your airfield, ask their engineers where they do their compass swings and use the same area. The chances are that this area will have been scanned and judged free from magnetic influences.

As it is the errors which are present when the glider is in the cruise which you are trying to ascertain, and if possible remove, all compass swings should be carried out, as far as is practical, with the glider in the cruise configuration. All electrics normally used in flight should be switched on. The glider's attitude should be close to that which it assumes in the cruise. This is not easy to accomplish, but as it is probably impractical to have the main wheel retracted when carrying out a compass swing, leaving the tail dolly on allows easier movement of the glider and goes some way to achieving a less nose-up attitude without adding to compass errors. The wings should be held level and the control column placed in a fairly central position. Lock the airbrakes and close the canopy. Move any metal objects well away from the glider. This includes the belly dolly, towing aids, trestles and your car.

As the glider will have to be manoeuvred, and compass readings will have to be compared with a remote hand-held compass, you will need to enlist the help of an assistant.

Standing at a reasonable distance from the glider (5 to 10 metres) and using a hand-held compass, align the glider's fuselage so that it points due north. Note any difference between the reading of the glider's compass and that of the remote compass.

It is helpful to mark the spot from which each remote compass reading was made with a wooden stick or some similar marker, as you will have to take another reading from that position later.

Now rotate the glider to point due south and note any difference between the two compasses. Repeat this exercise with the glider pointing due east and due west. Now you should have an idea of the error with which you are starting on each of the 4 cardinal points of the compass. If the errors are small (say less than 5 degrees), leave them be. If larger errors are found, then remove them in the following way.

Rotate the glider until it once again points due north.

Using a brass screwdriver or a brass key specifically designed for the purpose, turn the adjustment screw on the glider compass (marked either 'N–S' or simply 'C') until the difference which was previously observed between the two compasses is halved. For instance, if the glider compass reads 020° rather than north, adjust it until it reads 010°. Make a note of this new discrepancy.

Rotate the glider until it is again pointing due east and rotate the adjustment screw marked 'E–W', or 'B', until the error noted in this direction is halved. Note the new difference.

You can take new readings with the glider pointing towards south and west and make further corrections using the adjustment screws.

Errors when the glider is pointing east or west are always corrected using the adjustment screw marked 'E–W', or 'B', and those on a north or south heading using the 'N–S,' or 'C', adjustment screw.

Repeating the whole procedure two or three times should reduce any errors considerably. If any significant errors remain, these should be noted on a compass deviation card, which can be displayed in the cockpit.

Appendix 3

Motor Gliders and Turbo Gliders

Many glider pilots regard the installation of any kind of engine in a glider as incongruous – and because of this attitude treat the pilots of motor gliders and, to a lesser extent, turbo gliders with less respect than pilots of pure sailplanes.

Whether you fall into this purist category or not (and I fear the majority of glider pilots tend to), it is important to realise that, used properly, turbo gliders and motor gliders can prove a useful extension of our sport.

Before going further, let me define the type of motorised glider to which this appendix refers.

Self-launching motor gliders (as opposed to self-sustaining turbo gliders) come in two varieties. The first is basically a glider with an engine stuck on the front of the nose (or occasionally in the rear of the wing) with a propeller which, while often possessing the ability to be feathered to reduce drag, cannot be retracted. These motor gliders, which include aircraft such as the Scheibe Falkes, Grob 109s and the SZD Ogar, are more like light aircraft than sailplanes by appearance, if not by definition. Some of them are excellent tools for glider pilot training – in particular the teaching of field selection and approach techniques. However, their glide performance does not match that of modern sailplanes.

Then there are the motor gliders which have an engine and/or a propeller which can be retracted, converting the motor glider into a sailplane every bit as good at soaring as the equivalent non-engined version of the same sailplane (Pik 20E, DG400, ASH 26M, Stemme 10 and the like).

Lastly, there are the non-self-launching turbo gliders which have a small extendable (and retractable) engine. Although incapable of launching the glider, this engine is capable of supplying enough thrust to sustain flight or even give a small rate of climb. In their engine-stowed configuration, the soaring performance of these is also as good as the non-engined sailplanes from which they are derived.

It is the latter two types of motorised sailplane to which the following comments refer; that is, those essentially designed to soar, rather than cruise under engine power.

The following are some of the advantages of possessing such a motor glider or turbo glider.

Freedom to launch without reliance on external facilities

Possessing a self-launching sailplane allows a pilot to launch independently of other organisations or persons. While this is not all that great an advantage at an active gliding site where winches and aerotow facilities exist, there will be days at most clubs when there is no operation taking place. In addition, if a pilot wishes to operate from a non-gliding airstrip, then the ability to self-launch is essential. Expeditions to soar remote or unexplored areas or countries become a possibility.

At most airports, traditional glider launch methods are not permitted. Only by possessing the ability to taxi, self-launch and even comply with instructions from air traffic control, may operation of a glider be tolerated. In such cases, the need to self-launch and get clear of the controlled area, before beginning normal soaring flight, may make it necessary to use a motor glider.

Reaching soaring conditions

The ability to take off, climb and cruise initially under engine power allows a pilot to reach a soaring area which may be unattainable from even a reasonably long aerotow, and to do so at a cost which is not prohibitive. (One DG 400 pilot from my own 'flatland' club often takes off and flies to the mountains almost 70 miles away, where he enjoys some great wave soaring while the rest of us complain about winter's lack of thermals.)

Flying in inhospitable parts of the world

Many of you will have been on a vacation to some place where, from the ground, soaring conditions looked fantastic – and yet if you did have to land out, the problems of retrieving the glider would be considerable. While motor gliders or turbo gliders should never be flown where one becomes totally reliant on the engine starting successfully and not subsequently failing, these machines do offer the possibility of exploring what can appear to be epic soaring conditions which are too remote from a convenient landing area.

Guaranteed arrival back at base

Then there is the use to which most glider pilots envisage the engine would be put – getting home when the thermals finish.

Having the ability to 'get home on demand', as it were, apart from being appealing, can open up new possibilities for the soaring pilot. For instance, you are working a night shift but would dearly love to use most of the day to fly cross-country. The last thing you can afford is to be sitting in a field waiting for a retrieve crew when you should be heading for work. If a motor glider or turbo glider would mean that you could go soaring on more days, then surely this must be to their credit. Some of their critics would argue that knowing you can get home takes the challenge out of cross-country soaring. I can see their case, but if your circumstances mean that you can squeeze more soaring days into the year by having a motor glider or a turbo glider, then that is an important point in their favour.

Soaring tours

Each year the various soaring magazines tell of exceptional touring flights over great distances in motor gliders, where each day the pilot has set off soaring cross-country and at the end of each day landed at another place on the tour. Usually these flights have been made with the knowledge that, if soaring conditions collapse before the intended airport or gliding site is reached, the engine can be started and the leg safely continued. The ability to self-launch is essential if the tour is to be continued the next day.

Such tours have included crossing Australia, flying the length of Japan and touring parts of the USA and Europe. They, and the fantastic soaring and sights that they include, are all made possible thanks to the motor glider.

Motor/turbo glider competitions

In recent years, soaring competitions designed specifically for motor gliders and turbo gliders have been introduced. With suitable equipment, the position of the glider at the time its engine was started can be recorded. The glider can then be scored as if it had made a landing in a field at that position. Thus, normal competition scoring rules can be applied while everyone relaxes back at the base airfield, having avoided the possible hassle of retrieving the glider from a field. With a GPS/datalogger and an engine, competition outlandings can be a thing of the past.

For the record

In case, from the foregoing, you get the wrong idea and think that I am an ardent admirer of motorised gliders, let me say that for many years I have cringed at friends who have given me a strange look and said, 'Would you not be better sticking a little engine on it?' Throughout this appendix, I hope it is clear that apart from their use in the training of soaring pilots, my only support for any form of motorised glider is on the proviso that it offers the individual the potential for soaring or improved soaring which does not exist with a traditional, non-powered sailplane.

If pilots wish to cruise around in powered aeroplanes, then there are many more efficient, more comfortable and probably less expensive aeroplanes than most of the motor gliders on the market. Many gliding clubs have problems with noise complaints that restrict or make it difficult to use tow-planes for glider launching. It seems ludicrous to add to this problem the engine noise of motor gliders which are not being used specifically to further soaring (either directly or for the training of soaring pilots). Sadly, such noise problems have forced many clubs to restrict or prohibit the use of motor gliders which could have been used for genuine soaring purposes.

Appendix 4

Useful Addresses

BRITISH GLIDING ASSOCIATION
Kimberley House
Vaughan Way
Leicester LE1 4SE

Telephone 0116 253 1051
Website: www.gliding.co.uk

GLIDING FEDERATION OF AUSTRALIA
130 Wirraway Road
Essendon Airport
Victoria 3041
Australia

Telephone 03 379 4629/7411
Website: www.gfa.org.au

SOARING SOCIETY OF AMERICA
PO Box 2100
Hobbs
New Mexico 88241-2100

Telephone (505) 392-1177
Website: www.ssa.org

A comprehensive list of national gliding associations is available from the above sources.

Appendix 5

Conversion Factors

To convert from a unit in the first column to one in the second column, multiply by the conversion factor given in the third column.

DISTANCES AND HEIGHTS		
Nautical Miles	Statute Miles	1.151
Statute Miles	Nautical Miles	0.869
Nautical Miles	Feet	6080
Kilometres	Nautical Miles	0.54
Nautical Miles	Kilometres	1.853
Metres	Feet	3.281
Feet	Metres	0.305

SPEEDS		
Miles/hour	Knots	0.869
Knots	Miles/hour	1.151
Kilometres/hour	Knots	0.54
Knots	Kilometres/hour	1.853
Metres/second	Feet/minute	196.85
Feet/minute	Metres/second	0.00508
Knots	Feet/minute	101
Feet/minute	Knots	0.0099

WEIGHTS		
Pounds	Kilograms	0.454
Kilograms	Pounds	2.205
Stones	Pounds	14
Pounds	Stones	0.0714

TEMPERATURE	
Celsius (˚C)	Fahrenheit (˚F)
-20	-4
-17.8	0
-10	14
0	32
10	50
20	68
30	86
40	104
50	122

WATER BALLAST		
(Note: This conversion table applies only to water and not to other liquids or combinations of liquids.)		
Litres	Kilograms	1.0
	Pounds	2.2
	Imperial gallons	0.22
	US gallons	0.264
Imperial gallons	Kilograms	4.546
	Pounds	10.0
	Litres	4.546
	US gallons	1.2
US gallons	Kilograms	3.785
	Pounds	8.33
	Litres	3.785
	Imperial gallons	0.833
Kilograms	Litres	1.0
	Imperial gallons	0.22
	US gallons	0.264
	Pounds	2.2
Pounds	Litres	0.454
	Imperial gallons	0.1
	US gallons	0.12
	Kilograms	0.454

End Note

I am not sure that a book of this nature deserves an epilogue, but in case anyone has gained some wrong impressions from the preceding text, I feel the need to emphasise some points about the sport in which we are so fortunate to participate.

Much of this book has been about performance flying and achieving speed or awards. Many pilots crave these aspects of the sport. Many achieve very high standards in a sport which is continually seeing new records being set – which are not only the result of sailplane design, but equally the result of healthy ambition among its participants.

All this tends to put pressure on those whose achievements are less or slower in manifesting themselves, either because they are slower at learning new tricks, cannot afford to practise a lot, or perhaps are less ambitious. However, we should all remember that soaring is a sport which can be enjoyed in many ways, and by pilots possessing different levels of ability.

Some pilots are determined to be World Champion, others to win a national championship. Some take pleasure in flying cross-country for hours, enjoying the freedom and the challenge, with no interest in speed. The enjoyment of staying airborne while local hill soaring may be the peak of another pilot's ambition. I can remember one student who, having reached solo standard and proved to himself that he could safely fly solo, had no further ambition and was thereafter quite content to fly with instructors. This was not a result of any obvious lack of confidence or fear on his part, or any doubt by his instructors of his ability.

With all the different characters involved in gliding and soaring, it is hardly surprising that individuals find different elements of the sport enjoyable. The ideal instructor would recognise and accept the different goals of individual students and, while showing what else the sport has to offer, be able to teach, to a high standard, whatever facet a pilot wishes to learn.

Since no one instructor can be expected to be expert in all aspects of the sport, ideally a gliding club would have several instructors who, between them, can offer a pilot tuition towards whatever level he or she wishes to aspire. Ideally, our sport should benefit from a variety of instructors – some of whom are capable of training students to become safe, competent, solo pilots; some who can teach aerobatics; and others who can improve a pilot's speed flying, and so on.

Students should be encouraged to reach a level of soaring which they will enjoy, but should never be pushed or ignored because their ambitions do not match those of the club's Chief Flying Instructor.

Everyone from instructors to students should always remember that GLIDING IS WHAT YOU WANT IT TO BE – AND WHAT YOU MAKE IT.

Index